Astronomy at High Angular Resolution

Astrophysics and Space Science Library

More information about this series at http://www.springer.com/series/5664

Henri M.J. Boffin • Gaitee Hussain •
Jean-Philippe Berger • Linda Schmidtobreick
Editors

Astronomy at High Angular Resolution

A Compendium of Techniques in the Visible and Near-Infrared

Springer

Editors
Henri M.J. Boffin
ESO Vitacura
Santiago de Chile, Chile

Jean-Philippe Berger
ESO Vitacura
Santiago de Chile, Chile

Gaitee Hussain
ESO
Garching, Germany

Linda Schmidtobreick
ESO Vitacura
Santiago de Chile, Chile

ISSN 0067-0057 ISSN 2214-7985 (electronic)
Astrophysics and Space Science Library
ISBN 978-3-319-39737-5 ISBN 978-3-319-39739-9 (eBook)
DOI 10.1007/978-3-319-39739-9

Library of Congress Control Number: 2016950588

Cover illustration: ESO's Very Large Telescope is equipped with many instruments for high-resolution astronomy, including a laser for adaptive optics and an interferometer. The insert shows several applications of high angular resolution astronomy. Credit: N. Boffin/ESO/S. Brunier.

Printed on acid-free paper

This Springer imprint is published by Springer Nature
The registered company is Springer International Publishing AG Switzerland

Men who wish to know about the world must learn about it in its particular details.

(Heraclitus)

Preface

Recent years have seen a huge development in high-resolution astronomical techniques, which are critical to progress in many different areas of astronomy. These techniques can be divided in direct methods, interferometry, and reconstruction methods.

In 2001, a very successful book was produced in the LNP Springer Series about "astrotomography", a generic term for indirect mapping (or reconstruction) techniques that can be applied to a huge variety of astrophysical systems, ranging from planets via single stars and binaries to active galactic nuclei. Given the many past and coming improvements in instrumentation and telescopes, it was deemed necessary to revisit the landscape. It is also useful to put this in a broader context of high angular resolution astronomy, as many other techniques, from lucky and speckle imaging, to adaptive optics, to interferometry, are now becoming more and more used and have led to an amazing number of new discoveries. Moreover, it is indeed often necessary to combine all these techniques together in order to have a coherent and comprehensive idea of all the processes at work in a given astronomical environment.

In November 2014, a workshop was therefore held at the ESO headquarters in Garching, Germany, bringing together people from different communities who use various techniques to construct images at very high angular resolution. The workshop included many invited talks, and the scientific organisers thought that given the quality of the talks, it would be very useful to publish a fully edited book, where each of the invited speakers would write a chapter on a given technique or its application. The book you are holding in your hand is the result of this, and it is thus not only an update of the *Astrotomography* book but is clearly much wider in scope, given that it covers more methods and techniques, as well as more scientific fields. The book thus reviews these methods, the progress in the field, and the new harvest of results that were collected, as well as aims at preparing the next generation of astronomers to use these tools and techniques. It is the hope of the editors that this book will fulfil all these goals.

As is clear from the table of contents, high-resolution techniques cover quite a range of astronomical topics, from stellar surfaces and binary stars to active galactic nuclei. We should note, however, that this book only covers techniques used in the optical and infrared domain – we have on purpose not addressed high-resolution radio observations, for which specialised reviews and books are already available.

ESO Vitacura, Chile
April 2016

Henri M.J. Boffin

Contents

List of Contributors

Raymundo Baptista Departamento de Física, UFSC, Campus Trindade, 88040-900, Florianópolis, Brazil

Fabien Baron Georgia State University, Atlanta, GA, USA

Misty C. Bentz Department of Physics and Astronomy, Georgia State University, Atlanta, GA, USA

Wolfgang Brandner Max-Planck-Institut für Astronomie, Königstuhl 17, Heidelberg, Germany

Julien H. Girard ESO, Alonso de Córdova 3107, Vitacura, Santiago, Chile

Petr Hadrava Astronomical Institute, Academy of Sciences, Boční II 1401, Praha 4, Czech Republic

Colin Alastair Hill Astrophysics Research Centre, Queen's University Belfast, Belfast, UK

Sebastian F. Hönig Department of Physics & Astronomy, University of Southampton, Southampton, UK

Felix Hormuth Max-Planck-Institut für Astronomie, Königstuhl 17, Heidelberg, Germany

Michael J. Ireland Research School of Astronomy and Astrophysics, Australian National University, Canberra, ACT, Australia

Alain Jorissen Institut d'Astronomie et d'Astrophysique, Université libre de Bruxelles, Brussel, Belgium

Markus Kasper ESO, Karl-Schwarzschild-Straße 2, Garching, Germany

Kateryna Kravchenko Institut d'Astronomie et d'Astrophysique, Université libre de Bruxelles, Brussel, Belgium

Jean-Baptiste Le Bouquin Université Grenoble Alpes and CNRS, IPAG, F-38000 Grenoble, France

Thomas R. Marsh Department of Physics, University of Warwick, Coventry, UK

Dimitri Mawet Caltech, Pasadena, CA, USA

Julien Milli ESO, Alonso de Córdova 3107, Vitacura, Santiago, Chile

Julien Morin LUPM, Université de Montpellier & CNRS, Place Eugène Bataillon, 34095 Montpellier Cedex 05, France

David Mouillet IPAG, BP 53, F-38041, Grenoble Cédex 9, France

Stephen B. Potter South African Astronomical Observatory, Cape Town, South Africa

Axel D. Schwope Leibniz-Institute for Astrophysics Potsdam (AIP), Potsdam, Germany

Sophie Van Eck Institut d'Astronomie et d'Astrophysique, Université libre de Bruxelles, Brussel, Belgium

Christopher Allan Watson Astrophysics Research Centre, Queen's University Belfast, Belfast, UK

Chapter 1
Lucky Imaging in Astronomy

Wolfgang Brandner and Felix Hormuth

1.1 Motivations for Employing Lucky Imaging Techniques

While angular resolution is of lesser importance when studying extended objects with a flat surface brightness distribution, like, e.g., smooth emission line nebulae or reflection nebulosities, it is of crucial importance for the analysis and understanding of structured objects, such as close binary stars, sources embedded in crowded fields, or discerning details in shock-fronts of all kinds.

Even at very good astronomical observing sites, long integrations in the optical and near-infrared are typically limited to (at best!) an angular resolution $\geq 0.4''$ (see Fig. 1.1). Finer details at smaller angular scale are only revealed at very rare moments of exceptionally good astronomical seeing, and only for very brief periods of time. In long exposures, these moments of exceptional seeing get mixed with periods of worse seeing, and hence cannot be recovered.

Half of all binaries among solar type stars have separation of less than 30 A.U. [10], while for early to mid M-dwarfs this value decreases to 10 A.U. [20] and even smaller values for late-M-dwarfs and brown dwarfs [19]. Thus even for an exceptionally good seeing of $0.4''$, half of the solar type binaries more distant than 75 pc, and more than half of the M-dwarf binaries more distant than 25 pc remain unresolved. Circumstellar discs around young stellar objects have typical sizes of several 100 A.U., and thus remain largely unresolved in seeing limited observations of even the most nearby star forming regions. Resolving the cores of Milky Way open and globular clusters into individual stars in general also requires better than seeing limited observations. The same is true for detailed studies of the Galactic Centre region, or resolved studies of even the most nearby extragalactic stellar populations.

W. Brandner (✉) • F. Hormuth
Max-Planck-Institut für Astronomie, Königstuhl 17, 69117 Heidelberg, Germany
e-mail: brandner@mpia.de; hormuth@mpia.de

H.M.J. Boffin et al. (eds.), *Astronomy at High Angular Resolution*, Astrophysics and Space Science Library 439, DOI 10.1007/978-3-319-39739-9_1

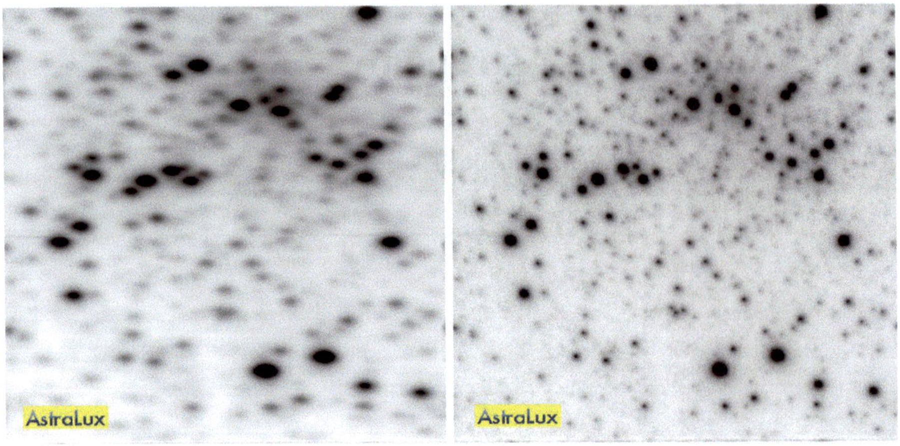

Fig. 1.1 (*left*) Seeing limited *i*-band observations of M15 at 0.4″ seeing with additional elongation due to telescope tracking errors: 5000 single frames with individual exposure times of 60 ms each were co-added for a total integration time of 300 s. (*right*) Lucky Imaging 5 % selection of M15 data set at 0.1″ resolution

This alone provides a strong scientific incentive to investigate and deploy modes for high-angular resolution to astronomical observatories. Among the wide variety of instrumental approaches and tools available, the relative simplicity in instrument development, deployment, and data analysis is what makes Lucky Imaging particularly attractive.

1.2 Atmospheric Seeing "101" and Lucky Imaging

Atmospheric turbulence is the driver behind the dynamically changing distribution of cold and warm air pockets of different density. The resulting localised variations in the refractive index of air optically distort incoming plane wavefronts originating at distant objects. This phenomenon is referred to as atmospheric seeing.

The atmospheric properties related to atmospheric seeing are characterised by two basic parameters:

Fried parameter or atmospheric *coherence length*: r_0 defines the diameter of the circular area with a time averaged wavefront variance $\sigma^2 = 1\,\text{rad}^2$. This corresponds to a wavefront error $\lambda_{\text{rms}} = \lambda_{\text{obs}}/6$.

Atmospheric *coherence time*: t_0 defines the time interval over which the variance of the observed wavefront changes by $\sigma^2 = 1\,\text{rad}^2$.

1.2.1 Angular Resolution and the Fried Parameter r_0

The Full Width at Half Maximum (FWHM) of the Point Spread Function (PSF) defines the angular resolution of a telescope. It is a function of the diameter of the entrance pupil of the telescope, the observing wavelength λ_{obs}, and the atmospheric seeing at the observing site.

Diffraction limit: the diffraction limited angular resolution of a telescope of diameter D is $\text{FWHM}_{\text{Diffraction}} = \lambda_{\text{obs}}/\text{D}$.
Seeing: the long exposure PSF has a $\text{FWHM}_{\text{Seeing}} = \lambda_{\text{obs}}/r_0$.

Good observing sites have typical values of $\text{r}_0 = 0.2$ to $0.4\,\text{m}$ in the *I*-band. As an example, let us assume that we are using a telescope with a $\text{D} = 3.5\,\text{m}$ primary mirror, are observing at a wavelength of $\lambda_{\text{obs}} = 850\,\text{nm}$, and at atmospheric conditions characterised by $\text{r}_0 = 0.2\,\text{m}$. While the telescope is capable of a diffraction limited angular resolution of $\text{FWHM}_{\text{Diffraction}} = 0.05''$, the atmospheric seeing would limit the angular resolution of a long exposure to $\text{FWHM}_{\text{Seeing}} = 0.9''$, i.e, $\text{FWHM}_{\text{Seeing}} \gg \text{FWHM}_{\text{Diffraction}}$.

As the Fried parameter scales with the wavelength as $\text{r}_0 \propto \lambda_{\text{obs}}^{6/5}$, the atmospheric seeing is getting better with longer observing wavelength: $\text{FWHM}_{\text{Seeing}} \propto \lambda_{\text{obs}}^{-1/5}$.

1.2.2 Strehl Ratio

The Strehl ratio (SR) is defined as the observed peak count of a PSF divided by the peak count of a perfectly diffraction limited PSF normalised to the same total count rate. Figure 1.2 shows the SR as a function of the wavefront variance σ^2 (or rms wavefront error) according to the Maréchal approximation (strictly valid only for random wavefront errors, see [35]):

$$SR = \exp(-\sigma^2) \tag{1.1}$$

For $\sigma^2 = 1\,\text{rad}^2$, the Strehl ratio is thus equal to 0.37. *Fully diffraction limited* is generally defined as $\text{SR} \geq 0.8$, i.e. $\sigma^2 \leq 0.223\,\text{rad}^2$. This corresponds to an rms wavefront error $\lambda_{\text{rms}} = \lambda_{\text{obs}}\sqrt{0.223}/(2\pi) = \lambda_{\text{obs}}/14$. Table 1.1 summarises the typical long-exposure seeing limited SR achievable at large optical telescopes.

1.2.3 Coherence Time τ_0

Taylor's hypothesis of frozen turbulence assumes that the time scale for the evolution of eddies embedded in an atmospheric layer is much longer than the time it takes the eddy pattern to move over the telescope (sub-)aperture [37]. For

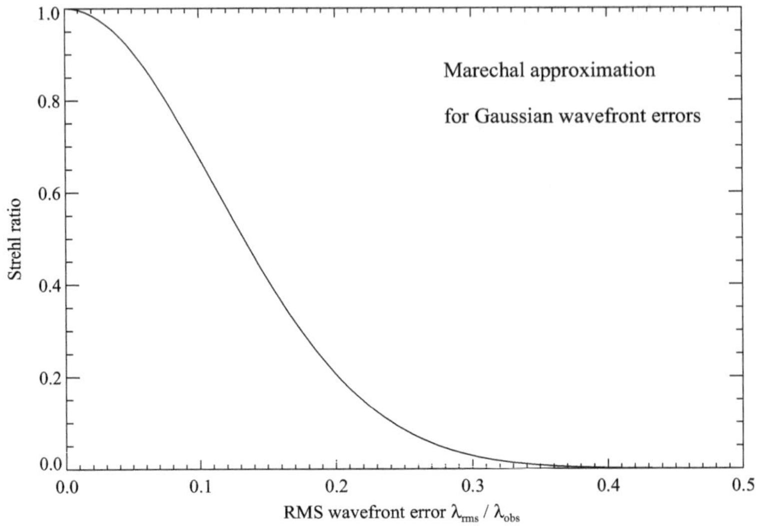

Fig. 1.2 Strehl ratio as a function of the rms wavelength error

Table 1.1 Typical long exposure SR for $r_0 = 0.2$ m at telescopes without Adaptive Optics in the *I*-band

D (m)	2.2	3.5	8.0	38
SR	0.01	0.003	6×10^{-4}	3×10^{-5}

a coherence length r_0 and wind speed v, the *coherence time* is defined as:

$$\tau_0 = 0.31 * r_0/v \quad (1.2)$$

Assuming typical values of $r_0 = 0.2$ m and $v = 10$ m/s, we get $\tau_0 = 0.02$ s. Thus short exposure times of the order of 10 ms are required to "freeze" turbulence in single exposures.

Figure 1.3 shows annual statistics of coherence time measurements for Paranal in two consecutive years. While in a very good year like 2003, τ_0 was equal or larger than 4.8 ms for half of the photometric observing time, in a bad year like 2002, the mean value was 2.6 ms, and τ_0 was larger than 4.8 ms only about a quarter of the time.

1.2.4 Probability of Lucky Imaging

The dynamical nature of atmospheric optical turbulence provokes varying magnitudes of wavefront distortions. Thus a certain fraction of short exposure images are

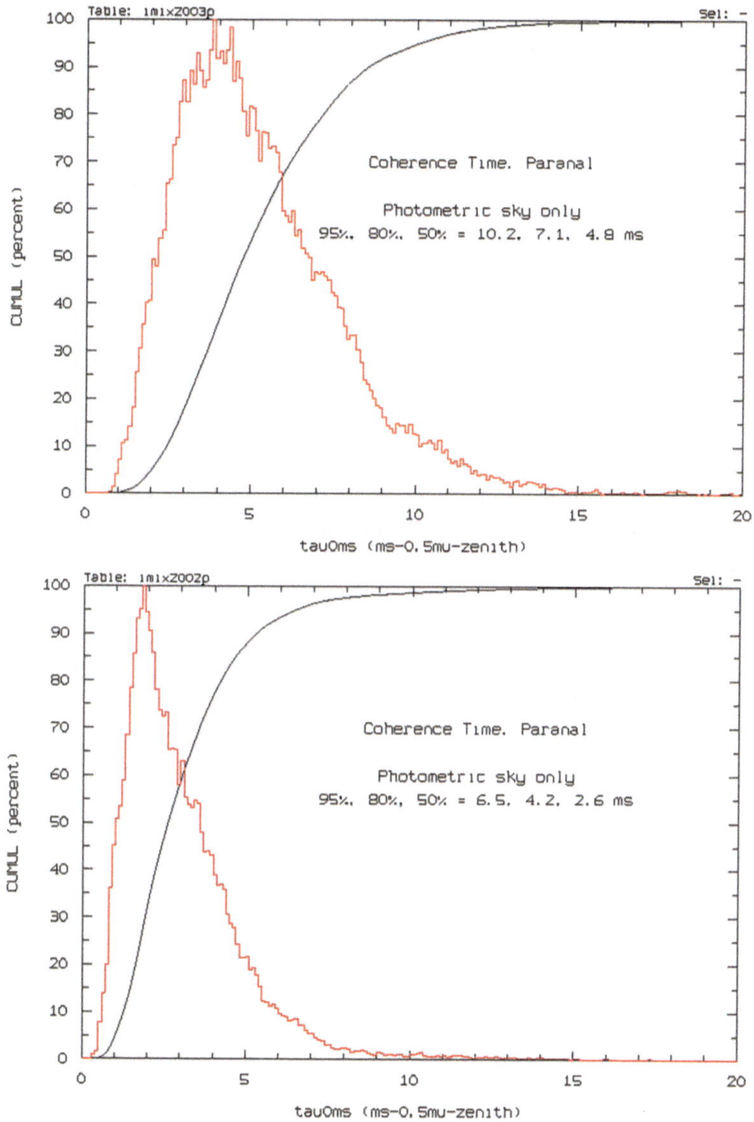

Fig. 1.3 Annual coherence time statistics for the Paranal observatory contrasting a very good year (2003, *top*) with a below average year (2002, *bottom*). The *red curve* shows the distribution of the coherence time (in ms), while the *black curve* is the cumulated distribution (Data from the ESO web site)

of significantly better angular resolution than the average λ/r_0, while others will be significantly worse than average. The idea behind Lucky Imaging is to pick the best images in a series with individual exposure times $t_{exp} \propto \tau_0$.

The probability P to obtain short exposure images with Strehl ratio ≥ 0.37 on a telescope of diameter D is:

$$P \approx 5.6 \exp[-0.1557 (D/r_0)^2] , \tag{1.3}$$

for $D/r_0 > 3.5$ [12]

As an example, let us consider a telescope with a $D = 2.2$ m primary mirror, and the atmospheric Fried parameter $r_0 = 0.35$ m. The probability to obtain short-exposure images with $SR \geq 0.37$ is then $P = 0.012$, i.e. 120 out of 10,000 images.

1.3 Lucky Imaging Precursors

Increasing from faint to bright light levels, the Human eye has a time resolution of 10–50 Hz. Behind a telescope of good optical quality with a magnification of 20–30, this makes for a powerful ancestor of modern Lucky Imaging techniques. Among the first recorded examples are Galileo Galilei's observations of the Sun in June and July 1613 (Fig. 1.4).

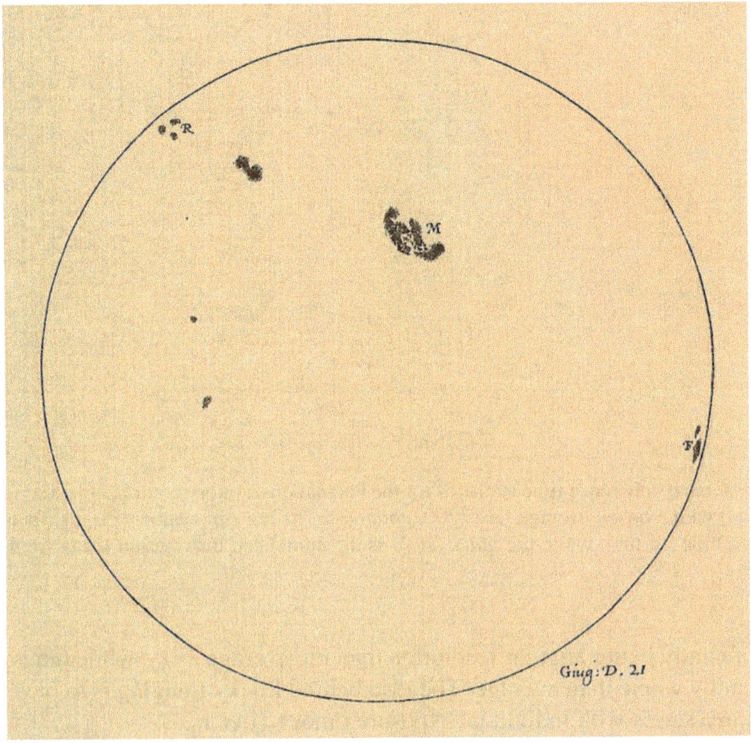

Fig. 1.4 The Sun on 21 June 1613 as recorded by Galileo Galilei

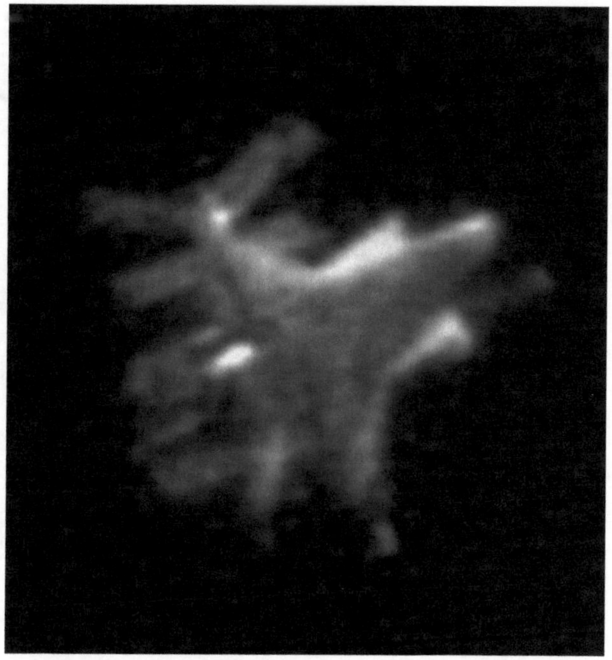

Fig. 1.5 Image of the visit of the space shuttle Atlantis to the space station MIR in 1996 as recovered by video astronomy based frame selection (Reproduced with permission from [8])

The theoretical foundation for Lucky Imaging was detailed by Fried in 1978 in a paper titled *Probability of getting a lucky short-exposure image through turbulence* [12]. The availability of affordable video cameras in the 1980s and 1990s with frame rates of 30 Hz brought the emergence of *Video astronomy* [28]. By using frame selection and combining the sharpest images, high resolution images of celestial objects and events became feasible. Examples are observations of the visit of space shuttle Atlantis to space station MIR in 1996 [8] (see Fig. 1.5), and high-resolution mapping of the surface of Mercury in 1998 [9], obtained by recording videos at frame rates of 30–60 Hz (using interlaced half-frames), and selecting, registering, and combining the sharpest (sub-sections of the) images.

Comparatively low quantum efficiency or high read noise restricted all these precursor techniques to relatively bright objects.

1.4 Technical Implementation

The early 2000s saw the emergence of *noise-free* array detector technology developed by Marconi Applied Technologies (now E2V). The devices are known as Low Light Level CCD (LLLCCD or L3CCD), and Electron Multiplication CCD (EM-CCD). L3CCDs were introduced in 2001 to the astronomical community in a paper discussing potential scientific applications [31].

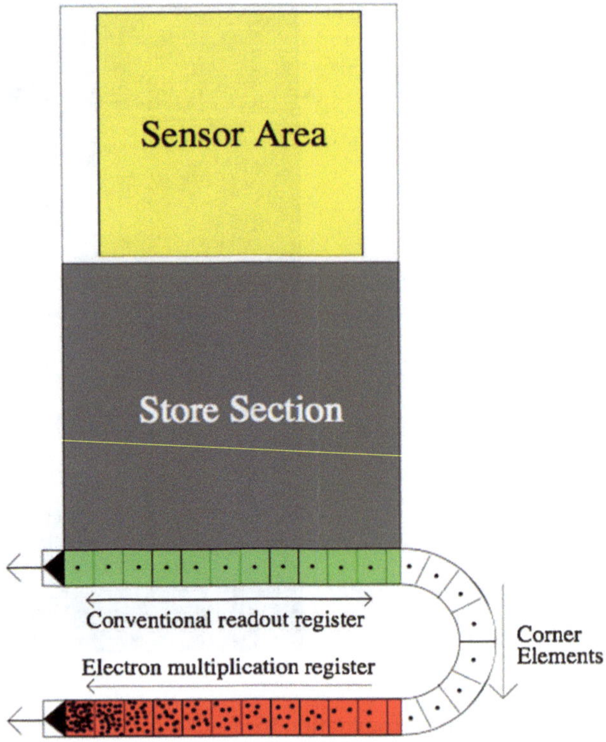

Fig. 1.6 Schematics of a frame transfer CCD with electron multiplication [16]

Figure 1.6 summarises the key features of an EM-CCD. Once an integration is complete, the entire frame is transferred to a storage section. A subsequent integration can start immediately, while the frame in storage is read out. The read-out can either be done using a conventional read-out register, or via an electron multiplication register. Individual electrons are "multiplied" by impact ionisation due to high clocking voltages in the electron multiplication register. The EM gain g for an EM register with s stages is $g = (1 + p)^s$, where p is the probability for "ionisation" by a single electron in one register. As an example for $p = 0.015$ and $s = 591$, we get an EM gain $g = 6629$. Figure 1.7 shows the required ionisation probability as a function of the desired EM gain for another implementation [31]. The high multiplication before read-out overcomes the conventional read-noise, resulting in an essentially *noise-free* detector (see Fig. 1.8).

Thanks to the extensive use of Commercial Off-The-Shelf technology (COTS), Lucky Imaging instruments like AstraLux Sur can be implemented and deployed on a short time scale. In the case of AstraLux Sur, the instrument was assembled within 6 months, which passed between approval of the observing project by the ESO Observing Programme Committee and the first light at the ESO NTT. In the following we briefly summarise the properties and components of the instrument (see Fig. 1.9).

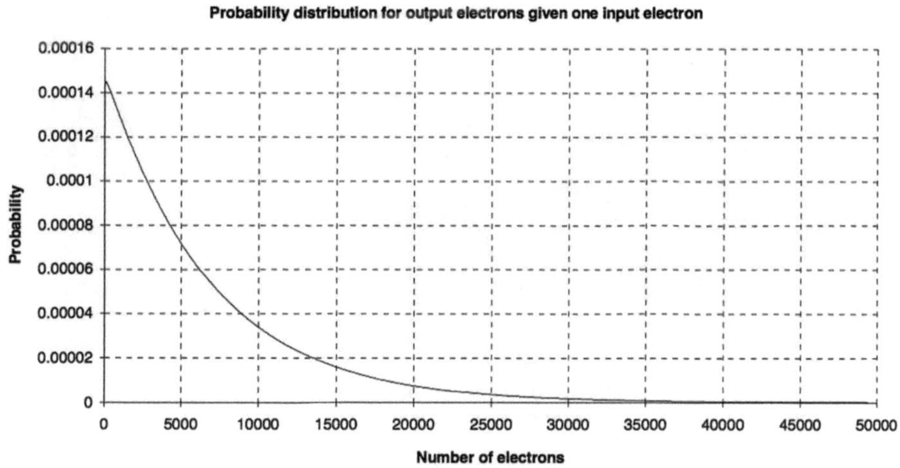

Fig. 1.7 Relation between desired Electron multiplication gain and required ionisation probability in an EM-CCD (Reproduced with permission from [31])

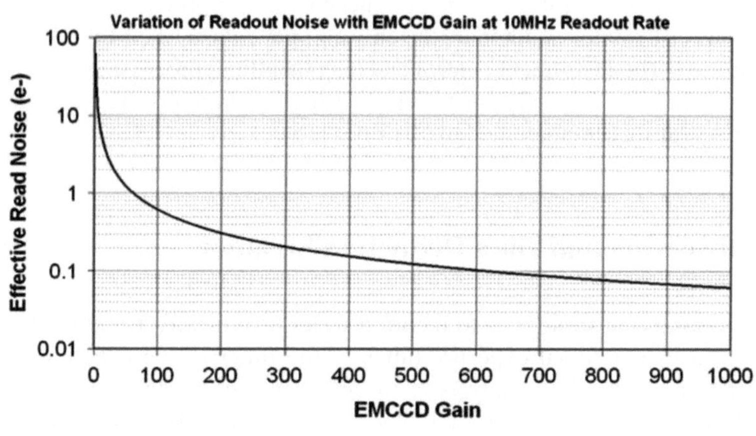

Fig. 1.8 Effective Readout Noise as a function of EM-CCD gain for an iXon+ DU-897 camera from Andor Technologies

AstraLux Sur is built around an iXon+ DU-897 camera from Andor Technologies. The back illuminated CCD with 512×512 pixel (16 μm pitch) has a quantum efficiency of QE > 90 % at 600 nm and >40 % at 900 nm. An air cooled Peltier lowers the CCD temperature to 190 K. Even lower temperatures are achievable by replacing the air cooling by water/glycol cooling, thus no liquid nitrogen is required for the operation. This greatly simplifies instrument design and operations. The camera has a full CCD frame rate of up to 34 Hz, and frame rates of several 100 Hz are possible using subarrays. The shortest exposure time is 20 μs. The cameras is equipped both with a conventional and an EM read-out register. EM-gain, read-out clock and voltages are adjustable.

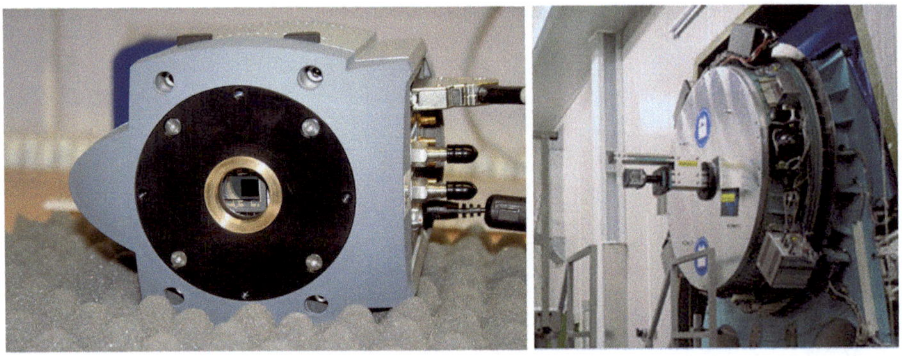

Fig. 1.9 AstraLux Sur: ANDOR iXon+ DU-897 camera (*left*), instrument attached to the adaptor rotator of the ESO 3.5 m NTT (*right*) (Reproduced with permission from [15])

A Barlow lens with ~3× magnification is used to match the pixel scale to the diffraction limit of the NTT in the *z*-band. Optical broad- and narrow-band filters are housed in an Oriel filter wheel. The mechanical support structure and adaptor plate to the NTT rotator-adaptor are the only components to be custom designed and manufactured. The instrument control and data handling, processing, and archiving is implemented in a local Gbit network with three computers. The total hardware costs for AstraLux Sur amounted to 50 kEuro [15].

1.5 Observing and Data Reduction Strategy

The basic observing strategy is to record a series of typically 10,000–20,000 short exposures with individual exposure times of the order of the coherence time τ_0. The first step in the data analysis is to determine the Strehl ratio or peak flux count of the brightest speckle for the object of interest in all images. In the case of observing extended objects like solar system planets, brightness contrast at object or illumination boundaries can also serve as a proxy to the Strehl ratio. Alternatively selection could be based on the properties of the power spectrum (see, e.g., [13]). In a second step the 1, 5, 10 %, etc. highest Strehl ratio images are registered and combined using the shift-and-add technique. The resulting Lucky Image has a Strehl ratio in the *I*-band of typically ~0.1, which is a factor of ~10 higher than the typical long exposure (see Table 1.1). The data volume created, processed, and archived within a 10 h night is typically 0.2–0.5 TB.

Figure 1.10 shows the triple system 2MASS J02490-1029 discovered with AstraLux Norte. The integrated brightness of the components is V ~ 14.5 mag. The observations were obtained in the *I*-band with individual integration times of 25 ms [20]. Intermediate and final results of the reduction steps are highlighted.

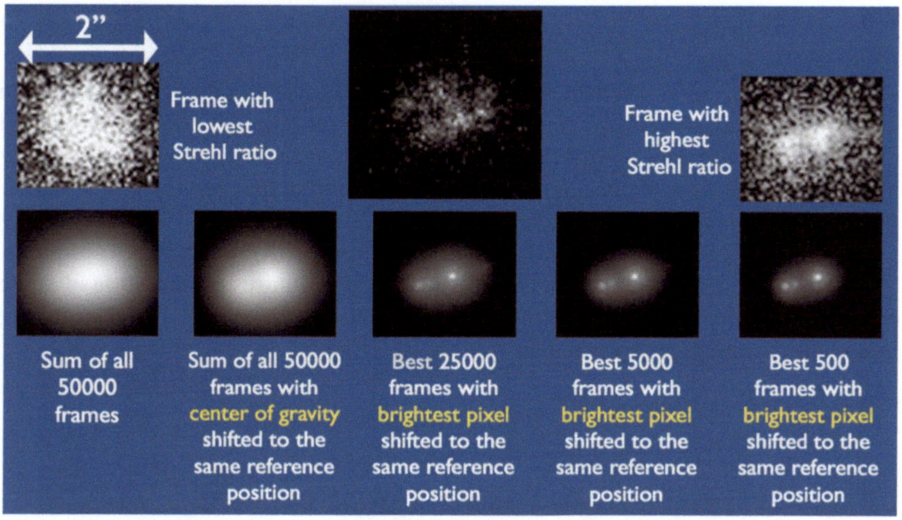

Fig. 1.10 Lucky Imaging view and data reduction strategy of the triple system 2MASS J02490-1029 discovered with AstraLux Norte at the CAHA 2.2-m telescope (Reproduced with permission from [20])

1.6 Lucky Imaging Compared to Other High-Angular Resolution Techniques

Astronomers have a large variety of passive and active high-angular resolution techniques at their disposal (see Table 1.2). While the passive techniques restrict themselves to simply recording data or monitoring events, the active techniques monitor wavefront aberrations and phase, and apply real time corrections to the optical light path. In the following we summarise the advantages and disadvantages of the various techniques.

Table 1.2 Overview of high-angular resolution techniques

Passive	Active
Simple shift-and-add	Tip-tilt stabilisation
Lucky Imaging	Higher order adaptive optics
Speckle interferometry	Interferometry with fringe tracking
Lunar occultation	...
...	

1.6.1 Passive Techniques

Shift-and-add requires the detection of several photons within the brightest speckle during one atmospheric coherence time [3]. As shift-and-add uses all individual exposures, it requires a brighter reference source than Lucky Imaging, which only uses the 1–10 % best images, and thus can afford individual exposure times up to 10× longer than the (average) coherence time [39].

Speckle interferometry is based on the Fourier analysis of the "specklegrams" of individual exposures. The image reconstruction thus requires preservation of some phase information [23, 34, 41]. Lucky Imaging has no such requirement, resulting in a fainter limiting magnitude than speckle interferometry.

1.6.2 Active Techniques

Tip-tilt stabilisation is capable of providing diffraction limited resolution only for telescope diameters $D \leq 3.4 \times r_0$ [32]. In contrast, Lucky Imaging can achieve diffraction limited resolution for telescope diameters $D \leq 7 \times r_0$ [38].

Higher-order Adaptive Optics, in general, samples telescope sub-pupils. For a Shack-Hartman wavefront sensor, the sub-aperture diameter equals r_0, i.e., is seven times smaller than for Lucky Imaging. Lucky Imaging has fainter limiting magnitude. We note that Pyramid wavefront sensors potentially gain up to 1–1.5 in terms of limiting magnitude, though in the optical this would require an extreme AO system with high-order corrections (see [38]).

1.6.3 Combining Passive and Active Techniques

In case partially or fully corrected adaptive optics (AO) observations of a target are feasible, the best results – in particular at short wavelengths – can be achieved by a combination of AO and Lucky Imaging.

Speckle stabilisation aims to identify the brightest speckle in real time and then stabilise it using a fast steering mirror. In terms of signal-to-noise ratio it is about a factor of three more time efficient than Lucky Imaging with 1 % image selection [22].

Adaptive Optics assisted Lucky Imaging includes higher order corrections going beyond tip-tilt or Speckle stabilisation, and is in particular successful in providing close-to-diffraction limited observations in the optical wavelength region [4, 26, 29, 30, 40].

1.6.4 Additional Advantages and Limitations of Lucky Imaging

While Lucky Imaging is capable of providing high angular resolution, its two main limitations are the achievable contrast and the limiting magnitudes towards the detection of faint sources (since only a fraction of the photons collected by the telescope passes the image selection process).

Lucky Imaging has also a number of technical and procedural advantages, which are relevant when considering project costs and time frames:

- low hardware complexity;
- low hardware cost (e.g., LuckyCam $\sim$ 20 k£. AstraLux Sur $\sim$ 50 kEuro);
- good limiting magnitude for reference source: $I_{\rm limit} \sim 16$ mag, which combined with the relatively large isoplanatic angle results in a sky coverage of about 25 % [38];
- fast implementation thanks to COTS components: AstraLux Norte took 12 months from project start to First light at CAHA 2.2-m, while AstraLux Sur took 6 months from approval by the Observing Programme Committee to First light at ESO's NTT.

> In summary, Lucky Imaging is capably of achieving a larger isoplanatic patch size of up to $60''$ in the I-band compared to single conjugated adaptive optics. Multi-conjugated AO can also deliver large corrected field sizes, though this comes at the cost of additional hardware complexity. Therefore "Lucky Imaging is (…) appealing for medium-sized astronomical observatories" [38].

1.7 Examples for Instrumentation and Science

Over the past decade Lucky Imaging cameras have been deployed at a large variety of small to mid-size telescopes (Fig. 1.11).

Some of the noteworthy Lucky Imaging instruments and the telescopes they have been operated with are summarised in the following[1]:

- LuckyCam at NOT, WHT, NTT, Palomar 5-m, …
- FastCam at 1.52-m TCS, NOT, WHT
- AstraLux Norte at CAHA 2.2-m
- AstraLux Sur at the ESO NTT
- Hungarian 1×1 k Andor camera at 1-m RCC

[1]This list is most likely incomplete.

Fig. 1.11 Lucky Imaging cameras: LuckyCam at the 2.5 m NOT (*left*) and AstraLux Norte at the CAHA 2.2-m (*right*)

- Adaptive Optics Lucky Imager at WHT.

LuckyCam was the first camera to use L3CCDs for Lucky Imaging. Starting from 2005, it has been operated at various telescopes, including NOT, WHT, NTT, and the Palomar 5-m Hale telescope. The initial hardware cost amounted to ~20 k£.

AstraLux Sur has been in operation at the ESO NTT since 2008. By 2015, eight different science programmes have been pursued distributed over 13 observing runs with a total of 54 telescope nights (including set-up nights).

Scientific programmes pursued with Lucky Imaging instruments range from companions to exoplanet host stars [1, 7, 11, 14, 27, 42], multiplicity study of M-dwarfs [2, 18–21, 25], the calibration of evolutionary and atmospheric models for very-low mass stars (Bergfors et al., in prep.), to orbital monitoring of binary stars [6, 17, 36, 43], multiplicity statistics of massive stars [33] or surveys for companions to white dwarfs.

1.8 Summary and Outlook

Lucky Imaging on its own is a powerful high-angular observing technique for mid- to large telescopes. Combined with adaptive optics it has the potential to provide diffraction limited observations in the visual wavelength range from the ground, matching – and potentially surpassing – the angular resolution of the Hubble Space Telescope [5, 13].

For more details on the principles of Lucky Imaging, instrumentation, and data analysis, we refer the reader to the theses by

- Tubbs on *Lucky Exposures: Diffraction Limited Astronomical Imaging through the Atmosphere* [38];

- Law on *Lucky Imaging: Diffraction-limited Astronomy from the Ground in the Visible* [24]; and
- Hormuth on *High Angular Resolution Astronomy with an Electron Multiplying CCD* [16].

References

1. Bergfors, C., Brandner, W., Daemgen, S., et al.: Stellar companions to exoplanet host stars: Lucky Imaging of transiting planet hosts. MNRAS **428**, 182 (2013)
2. Bergfors, C., Brandner, W., Janson, M., et al.: Lucky Imaging survey for southern M dwarf binaries. A&A **520**, A54 (2010)
3. Christou, J.C.: Image quality, tip-tilt correction, and shift-and-add infrared imaging. PASP **103**, 1040 (1991)
4. Crass, J., King, D., MacKay, C.: The AOLI low-order non-linear curvature wavefront sensor: laboratory and on-sky results. In: Society of Photo-Optical Instrumentation Engineers (SPIE) Conference Series, vol. 9148, p. 2. SPIE, Bellingham (2014)
5. Crass, J., Mackay, C., King, D., et al.: The adaptive optics Lucky Imager: diffraction limited imaging at visible wavelengths with large ground-based telescopes. In: American Astronomical Society Meeting Abstracts, vol. 225, p. #413.06. American Astronomical Society, Washington, DC (2015)
6. Csépány, G., Ábrahám, P., Regály, Z., Mező, G., Brandner, W., Hormuth, F.: Examining young stellar systems in birth by high angular resolution observations. Contrib. Astron. Obs. Skalnate Pleso **43**, 425 (2014)
7. Daemgen, S., Hormuth, F., Brandner, W., Bergfors, C., Janson, M., Hippler, S., Henning, T.: Binarity of transit host stars. Implications for planetary parameters. A&A **498**, 567 (2009)
8. Dantowitz, R.F.: Keeping an eye piece on satellites. Sky & Telesc. **92**(2), 86 (1996)
9. Dantowitz, R.F., Teare, S.W., Kozubal, M.J.: Ground-based high-resolution imaging of Mercury. AJ **119**, 2455 (2000)
10. Duquennoy, A., Mayor, M.: Multiplicity among solar-type stars in the solar neighbourhood. II – distribution of the orbital elements in an unbiased sample. A&A **248**, 485 (1991)
11. Faedi, F., Staley, T., Gómez Maqueo Chew, Y., Pollacco, D., Dhital, S., Barros, S.C.C., Skillen, I., Hebb, L., Mackay, C., Watson, C.A.: Lucky imaging of transiting planet host stars with LuckyCam. MNRAS **433**, 2097 (2013)
12. Fried, D.L.: Probability of getting a lucky short-exposure image through turbulence. J. Opt. Soc. Am. **68**(12), 1651 (1978)
13. Garrel, V., Guyon, O., Baudoz, P.: A highly efficient Lucky Imaging algorithm: image synthesis based on Fourier amplitude selection. PASP **124**, 861 (2012)
14. Ginski, C., Mugrauer, M., Neuhäuser, R.: Lucky imaging multiplicity studies of exoplanet host stars. Contrib. Astron. Obs. Skalnate Pleso **43**, 410 (2014)
15. Hippler, S., Bergfors, C., Brandner, W., Daemgen, S., Henning, T., Hormuth, F., Huber, A., Janson, M., Rochau, B., Rohloff, R.R., Wagner, K.: The AstraLux Sur Lucky Imaging instrument at the NTT. The Messenger **137**, 14 (2009)
16. Hormuth, F.: AstraLux: high angular resolution astronomy with an electron multiplying CCD, Ph.D. thesis, Faculty of Physics and Astronomy, University of Heidelberg (2007)
17. Hormuth, F., Brandner, W., Hippler, S., Janson, M., Henning, T.: Direct imaging of the young spectroscopic binary HD 160934. A&A **463**, 707 (2007)
18. Janson, M., Bergfors, C., Brandner, W., Bonnefoy, M., Schlieder, J., Köhler, R., Hormuth, F., Henning, T., Hippler, S.: Orbital monitoring of the AstraLux large M-dwarf multiplicity sample. ApJ Suppl. **214**, 17 (2014)

19. Janson, M., Bergfors, C., Brandner, W., Kudryavtseva, N., Hormuth, F., Hippler, S., Henning, T.: The AstraLux multiplicity survey: extension to late M-dwarfs. ApJ **789**, 102 (2014)
20. Janson, M., Hormuth, F., Bergfors, C., Brandner, W., Hippler, S., Daemgen, S., Kudryavtseva, N., Schmalzl, E., Schnupp, C., Henning, T.: The AstraLux large M-dwarf multiplicity survey. ApJ **754**, 44 (2012)
21. Jódar, E., Pérez-Garrido, A., Díaz-Sánchez, A., Villó, I., Rebolo, R., Pérez-Prieto, J.A.: New companions to nearby low-mass stars. MNRAS **429**, 859 (2013)
22. Keremedjiev, M., Eikenberry, S.S.: A comparison between Lucky Imaging and speckle stabilization for astronomical imaging. PASP **123**, 213 (2011)
23. Labeyrie, A.: Attainment of diffraction limited resolution in large telescopes by Fourier analysing speckle patterns in star images. A&A **6**, 85 (1970)
24. Law, N.M.: Lucky imaging: diffraction-limited astronomy from the ground in the visible, Ph.D. thesis, St. Johns College, University of Cambridge (2006)
25. Law, N.M., Hodgkin, S.T., Mackay, C.D.: Discovery of five very low mass close binaries, resolved in the visible with lucky imaging*. MNRAS **368**, 1917 (2006)
26. Law, N.M., Mackay, C.D., Dekany, R.G., Ireland, M., Lloyd, J.P., Moore, A.M., Robertson, J.G., Tuthill, P., Woodruff, H.C.: Getting lucky with adaptive optics: fast adaptive optics image selection in the visible with a large telescope. ApJ **692**, 924 (2009)
27. Lillo-Box, J., Barrado, D., Bouy, H.: High-resolution imaging of Kepler planet host candidates. A comprehensive comparison of different techniques. A&A **566**, A103 (2014)
28. MacFarlane, A.W.: A primer for video astronomy. Sky & Telesc. **79**(2), 226 (1990)
29. Mackay, C.: High-efficiency lucky imaging. MNRAS **432**, 702 (2013)
30. MacKay, C., Rebolo, R., Crass, J., et al.: High-resolution imaging in the visible on large ground-based telescopes. In: Society Of Photo-Optical Instrumentation Engineers (SPIE) Conference Series, vol. 9147, p. 1. SPIE, Bellingham (2014)
31. Mackay, C.D., Tubbs, R.N., Bell, R., Burt, D.J., Jerram, P., Moody, I.: Subelectron read noise at MHz pixel rates. In: Blouke, M.M., Canosa, J., Sampat, N. (eds.) Sensors and Camera Systems for Scientific, Industrial, and Digital Photography Applications II. Society of Photo-Optical Instrumentation Engineers (SPIE) Conference Series, vol. 4306, pp. 289. SPIE, Bellingham (2001)
32. Noll, R.J.: Zernike polynomials and atmospheric turbulence. J. Opt. Soc. Am. **66**, 207 (1976)
33. Peter, D., Feldt, M., Henning, T., Hormuth, F.: Massive binaries in the Cepheus OB2/3 region. Constraining the formation mechanism of massive stars. A&A **538**, A74 (2012)
34. Roddier, F.: Interferometric imaging in optical astronomy, Physics Reports **170**, 97 (1988)
35. Ross, T.S.: Limitations and applicability of the maréchal approximation. Appl. Opt. **48**(10), 1812 (2009)
36. Scardia, M., Argyle, R.W., Prieur, J.L., Pansecchi, L., Basso, S., Law, N.M., Mackay, C.D.: The orbit of the visual binary ADS 8630 (γ Vir). Astronomische Nachrichten **328**, 146 (2007)
37. Taylor, G.I.: The spectrum of turbulence. Proc. R. Soc. Lond. **164**, 476 (1938)
38. Tubbs, R.N.: Lucky exposures: diffraction limited astronomical imaging through the atmosphere, Ph.D. thesis, St. Johns College, University of Cambridge (2003)
39. Tubbs, R.N., Baldwin, J.E., Mackay, C.D., Cox, G.C.: Diffraction-limited CCD imaging with faint reference stars. A&A **387**, L21 (2002)
40. Velasco, S., Rebolo, R., Mackay, C., et al.: Adaptive optics and lucky imager (AOLI): presentation and first light. Highlights Span. Astrophys. **VIII**(8), 850 (2015)
41. Weigelt, G.P.: Modified astronomical speckle interferometry 'speckle masking'. Opt. Commun. **21**, 55 (1977)
42. Wöllert, M., Brandner, W., Bergfors, C., Henning, T.: A Lucky Imaging search for stellar companions to transiting planet host stars. A&A **575**, A23 (2015)
43. Wöllert, M., Brandner, W., Reffert, S., Schlieder, J.E., Mohler-Fischer, M., Köhler, R., Henning, T.: The young binary HD 102077: orbit, spectral type, kinematics, and moving group membership. A&A **564**, A10 (2014)

Chapter 2
Adaptive Optics in High-Contrast Imaging

Julien Milli, Dimitri Mawet, David Mouillet, Markus Kasper, and Julien H. Girard

2.1 Introduction

2.1.1 Science Case

While most confirmed exoplanets were discovered by indirect techniques such as radial velocities or transits, adaptive optics (AO) assisted direct imaging is a very rich and complementary method that can reveal the orbital motion of the planet, the spectrophotometry of its atmosphere but also the architecture and properties of its circumstellar environment. It can unveil possible interactions with a disc, whether a proto-planetary disc in case of on-going planetary accretion, or a debris disc for more evolved, gas-poor systems. From a statistical point of view, it probes a region, in the mass versus semi-major axis discovery space, different from other techniques, as illustrated in Fig. 2.1. Reaching a uniform sampling of such a parameter space is essential to derive the frequency of planets as a function of mass and semi-major axis, and therefore constrain the planet formation mechanisms. For instance, theories of planet formation predict a higher efficiency of giant planet

J. Milli (✉) • J.H. Girard
ESO, Alonso de Córdova 3107, Vitacura, Santiago, Chile
e-mail: jmilli@eso.org

D. Mawet
Caltech, 1200 E. California Blvd, Pasadena, CA 91125, USA
e-mail: dmawet@astro.caltech.edu

D. Mouillet
IPAG, BP 53, F-38041, Grenoble Cédex 9, France
e-mail: david.mouillet@obs-ujf.grenoble.fr

M. Kasper
ESO, Karl-Schwarzschild-Straße 2, 85748 Garching, Germany
e-mail: mkasper@eso.org

H.M.J. Boffin et al. (eds.), *Astronomy at High Angular Resolution*, Astrophysics and Space Science Library 439, DOI 10.1007/978-3-319-39739-9_2

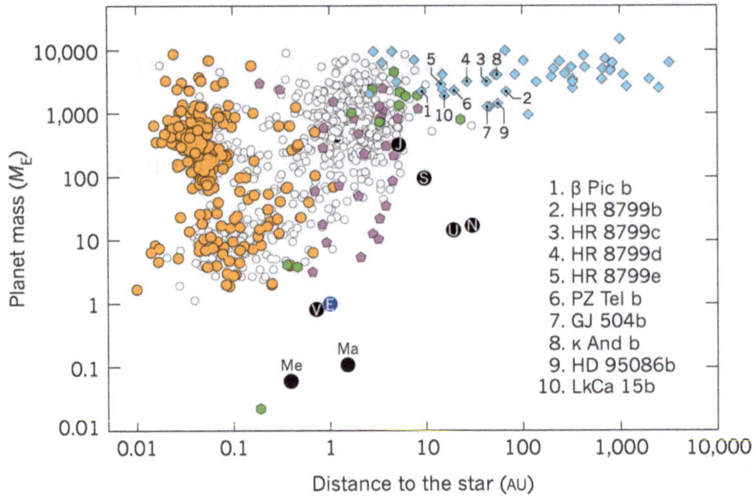

Fig. 2.1 Mass (in units of the Earth's mass) as a function of semi-major axis for solar system planets (*filled black circles*) and detected extrasolar planets (*other symbols*). The colours indicate the detection technique: *light blue* for direct imaging, *white* for radial velocities, *yellow* for transits (planets with measured mass), *pink* for microlensing, *green* for pulsation timing. The 10 labelled planets are giant planets detected with AO-assisted direct imaging within 100 au of their host star and with a mass ratio to their host star below 0.02 (Reprinted by permission from Macmillan Publishers Ltd: Nature [49], (c) 2014)

formation close to the snow line, where radial velocity and transit techniques are poorly sensitive. They require additional ingredients such as migrations and orbital instabilities to explain the current view depicted in Fig. 2.1.

2.1.2 *Requirements*

Imaging extrasolar planets and discs requires dedicated instruments and strategies to overcome two main challenges:

1. the tiny angular separation between the star and the planet or disc. The projected separation is below $0.1''$ for a planet orbiting at 10 au from a star distant of 100 pc.
2. the contrast between a star and its planet ranges between 10^{-4} for a young giant planet to 10^{-10} for an Earth shining in reflected light. Discs are also very tenuous, with contrasts[1] ranging from 10^{-4} for the brightest debris discs down to 10^{-10} for an analog of our zodiacal belt at 10 pc [63].

[1]For extended structures, the contrast is defined per resolution element.

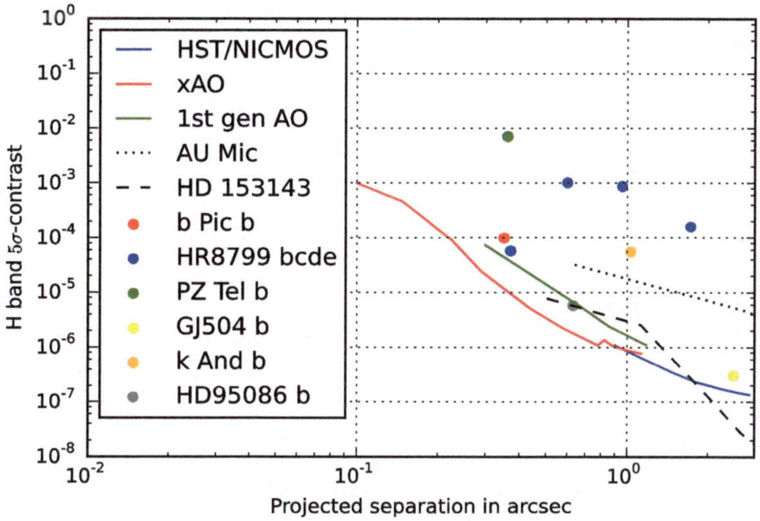

Fig. 2.2 Typical contrast obtained on sky in the H-band from the first generation of AO systems (NaCo, NIRC2, NICI, HiCIAO, *green curve*), the second generation of extreme AO system (SPHERE/GPI, *red curve*), compared to the space-based instrument HST/NICMOS (*blue curve*). These detection curves are at 5σ for a typical 30 min observations. We overplotted the typical contrast of confirmed exoplanets as well as two disc contrast expressed per resolution element (*black dotted and dashed curves*)

These two requirements are summarised in Fig. 2.2 that shows the separation and contrast of a few planets detected in direct imaging (dots), and two debris disc surface brightnesses (black lines). These two requirements are indissociable. From the ground, adaptive optics is one answer to the first requirement. Other techniques exist[2] such as speckle imaging [26], Lucky Imaging [29], sparse aperture masking [4], interferometry [30], but AO-assisted imaging is currently the only option to reach both the diffraction limit of the telescope and provide a contrast below 10^{-6} at a few resolution elements. The size of a resolution element, e.g., the angular resolution, is given by the size of the telescope pupil. For of a circular aperture, the point spread function (hereafter PSF) is an Airy function of full width at half maximum $\sim \lambda/D$ where D is the diameter of the telescope and λ the wavelength. The first three lines of Table 2.1 summarise the angular resolution of an 8-m diffraction-limited telescope (in milli-arcseconds, mas) in the main optical and near-infrared filters. In the optical, this corresponds to the angular diameter of the more massive nearby stars.

To reach the contrast requirements, AO alone is however not sufficient because residual starlight still contaminates the region of interest within 1″, as illustrated in Fig. 2.3. These residuals come from both the diffracted light of the telescope

[2] Some of them being covered in this book.

Table 2.1 Trade-off between angular resolution, AO performance and sensitivity for planets in the different optical and near-infrared filters

Band		V	R	I	J	H	Ks	Lp
λ (μm)		0.55	0.65	0.82	1.22	1.63	2.2	3.8
Angular resolution (mas)		14	17	21	31	42	57	98
First	Typical Strehl ratio[a] (%)	< 5	< 5	< 5	5	19	40	73
gen.	Typical contrast[b] at 0.5″ ($\times 10^{-4}$)	N/A	N/A	N/A	1.6	1.4	1.0	0.44
AO[a]	Corresponding sensitivity[c] (M_{Jup})	N/A	N/A	N/A	14/43	12/37	10/32	5/12
	Typical Strehl ratio[a] (%)	26	38	55	76	86	92	97
xAO[d]	Typical contrast[e] at 0.5″ ($\times 10^{-5}$)	5	4	3	1.7	1	0.5	0.2
	Corresponding sensitivity[c] (M_{Jup})	N/A	N/A	N/A	6/13	5/12	4/11	2/7

[a] 40 % Strehl ratio (SR; see Sect. 2.3.1) was assumed in K-band, and the Strehl ratio scales as $SR_{\lambda_2} = SR_{\lambda_1}^{(\lambda_1/\lambda_2)^2}$

[b] A contrast of 1×10^{-4} was assumed in H-band. The scaling in wavelength follows Eq. 2.2

[c] The two values refer respectively to a 10 and 100 Myr-old self-luminous planet, orbiting an A0V star. The luminosity to mass conversion used the AMES-COND evolutionary tracks [5]

[d] 92 % Strehl ratio was assumed in K-band

[e] 1×10^{-5} was assumed in H-band

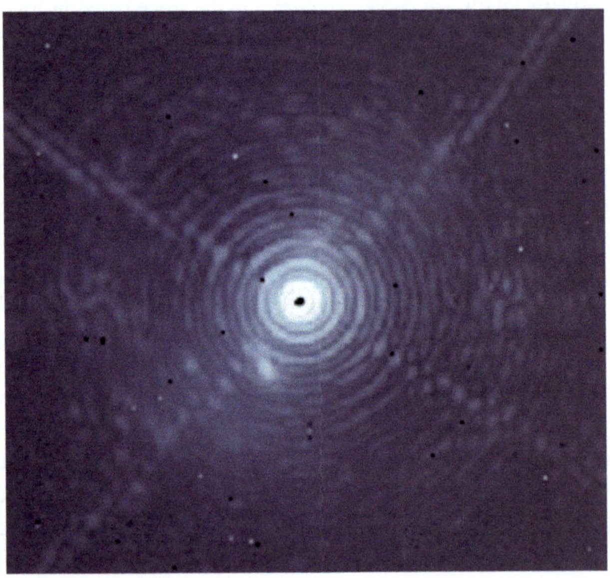

Fig. 2.3 Typical high-Strehl ratio (∼90 %) raw PSF, as seen from the real-time display during the standard operations of the xAO instrument VLT/SPHERE on a very bright star in the $H2$-band. The Airy rings are clearly visible, along with the diffraction spikes of the spiders of the telescope and many bright speckles, usually pinned to the diffraction pattern

entrance pupil, and the residual wavefront error due to uncorrected atmospheric perturbations and imperfect optics within the telescope and instrument. At four resolution elements, the Airy pattern still reaches an intensity of 3×10^{-4} the peak value. Therefore, detecting a signal at this level of contrast without any other high-contrast technique is highly ineffective because significant time must be spent to get enough signal. One must rely on additional diffraction light suppression technique, such as coronagraphy. The second source of residual starlight is more difficult to address. Wavefront errors create speckles in the image plane and mimic point-sources, degrading the contrast [52]. Unlike the diffraction pattern, speckles vary temporally on different timescales: from a fraction of milliseconds for non-corrected atmospheric speckles to hours or days for quasi-static speckles slowly variable with temperature, mechanical flexions or the rotation of optical parts. Moreover, they can interfere with the diffraction pattern of the pupil and be reinforced to create pinned speckles [1]. Adequate observing strategies and post-processing techniques based on differential imaging can remove the remaining starlight residual to the necessary level. Although coronagraphy and differential imaging are not the focus of this review, they put stringent constraints on the AO system that will be discussed here.

2.1.3 From Pioneering Adaptive Optics Experiments to Extreme Adaptive Optics System

AO developments have seen a tremendous progress over the past 30 years, from the first experiments to the most advanced systems now in operations and known as extreme AO systems. The concept of compensating astronomical seeing was first proposed by Babcock in 1953 [3]. Military research further developed the concept, and the first prototype for astronomical observations, called COME-ON was installed in 1989 at the ESO La Silla Observatory [60]. ADONIS, an upgraded version of COME-ON and COME-ON+ became the first user-facility instrument equipped with an AO system [7]. Ten years later, VLT/Naos and Keck-AO were the workhorses of the first generation of adaptive optics systems, integrating the lessons learnt from the first pioneering experiments. Combined to the science camera Conica and NIRC2, respectively, these instruments led to many science breakthroughs, such as the first discovery of an exoplanet with the VLT in 2004 [9], at a projected separation of 0.8″, or the discovery of a system of four giant planets orbiting the star HR 8799 in 2008 [37]. In parallel, the design choices of a second generation of AO systems were made, and these extreme AO (xAO) instruments saw their first light in the past few years. The high contrast requirements set new constraints on the AO systems, in order to control the wavefront to an exquisite level, to feed high-rejection coronagraph and to allow differential imaging and maintain the temporal evolution of aberrations as slow as possible. These new systems benefited from the lessons learnt from the first generations of instruments and will in turn provide valuable feedback for the on-going design of AO systems for extremely large telescopes. The first xAO systems on sky were the PALM-3000/P1640 [13, 47], followed by MagAO [12]. Now three systems had their first light in the last two years: GPI at Gemini South [31], SCExAO at Subaru [18] and SPHERE at the VLT [6]. Table 2.2 summarises the main instruments and AO systems contributing or having contributed to the field. The design of those

Table 2.2 Main instruments benefiting from AO and high-contrast imaging capabilities. This non exhaustive list groups the instruments by generation

Instrument	Telescope	Wavelength (μm)	Operations[a]
ADONIS	La Silla 3.6	1–5	1996–2001
PUEO	CFHT	0.7–2.5	1996–2013
NaCo	VLT	1–5	2002
Lyot Project	AEOS	0.8–2.5	2003–2007
ALTAIR-NIRI	Gemini N.	1.1–2.5	2003
NIRC2	Keck	1–5	2004
NICI	Gemini S.	1.1–2.5	2007
HiCIAO	Subaru	1.1–2.5	2009
PALM-3000/P1640	Palomar 200″	1.1–1.65	2009
FLAO/LMIRCam	LBT	3–5	2012
GPI	Gemini S.	1.0–2.3	2013
MagAO/VisAO	Clay	0.5–5	2014
SPHERE	VLT	0.5–2.3	2014
SCExAO	Subaru	0.5–2.2	2015

[a] Instruments without end date are still in operation in 2015

instruments are a trade-off between angular resolution, AO performance and planet sensitivity, as illustrated by the last rows of Table 2.1, that compare the typical AO performance, contrast and sensitivity for both first generation AO systems and xAO systems in different filters.

2.2 Fundamentals of High-Contrast Adaptive Optics Systems

2.2.1 Characteristics of Images Distorted by the Atmospheric Turbulence

The limit of angular resolution set by the atmospheric turbulence in the absence of AO correction is λ/r_0 where r_0 is the Fried parameter, scaling as $\lambda^{6/5}$. To evaluate the level of performance of an AO system, the correlation time, also known as the Greenwood time delay, is the most relevant parameter. It is defined as $\tau_0 = 0.314\ r_0/v$ where v is the mean wind speed weighted by the turbulence profile along the line of sight [55]. This parameter sets the speed at which an AO system has to react to correct for the atmospheric turbulence. It is also proportional to $\lambda^{6/5}$, therefore correcting in the near-infrared is easier than in the visible where the turbulence evolves faster.

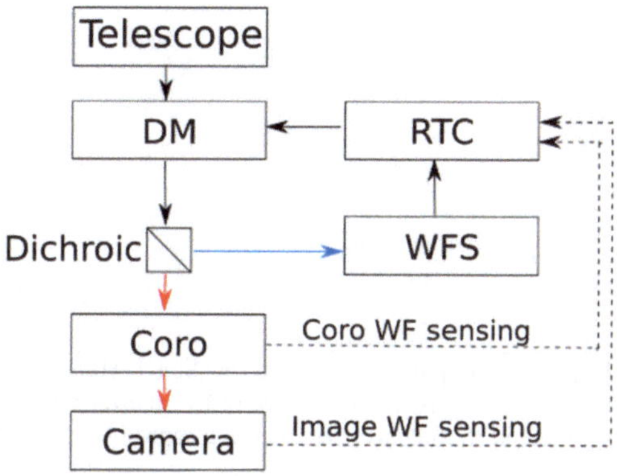

Fig. 2.4 Architecture of a standard AO system (*plain lines and rectangles*). The *dashed lines* represent the additional features implemented in xAO systems (described in Sect. 2.3.3). *Coro* indicates a coronagraph

2.2.1.1 Architecture of an AO System

We review here briefly the architecture and key parameters of an AO system before presenting the specific constraints set by high-contrast observations. We refer the reader to the review by Roddier [57] for further details on general AO systems.

The architecture of the first generations of AO systems is composed of three key elements represented in Fig. 2.4:

- a wavefront sensor (WFS), whose role is to measure the optical disturbance in quasi real time; the measurements are sent to
- a real-time controller (RTC) that reconstructs the wavefront and sends the command to
- a deformable mirror (DM) that corrects for the distorted wavefront.

2.2.2 *Wavefront Sensing*

The wavefront sensor is the element measuring in real time the distorsion of the wavefront. Spatial resolution, speed and sensitivity are the three parameters driving the design. An overall review of WFS is given in [59]. In the visible and near-infrared, there exists no sensor to measure directly the phase, therefore it must be encoded in intensity variations, and different techniques have been developed. The main challenge consists in getting enough spatial resolution to drive as many modes as the deformable mirror can correct, as fast as the turbulence evolves and without amplifying and propagating the measurement noise. The fundamental limit is set by photon noise.

The two most common implementations measure the slope of the wavefront. For the Shack-Hartmann WFS (SHWFS), a lenslet array placed in a conjugated pupil plane samples the incoming wavefront (Fig. 2.5 left). Each lenslet creates an image of the source, called a spot, at its focus. In presence of a non-planar incident wave,

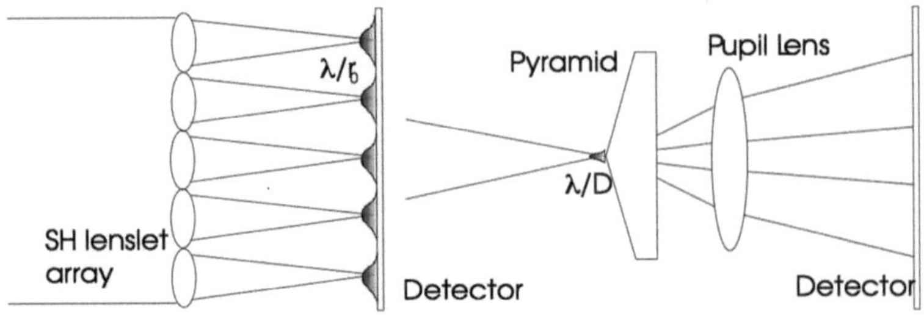

Fig. 2.5 Comparison between the Shack-Hartmann (*left*) and pyramid (*right*) WFS, illustrating the highest sensitivity of low order aberrations of the pyramid WFS when a tilt by λ/D is introduced in the wavefront (Figure reproduced with permission from [54])

the lenslet receives a tilted wavefront and the spot is shifted. Therefore, measuring the spot displacement enables to derive the local slope of the wavefront in each lenslet. This requires many pixels to locate accurately the center position of each spot, these are subsequently operating in a low flux regime which requires high sensitivity- and low noise detectors.[3] Because this is a relative measurement, it requires also an accurate and regular calibration with a flat wavefront, to avoid any drift. It is a well-proven and mature technology, implemented both by Gemini/GPI and VLT/SPHERE. The pyramid wavefront sensor (PWFS) is a more recent development proposed in 1995 [53]. A pyramidal prism is inserted in a focal plane. Each face of the prism deflects the light in a different direction and a lens relay conjugates the four apparent exit pupils onto four pupil images on the detector (Fig. 2.5 right). The prism can be fixed or oscillating. The measured flux in each quadrant can be related to the slope of the wavefront. The PWFS is more sensitive to low-order modes than the SHWFS [54]. This drawback of the SHWFS is of critical importance because many high-rejection coronagraphs require an excellent correction of low-order aberrations to provide a high contrast. The PWFS is also less prone to aliasing effects than the SHWFS, although this can be mitigated by the use of an adjustable spatial filter, as in Gemini/GPI and VLT/SPHERE. The curvature WFS (CWFS) uses two out of focus measurements to derive the curvature (second derivative) of the wavefront [56]. It can be implemented by an oscillating membrane. The error propagation is worse than for the SHWFS for equivalent photon noise properties, which is why it is not commonly used on high-order AO systems.

The Zernike phase contrast WFS implements an idea developed by Zernike to convert the phase variation in the entrance pupil in an intensity variation in a re-imaged pupil by inducing a phase shift of $\pi/2$ over a diffraction-limited spot in the focal plane [8]. It is very attractive because it directly converts the phase into intensity instead of measuring the first or secondary derivative of the phase as in the techniques described above, reducing therefore the computation cost and the error

[3]Hence the use of Electron Multiplying Charge Coupled Device (EM-CCD), for instance in VLT/SPHERE.

propagation. Moreover, the SHWFS is insensitive to certain phase aberration pattern (waffle mode, differential piston), whereas the ZWFS is free from these artefacts.

2.2.3 Deformable Mirror Technologies

The correction of the wavefront is generally achieved in two or more stages. The tip/tilt mirror corrects for overall shifts of the PSF due either to atmospheric variations, wind or vibrations in the instrument. The correction of higher-order modes of the turbulence is done with a deformable mirror (DM), having up to several thousands actuators. The design of a DM is a trade-off between fast response, density of the actuators, and amplitude and accuracy of their stroke. They are several technologies available, reviewed in [32]. Some systems even combine different technologies in multi-stage DMs.

Stacked array DMs are made of discrete actuators in ferroelectric material (either piezoelectric or electrostrictive). Lead-zirconate titanate (PZT) and lead magnesium niobate (PMN) are the most common form. When an electric field is applied, it elongates in the direction of the field (longitudinal effect). Although they require high voltages, they provide a large stroke, a high stiffness, reliability and accuracy. This technology was implemented in SPHERE. To increase further the density of actuators or reduce the DM size, MEMS (Micro-Electro Mechanical Systems) are an appealing alternative that use a thin mirror membrane attached to an intermediate flexible support actuated by electrostatic or electromagnetic fields. The use of surface micro-machining technologies to etch the electrodes enables to make very compact, high-density and cost-effective DMs. This solution was preferred for Gemini/GPI.

Bimorph DMs exploit the transverse effect of a piezoelectric material. Only two wafers of piezoelectric materials separated by an electrode are needed, which makes manufacturing easier than stacked arrays. They also need high voltages but provide large stroke at a reasonable price. At the VLT, the instruments SINFONI, CRIRES and the Unit Telescopes in interferometric mode all implement this technology. To provide the large strokes required to drive adaptive secondary mirrors (ASM), voice coil is the preferred solution. The actuators are made of a dense array of voice calls that create a magnetic field to drive the magnets attached to the mirror thin shell. It provides a fast response but dissipates much heat. It currently equips the LBT and MagAO.

2.3 The Transition to Extreme AO Systems

Extreme AO systems is an evolution of the concept of single-conjugated AO systems to reach the requirements described in Sect. 2.1.2 for the detection of exoplanets: high contrast at short separations below $1''$. This translates into a requirement for

Table 2.3 Extreme AO characteristics [13, 24]

Instruments	DM technology	WFS	Frame rate (kHz)
PALM-3000/P1640	Electrostrictive, 241 + 3388 act.	SHWFS	2
FLAO/LMIRCam	Voice coil 672 act.	PWFS	0.9
MagAO	Voice coil 585 act.	PWFS	1
GPI	Piezoelectric 100 act. + MEMS 1500 act.	SHWFS	1.2
SPHERE	Piezoelectric 1377 act.	SHWFS	1.2
SCExAO	Bimorph 188 act. + MEMS 1000 act.	PWFS	1.5 to 3.6

very low wavefront errors, described in Sect. 2.3.1. Two approaches are explored in parallel. On the one hand, a lot of effort is devoted to measure and control the wavefront to an exquisite level. To do so, xAO systems use advanced sensing schemes combined with fast, high-density DMs to correct for more than a thousand modes (Sects. 2.3.2 and 2.3.3), filtered by the real-time calculator to match the WFS sensitivity and DM response. Table 2.3 summarises the main DM and WFS technologies implemented in current xAO systems. On the other hand, a calibration of the static and slowly variable aberrations (Sect. 2.3.4) is implemented through specific observing strategies and data reduction algorithms.

2.3.1 Wavefront Error Requirement for High Contrast

The performance of an AO system is quantified using the Strehl ratio, *SR*, which is the ratio of the peak on-axis intensity of an aberrated wave, to that of a reference unaberrated wave. It can be approximated by the Marechal expression [33, 35]:

$$SR = e^{-\sigma_\phi^2} \tag{2.1}$$

where σ_ϕ^2 is the variance of the phase aberration across the pupil and can be decomposed as $\sigma_\phi = \frac{2\pi\delta}{\lambda}$ with δ the root-mean-square (rms) wavefront error in nm. As an example, for a wavefront error of 1 rad rms, which corresponds to $\delta = 100$ nm at $\lambda = 630$ nm, the Strehl ratio is 37 %, meaning that 37 % of the PSF energy is in the core of the PSF. The fraction of the PSF energy which is not controlled is critical in high-contrast imaging. Extreme AO systems are systems able to achieve Strehl ratios above 90 %, while standard first generation AO systems typically reach 40–60 % at 1.6 μm, corresponding to ~50 nm rms in the first case and 200 nm rms in the second case. Not only is the total variance of the phase σ_ϕ^2 critical but the structure of the wavefront error is also important. Wavefront errors at low spatial frequencies need to be controlled at the best level because they correspond to short-separation aberrations in the focal plane.

As an illustration, let us consider a single mode of the aberrated wavefront in the form of a pure sinusoidal wave wavefront error of amplitude *h* (expressed in m) and spatial frequency f. It can be shown that if the amplitude of the aberration is small with respect to the wavelength, e.g. $h \ll \lambda$, such a mode creates two symmetric replicas of the central PSF at an angular distance $f\lambda$ from the central PSF and with a brightness ratio of $\left(\frac{\pi h}{\lambda}\right)^2$ [17]. As an application, with $h = \frac{\lambda}{3140}$, the contrast between the PSF replicas and the central PSF is already 10^{-6}.

In practice, the spatial structure of the wavefront error, that is, its power density function, is unknown and σ_ϕ^2 corresponds to an average over many incoherent modes of the aberrated wavefront. Relating the level of aberration σ_ϕ^2 to the science requirements of exoplanet detection, e.g. the contrast, is not straightforward. Serabyn et al. [64] provide a rough estimate of the level of contrast reached in the well-corrected area of a high-order deformable mirror:

$$C = \frac{1 - SR}{N_{act}} \tag{2.2}$$

where N_{act} is the number of actuators of the deformable mirror. Combining Eq. 2.2 with Eq. 2.1 yields the following expression for the raw contrast as a function of the rms wavefront error:

$$C = \frac{1 - e^{-\left(\frac{2\pi\delta}{\lambda}\right)^2}}{N_{act}} \tag{2.3}$$

It is represented in Fig. 2.6 for typical wavelengths in the visible and near-infrared. Because the rms of the phase error σ_ϕ scales with λ/D, the control of the wavefront error is less critical for AO systems working in the near-infrared than in the visible.

In AO-assisted imagers with fast WFS, the major contributor to the error in the wavefront correction is the fitting error. It is due to the finite number of actuators of the DM, and scales with $N_{act}^{-5/6}\left(\frac{D}{r_0}\right)^{5/3}$. The cutoff frequency which corresponds to the largest spatial frequency that can be corrected by the DM, is given by $f_c = \frac{N\lambda}{2D}$ where N is the number of actuators on a diameter. In high contrast imaging, the goal is to reach the deepest in-band contrast, within the well-corrected area of the DM, i.e. for $f < f_c$. As detailed in [24] and [17], the breakdown of the error within this well-corrected region is strongly dependent on the design choices of the AO system, especially the WFS concerning the associated photon noise. We provide in Fig. 2.7 the typical error budget expressed in contrast as a function of separation for a 30-m telescope. Servolag is the significant contributor between 0.05″ and 0.3″. It is due to the finite temporal bandwidth of the AO, limited by the frequency of the WFS. It scales with $\left(\frac{\tau}{\tau_0}\right)^{5/3}$ where τ is the time lag in the AO loop and τ_0 the atmospheric correlation time. Last, at very short separations below a few resolution elements, the chromaticity of the optical path length difference and the amplitude aberrations induced by scintillation start to dominate the error budget.

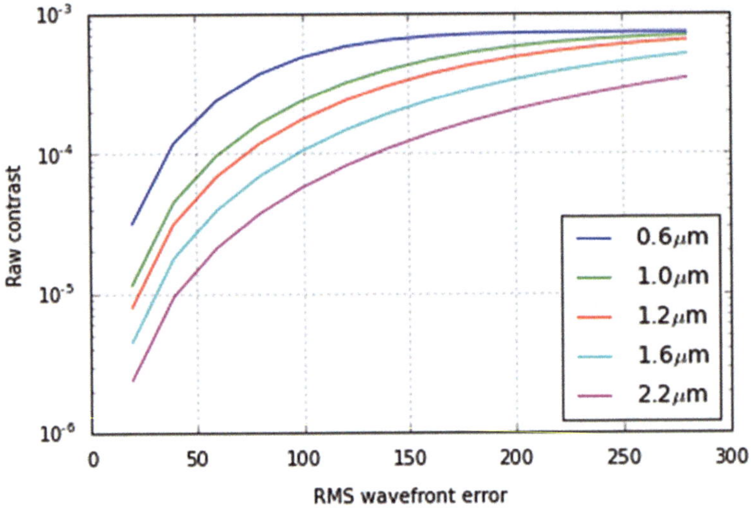

Fig. 2.6 Conversion between wavefront error and contrast at different wavelengths in the visible and near-infrared. The typical wavefront error of an xAO system is less than 100 nm

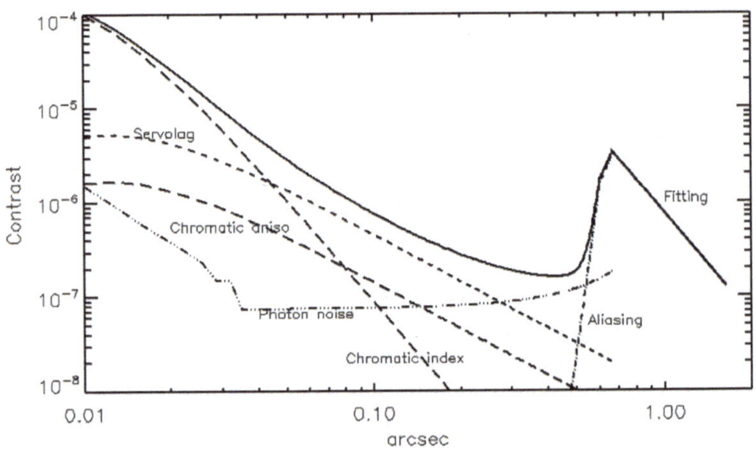

Fig. 2.7 Typical error budget as a function of angular separation for an xAO system running at 3 kHz on a bright ($I = 6$) star, at 1.6μm, with a fixed pyramid WFS (Figure reproduced with permission from [24])

2.3.2 Coronagraphy and Diffraction Control

The role of a coronagraph is to block the starlight and let as much off-axis signal (planet or circumstellar material emission) as possible through the system. It can only remove the coherent, static part of the diffraction pattern but cannot remove speckles due to wavefront errors. This is still greatly valuable because it reduces

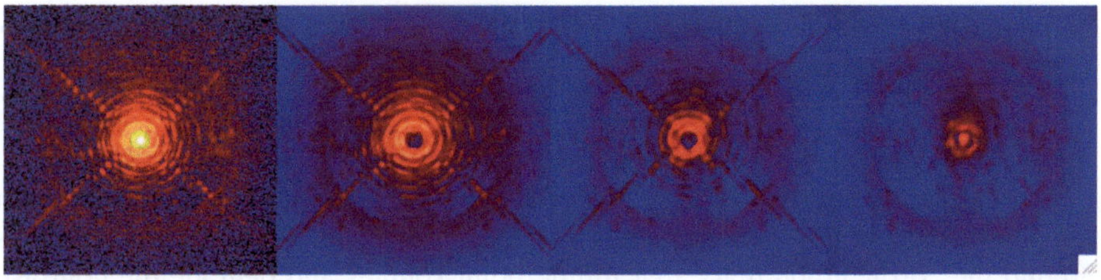

Fig. 2.8 Typical high Strehl ratio raw PSF in the presence of turbulence in the VLT pupil (circular aperture obscured by the secondary mirror held by the spiders): without coronagraph (*left*), with a Lyot stop (*middle left*), with a Lyot stop and apodiser (*middle right*), with apodiser, Lyot stop and pupil stop masking the telescope spiders (*right*) (These images were obtained with SPHERE [6]; the well-controlled radius measures 0.8″)

the photon noise of the diffraction pattern and the coherent amplification between the speckles and the diffraction pattern described in [1] and visible in Fig. 2.3. In practice, detectors have a limited dynamical range, therefore the use of coronagraphs avoids saturation and detrimental bleeding. It also limits scattering and parasitic reflections in the optical train, downstream of the coronagraph.

Most coronagraph designs are a trade-off between coronagraphic rejection, throughput, inner working angle (IWA) and angular resolution. The state-of-the-art designs are reviewed in [19] and more recently in [38]. They can be sorted between amplitude masks and phase masks, whether they act on the amplitude or the phase of the wavefront, or between focal and pupil masks, wether they are located in a pupil or in a focal plane.

In VLT/SPHERE, Gemini/GPI and Subaru/HiCIAO, apodised Lyot coronagraphs are mostly used. They represent an evolution of the Lyot coronagraph to include an apodised entrance pupil to further improve the achievable contrast [65] by removing the diffraction pattern. Figure 2.8 illustrates the level of light suppression and diffraction control that can be done by combining an apodiser, a Lyot stop and a pupil stop. They are typically limited to an IWA of $3-4\lambda/D$ in their current design.

To provide a smaller IWA than conventional Lyot coronagraphs, Guyon et al. [20] proposed the concept of a focal plane phase mask. The four-quadrant phase mask [58] and vortex coronagraph known as the AGPM [39] are evolutions of this concept and are used today on VLT/NaCo. The latter has been successfully commissioned on Keck/NIRC2, on VLT/VISIR in the N-band and is foreseen to equip VLT/SPHERE. They enable a smaller IWA, but chromaticity is hard to achieve. Pupil apodisation is another technique developed in order to smooth the pupil edges that create the Airy rings, implemented through a continuous amplitude mask or a binary mask known as a shaped pupil [23]. Loss-less pupil apodisation can be achieved through phase remapping. Successful implementations of this concept include the apodising phase plate (APP) on VLT/NaCo [25] or the phase induced amplitude apodisation coronagraph (PIAA) on Subaru/SCExAO [16].

2.3.3 Low-Order Wavefront Sensing and Non-common Path Aberrations

Most coronagraphs require the best possible correction of low-order aberrations, most importantly tip/tilt, focus, astigmatism and coma, in order to suppress light efficiently at their IWA. Uncorrected low-order wavefront error create light leaks around the coronagraph mask that mimic point-sources and degrade the contrast. This requirement is challenging in standard AO systems because there are non-common path aberrations (NCPA) between the WFS arm and the science arm and the WFS might not be sensitive enough to these low-order aberrations, as it is the case for the SHWFS. A number of practical solutions exist for current xAO systems and are reviewed in [38]. In short, the simple architecture of Fig. 2.4 (plain lines) is modified to include two additional feedbacks from the coronagraphic plane and image plane in order to sense the wavefront errors close to the coronagraph and at the same wavelength as the science camera. The first feedback loop requires a dedicated calibration system. In VLT/SPHERE, this is implemented through a beamsplitter close to the coronagraphic focus, sending a small fraction of the light to the differential tip/tilt sensor [51]. A distinct choice was made by Gemini/GPI and Subaru/SCExAO where a modified Mach-Zehnder interferometer combines the light reflected off the coronagraph spot with a reference wavefront. The second feedback uses directly the image recorded from the science camera to reconstruct the NCPA using a phase diversity algorithm [48, 61].

2.3.4 Observation Strategies for Improved Stability and Speckle Removal

2.3.4.1 Stability Considerations

Soon after the first generation of AO systems came online, it was realised that quasi-static residual speckles were the largest source of noise at short separations that prevent detection of faint companions [52]. Unlike atmospheric residuals that would average over time, these speckles are long-lived [22]. They come from mechanical flexures, imperfect optics, non-common path errors between the WFS arm and the science arm that are slowly varying with the tracking of the telescope and rotation of the optics. Because most of these slowly varying wavefront errors are fixed in a pupil frame rather than in a sky frame, high-contrast imaging is performed in pupil-stabilised mode. This way, the PSF is kept as stable as possible during the science observations and can be calibrated in the post-processing stage. On the Cassegrain focus of an alt/az telescope (e.g., Gemini/GPI), this means turning off the rotator, whereas on the Nasmyth platform (e.g., VLT/SPHERE), this means introducing a derotator as early as possible in the optical train. Around meridian passage, the apparent motion of a star on sky is the slowest, the altitude motion of the telescope

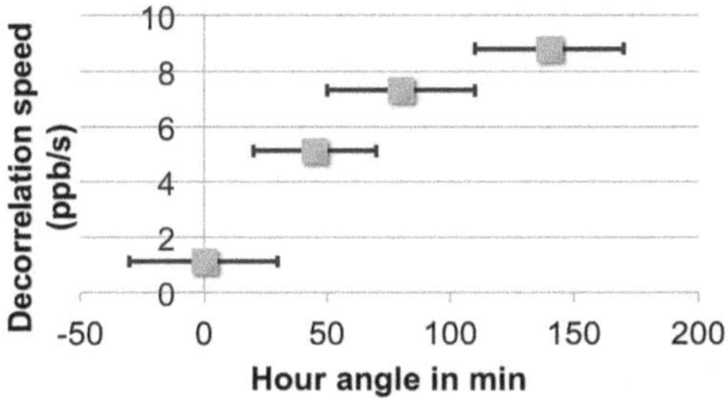

Fig. 2.9 Decorrelation speed (in units of 10^{-9}/s; part per billion per second – ppb/s) as a function of the hour angle. These data were obtained with NaCo in the *Lp*-band and averaged over one hour (horizontal error bar) to derive meaningful conclusions (Milli et al. in prep). Decorrelations is much slower around meridian than at larger hour angles

goes to zero. This stable configuration can be traced down to the science camera, as shown in Fig. 2.9, showing that deep coronagraphic images decorrelate slower around meridian passage.

It is interesting to note that the two xAO instruments VLT/SPHERE and Gemini/GPI have adopted different solutions to minimise the impact of slowly variable aberrations. GPI is a light and compact instrument attached to the moving Cassegrain focus, and implements a calibration unit (see Sect. 2.3.3) to probe in real time the evolution of the quasi-static speckle and control them. On the other hand, SPHERE is a heavy instrument that rests on the Nasmyth platform via an actively-controlled support to damp vibrations to stay as stable as possible.

2.3.4.2 Observing Strategies

Once all possible efforts have been made to keep the wavefront error as low as possible and the PSF as stable as possible, the post-processing stage aims at further enhancing the detection of astrophysical signal and removing instrumental speckles. This step relies on the introduction of diversity between the astrophysical signal and the speckles.

In pupil-stabilised observation – now the baseline for most high-contrast observations[4] – the relative motion of an on-sky signal with respect to the pupil is used to disentangle between a fixed speckle and a rotating companion, a strategy called Angular Differential Imaging (ADI; [36]). Advanced data reduction algorithms have been developed to further improve the efficiency of ADI [2, 14, 28, 44, 67]. However,

[4]Except for polarimetry where high polarimetric accuracy may drive the need for a different stabilisation scheme.

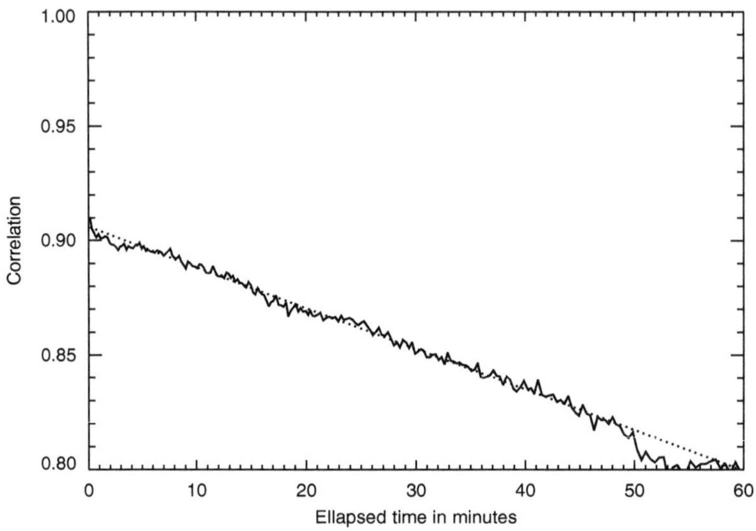

Fig. 2.10 Correlation over time for a set of deep *H*-band observations with VLT/NaCo. The correlation shows a sharp decrease within the first seconds and decreases linearly in the following hour

it is intrinsically limited by two considerations:

(1) the decorrelation of the PSF over time [22], illustrated in Fig. 2.10, linear on a timescale of several tens of minutes but very steep within the first seconds
(2) the very slow rotation of the field at very short separations where planet are expected, which implies a lot of self-subtraction of the astrophysical signal, especially for the case of extended sources such as discs [41].

Reference star differential imaging (RDI; see Fig. 2.11) is a solution to circumvent these intrinsic difficulties. It implies observing a reference star as close as possible in time and space, ideally with the same parallactic angle variation to keep the motion of the optics as close as possible to the science observations. In practice, fast switching between science targets and calibrators suffers from the overheads of the WFS acquisition, but solutions exist to close the loop after switching target without reacquiring on the WFS, as implemented on VLT/NaCo ("star hopping" [15, 27]). On the other hand, it is also possible to rely on a large library of PSF acquired as part of a survey of many stars along several nights or weeks to build an optimal PSF for the star subtraction. Such a strategy was proposed on the Hubble Space Telescope [11, 66] and observing programs are currently on-going to validate this strategy on ground-based instruments in the new high-Strehl ratio regime opened by xAO instruments.

Simultaneous differential imaging represents another solution, whether spectral (SDI) or polarimetic (PDI). For that, the target needs to have a spectral feature (usually a methane feature for gas giants) or must show linear polarisation (discs

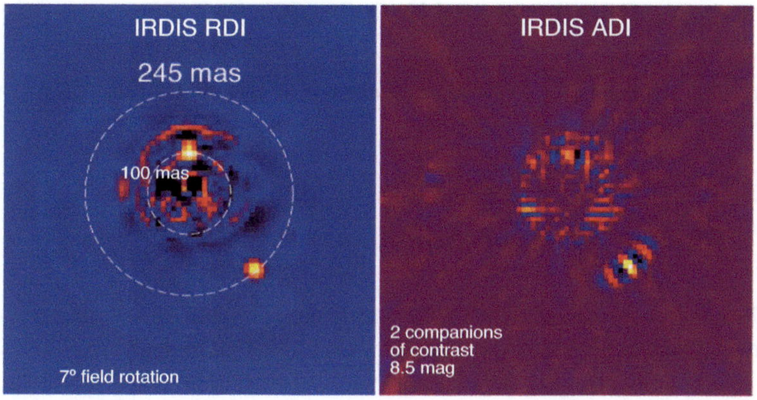

Fig. 2.11 Illustration of the power of RDI with the near-infrared subsystem IRDIS of the VLT/SPHERE instrument in the J-band. A sub-stellar companion with a contrast of 8.5 magnitudes is detected at $0.245''$ with a higher signal-to-noise ratio in RDI (*left image*) than ADI (*right image*). The same conclusion can be drawn for the fake companion injected at $0.1''$. The RDI does not suffer from the self-subtraction issue that create negative side lobes around the companion in ADI. The ADI image was scaled to account for the self-subtraction which decreases radially. The field rotation of 7° corresponds to only $0.4\ \lambda/D$ at $0.1''$

Table 2.4 Overview of the most common differential imaging techniques

Name	Diversity parameter	Strengths	Drawbacks
ADI	pupil/field relative rotation	easy implementation	PSF decorrelation over time, self-subtraction
RDI	science/ref. star	no self-subraction	rapid switching required or large library, subject to change in the PSF shape/AO correction, overheads due to the calibrator
SDI	wavelength	simultaneous difference	relies on spectral features, chromaticity
PDI	linear polarisation	simultaneous difference, achromatic	relies on linear polarisation feature, calibration of instrumental polarisation.

or planets in scattered light). SDI relies on the fact that speckles scale with the wavelength unlike on-sky signal, whereas PDI relies on the fact that the thermal emission from the central star is unpolarised unlike scattered light from circumstellar material. Table 2.4 summarises the benefits and drawbacks of each differential imaging technique. Note that ADI can be carried out on top of SDI or PDI.

2.4 Science Highlights and New Challenges

We provide in this section a few selected examples that showcase the new possibilities offered by xAO instruments.

2.4.1 Discs at Very Short Separations

HR 4796 is a prototypical debris disc that makes an ideal benchmark to compare instruments because it has been observed with almost all high-contrast instruments. The disc has a semi-major axis of $\sim 1''$ and a semi-minor axis of $\sim 0.2''$. It was not detected in early AO imaging with the pioneering AO system COME-ON+ [43]. First generation AO systems revealed the ring along its semi-major axis (see for instance the recent Subaru NICI image from [69] in Fig. 2.12 left). They were mostly blind to the semi-minor axis until the recent NaCo polarimetric image could reveal the disc almost entirely (Fig. 2.12, second image from [40]). An uncontrolled mode of the DM (waffle) creates a four-point pattern at the DM cutoff frequency ($7\ \lambda/D$). With the xAO instruments VLT/SPHERE and Gemini/GPI, the disc is detected directly in raw frames of a few seconds without any star subtraction. The disc was observed in ADI in the *H*-band during VLT/SPHERE commissioning (ESO press release 1417, third image in Fig. 2.12) under poor conditions. Light leak around the coronagraph below $0.2''$ and the small field rotation of only 22° prevent a clear detection of the semi-minor axis but the improvement in achieved separation and image quality is striking. With GPI, the disc is detected unambiguously in polarimetry in the *Ks*-band (right image, [50]). The concept of integral field polarimetry implemented in GPI minimises differential wavefront error between

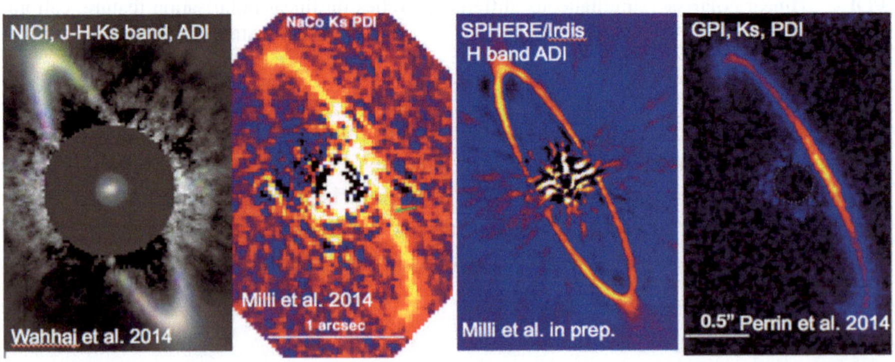

Fig. 2.12 Comparison between the first generation AO imagers (*left images*) and xAO imagers (*right images*). The first and third images are coronagraphic images obtained in ADI, while the second and fourth are polarised images (Figure adapted from [40, 50, 69], reproduced with permission)

the two polarisation channels, whereas the Wollaston prism of NaCo introduces differential aberrations.

2.4.2 Planets in the Visible

The new generation of xAO systems provides a very high wavefront correction that opens the door to visible-light AO-assisted imaging, much more demanding due to the shorter wavelengths. The visible range is appealing because the angular resolution is higher, and polarisation can be efficiently used both for starlight rejection and planet/disc characterisation. The first system on-sky was MagAO on the Magellan Clay telescope. MagAO is a 585-actuator adaptive secondary mirror with a pyramid WFS [12]. It can operate simultaneously in the visible with the camera VisAO and in the infrared with Clio2. The gas giant β Pictoris b was detected for the first time in the red (band Y_S, that still corresponds to the planet thermal emission) with VisAO in 2012 [34], with a signal-to-noise ratio of $\sim$4 (Fig. 2.13 left) and a Strehl ratio of 40 %.

Since then, other visible AO instruments arrived on-sky, such as VAMPIRES [45] fed by the SCExAO system, or SPHERE/Zimpol [62], both implementing differential polarimetry. Zimpol is an unique concept of a high-accuracy polarimeter and imager. The polarimeter is based on the concept of high-frequency modulation of the polarisation to freeze the non-corrected atmospheric residuals, using a ferroelectric liquid cristal operating at 1 kHz. Demodulation is carried out with a dedicated CCD camera synchronised with the modulator and switching the charges alternatively up and down. Combined with the SPHERE xAO system, Zimpol aims at detecting exoplanets in reflected light [42], due to its exquisite inner working angle of 30 mas (2 λ/D at 600 nm). During science verification, the Zimpol

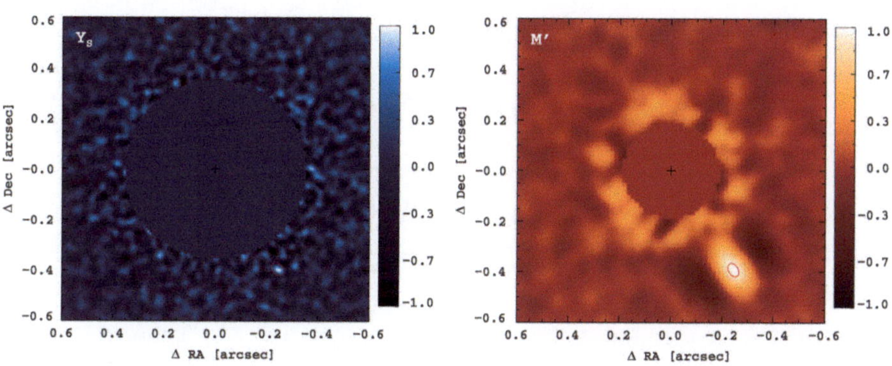

Fig. 2.13 Simultaneous detection of β Pictoris b in the *red* (Y_S, *left*) and in the near-infrared at M' (*right*). This illustrates the higher resolution provided by shorter wavelengths (Figure reproduced with permission from [34])

instrument in imaging mode resolved for the first time the surface of a star: the nearby red giant R Doradus, with a diameter of 57 mas.

2.4.3 Spectra of Exoplanets and Brown Dwarfs

Combined with integral field spectrographs (IFS), xAO provided the first spectra of exoplanets. The first IFS on-sky was P1640 at Palomar [47]. It revealed the spectra of the four known exoplanets HR 8799 b, c, d and e (Fig. 2.14) from 995 to 1769 nm, and identified the presence of CH_4 along with NH_3, C_2H_2, and possibly CO_2 or HCN in variable amounts in each component of the system. More recently, Gemini/GPI revealed the H-band spectra of β Picoris b [10] at only 436 mas of the central star. The spectrum, obtained with a resolving power of $R \sim 45$, shows a triangular shape, typical of cool low-gravity substellar objects. Extreme AO systems can also be combined with long-slit spectroscopy (LSS) to provide medium resolution spectra. The concept of coronagraphic LSS is implemented in VLT/SPHERE [68] to enhance starlight rejection. It was recently illustrated during science verification to reveal an $R \sim 350$ spectrum of the young substellar companion 2MASS 0122-2439B across the YJH-bands [21].

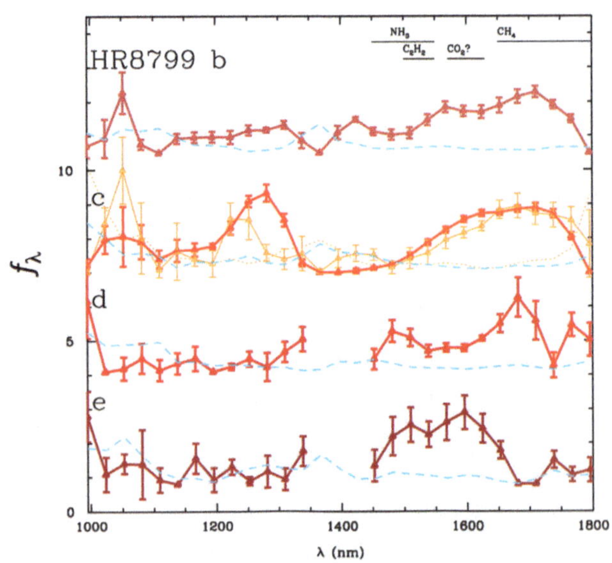

Fig. 2.14 Spectra of the four exoplanets detected in the system around HR8799, obtained with a resolution of $\sim$ 35 with the system PALM3000/P1640 (Figure reproduced with permission from [46])

2.5 Conclusions and Future Challenges

Today, several large-scale surveys of extrasolar planets using xAO systems are under way. These studies probe a new parameter space at smaller separations and higher contrast and will probably lead to tens of new detections in the coming years, that will help to understand the population and formation mechanisms of giant planets. On individual systems, these new capabilities also open up new perspectives. The photometric accuracy will enable to capture temporal variations of exoplanet emission due to non-uniform cloud coverage, or temporal evolution of disc asymmetries as a result of gravitational perturbation or stellar winds. By reaching shorter separations, direct imaging will bridge the gap with the radial velocity technique, to get access to the dynamical mass of giant planets and therefore test the evolutionary models.

From a technical point of view, the higher Strehl ratios and stability of xAO systems unexpectedly revealed fine instrumental effects that passed unnoticed in first generation systems because of the lack of sensitivity. Such effects include vibration issues, peculiar dome- or low-altitude atmospheric conditions or subtle polarisation mechanisms. It also triggered an intense activity on post-processing techniques and signal detection theory, for which other fields of physics were a major source of inspiration. Observing strategies and data extraction methods valid at large separation and widely used on first generation instruments are not necessarily relevant at very short separation of a few resolution elements and need to be tailored for this new regime.

These developments and the experience acquired through xAO instruments will contribute to the preparation of the first planet-finder instruments for 30m-class telescopes arriving in the 2020s. Such systems are expected to unveil even shorter separations below 10 mas, but they will also have to overcome new challenges such as chromatic effects or segmented mirrors. Complemented by space-based coronagraphic instruments aiming for deeper contrasts, they are expected to reveal the first images of rocky planets around nearby stars.

References

1. Aime, C., Soummer, R.: The usefulness and limits of coronagraphy in the presence of pinned speckles. ApJ **612**, L85 (2004)
2. Amara, A., Quanz, S.P.: PYNPOINT: an image processing package for finding exoplanets. MNRAS **427**, 948 (2012)
3. Babcock, H.W.: The possibility of compensating astronomical seeing. PASP **65**, 229 (1953)
4. Baldwin, J.E., Haniff, C.A., Mackay, C.D., Warner, P.J.: Closure phase in high-resolution optical imaging. Nature **320**, 595 (1986)
5. Baraffe, I., Chabrier, G., Barman, T.S., Allard, F., Hauschildt, P.H.: Evolutionary models for cool brown dwarfs and extrasolar giant planets. The case of HD 209458. A&A **402**, 701 (2003)

6. Beuzit, J.L., Feldt, M., Dohlen, K., et al.: SPHERE: a planet finder instrument for the VLT. In: McLean, I.S., Casali, M.M. (eds.) Ground-Based and Airborne Instrumentation for Astronomy II, Marseille, 23 June 2008. Society of Photo-Optical Instrumentation Engineers (SPIE) Conference Series, vol. 7014 (2008)
7. Beuzit, J.L., Hubin, N., Gendron, E., et al: ADONIS: a user-friendly adaptive optics system for the ESO 3.6-m telescope. In: Ealey, M.A., Merkle, F. (eds.) Adaptive Optics in Astronomy, Kailua, 13 Mar 1994. Society of Photo-Optical Instrumentation Engineers (SPIE) Conference Series, vol. 2201, p. 955 (1994)
8. Bloemhof, E.E., Wallace, J.K.: Phase contrast techniques for wavefront sensing and calibration in adaptive optics. In: Tyson, R.K., Lloyd-Hart, M. (eds.) Astronomical Adaptive Optics Systems and Applications, San Diego, 03 Aug 2003. Society of Photo-Optical Instrumentation Engineers (SPIE) Conference Series, vol. 5169, pp. 309 (2003)
9. Chauvin, G., Lagrange, A.M., Dumas, C., Zuckerman, B., Mouillet, D., Song, I., Beuzit, J.L., Lowrance, P.: A giant planet candidate near a young brown dwarf. Direct VLT/NACO observations using IR wavefront sensing. A&A **425**, L29 (2004)
10. Chilcote, J., Barman, T., Fitzgerald, M.P., et al.: The first H-band spectrum of the giant planet β pictoris b. ApJ **798**, L3 (2015)
11. Choquet, E., Hagan, J.B., Pueyo, L., et al.: Archival legacy investigation of circumstellar environments using KLIP algorithm on HST NICMOS coronagraphic data. In: Oschmann, J.M., Clampin, M., Fazio, G.G., MacEwen, H.A. (eds.) Space Telescopes and Instrumentation 2014: Optical, Infrared, and Millimeter Wave, Montréal, 22 June 2014 (2014)
12. Close, L.M., Males, J.R., Kopon, D.A., et al.: First closed-loop visible AO test results for the advanced adaptive secondary AO system for the Magellan Telescope: MagAO's performance and status. In: Ellerbroek, B.L., Marchetti, E., Véran, J.-P. (eds.) Adaptive Optics Systems III, Amsterdam, 1 July 2012. Society of Photo-Optical Instrumentation Engineers (SPIE) Conference Series, vol. 8447 (2012)
13. Dekany, R., Roberts, J., Burruss, R., et al.: PALM-3000: exoplanet adaptive optics for the 5 m Hale Telescope. ApJ **776**, 130 (2013)
14. Galicher, R., Marois, C., Macintosh, B., Barman, T., Konopacky, Q.: M-band imaging of the HR 8799 planetary system using an innovative LOCI-based background subtraction technique. ApJ **739**, L41 (2011)
15. Girard, J.H.V., O'Neal, J., Mawet, D., Kasper, M., Zins, G., Neichel, B., Kolb, J., Christiaens, V., Tourneboeuf, M.: Image quality and high contrast improvements on VLT/NACO. In: Ellerbroek, B.L., Marchetti, E., Véran, J.-P. (eds.) Adaptive Optics Systems III, Amsterdam, 1 July 2012. Society of Photo-Optical Instrumentation Engineers (SPIE) Conference Series, vol. 8447 (2012)
16. Guyon, O.: Phase-induced amplitude apodization of telescope pupils for extrasolar terrestrial planet imaging. A&A **404**, 379 (2003)
17. Guyon, O.: Limits of adaptive optics for high-contrast imaging. ApJ **629**, 592 (2005)
18. Guyon, O., Martinache, F., Garrel, V., Vogt, F., Yokochi, K., Yoshikawa, T.: The Subaru coronagraphic extreme AO (SCExAO) system: wavefront control and detection of exoplanets with coherent light modulation in the focal plane. In: Ellerbroek, B.L., Hart, M., Hubin, N., Wizinowich, P.L. (eds.) Adaptive Optics Systems II, San Diego, 27 June 2010. Society of Photo-Optical Instrumentation Engineers (SPIE) Conference Series, vol. 7736, p. 24 (2010)
19. Guyon, O., Pluzhnik, E.A., Kuchner, M.J., Collins, B., Ridgway, S.T.: Theoretical limits on extrasolar terrestrial planet detection with coronagraphs. ApJS **167**, 81 (2006)
20. Guyon, O., Roddier, C., Graves, J.E., Roddier, F., Cuevas, S., Espejo, C., Gonzalez, S., Martinez, A., Bisiacchi, G., Vuntesmeri, V.: The nulling stellar coronagraph: laboratory tests and performance evaluation. PASP **111**, 1321 (1999)
21. Hinkley, S., Bowler, B.P., Vigan, A., et al.: Early results from VLT-SPHERE: long-slit spectroscopy of 2MASS 0122-2439B, a young companion near the deuterium burning limit. ApJ **805**, 10 (2015)
22. Hinkley, S., Oppenheimer, B.R., Soummer, R., et al.: Temporal evolution of coronagraphic dynamic range and constraints on companions to Vega. ApJ **654**, 633 (2007)

23. Kasdin, N.J., Vanderbei, R.J., Spergel, D.N., Littman, M.G.: Extrasolar planet finding via optimal apodized-pupil and shaped-pupil coronagraphs. ApJ **582**, 1147 (2003)
24. Kasper, M.: Adaptive optics for high contrast imaging. In: Ellerbroek, B.L., Marchetti, E., Véran, J.-P. (eds.) Adaptive Optics Systems III, Amsterdam, 1 July 2012. Society of Photo-Optical Instrumentation Engineers (SPIE) Conference Series, vol. 8447 (2012)
25. Kenworthy, M.A., Codona, J.L., Hinz, P.M., Angel, J.R.P., Heinze, A., Sivanandam, S.: First on-sky high-contrast imaging with an apodizing phase plate. ApJ **660**, 762 (2007)
26. Labeyrie, A.: Attainment of diffraction limited resolution in large telescopes by fourier analysing speckle patterns in star images. A&A **6**, 85 (1970)
27. Lacour, S., Tuthill, P., Amico, P., Ireland, M., Ehrenreich, D., Huelamo, N., Lagrange, A.M.: Sparse aperture masking at the VLT. I. Faint companion detection limits for the two debris disk stars HD 92945 and HD 141569. A&A **532**, A72 (2011)
28. Lafrenière, D., Marois, C., Doyon, R., Nadeau, D., Artigau, É.: A new algorithm for point-spread function subtraction in high-contrast imaging: a demonstration with angular differential imaging. ApJ **660**, 770 (2007)
29. Law, N.M., Mackay, C.D., Baldwin, J.E.: Lucky imaging: high angular resolution imaging in the visible from the ground. A&A **446**, 739 (2006)
30. Le Bouquin, J.B. & Absil, O.: On the sensitivity of closure phases to faint companions in optical long baseline interferometry. A&A **541**, A89 (2012)
31. Macintosh, B., Graham, J.R., Ingraham, P., et al.: First light of the Gemini planet imager. Proc. Natl. Acad. Sci. **111**, 12 661 (2014)
32. Madec, P.Y.: Overview of deformable mirror technologies for adaptive optics and astronomy. In: Ellerbroek, B.L., Marchetti, E., Véran, J.-P. (eds.) Adaptive Optics Systems III, Amsterdam, 1 July 2012. Society of Photo-Optical Instrumentation Engineers (SPIE) Conference Series, vol. 8447, p. 5 (2012)
33. Mahajan, V.N.: Strehl ratio for primary aberrations in terms of their aberration variance. J. Opt. Soc. Am. (1917–1983) **73**, 860 (1983)
34. Males, J.R., Close, L.M., Morzinski, K.M., et al.: Magellan adaptive optics first-light observations of the exoplanet β Pic B. I. Direct imaging in the far-red optical with MagAO+VisAO and in the near-ir with NICI. ApJ **786**, 32 (2014)
35. Maréchal, A.: Etude des effets combinés de la diffraction et des aberrations géometriques sur l'image d'un point lumineux. Rev. Opt. Theor. Instrum. **26**, 257 (1947)
36. Marois, C., Lafrenière, D., Doyon, R., Macintosh, B., Nadeau, D.: Angular differential imaging: a powerful high-contrast imaging technique. ApJ **641**, 556 (2006)
37. Marois, C., Lafrenière, D., Macintosh, B., Doyon, R.: Confidence level and sensitivity limits in high-contrast imaging. ApJ **673**, 647 (2008)
38. Mawet, D., Pueyo, L., Lawson, P., et al.: Review of small-angle coronagraphic techniques in the wake of ground-based second-generation adaptive optics systems. In: Clampin, M.C., Fazio, G.G., MacEwen, H.A., Oschmann, J.M. (eds.) Space Telescopes and Instrumentation 2012: Optical, Infrared, and Millimeter Wave, Amsterdam, 1 July 2012. Society of Photo-Optical Instrumentation Engineers (SPIE) Conference Series, vol. 8442 (2012)
39. Mawet, D., Riaud, P., Absil, O., Surdej, J.: Annular groove phase mask coronagraph. ApJ **633**, 1191 (2005)
40. Milli, J., Mawet, D., Pinte, C., Lagrange, A.M., Mouillet, D., Girard, J.H., Augereau, J.C., de Boer, J., Pueyo, L., Choquet, E.: New constraints on the dust surrounding HR 4796 A. A&A **577**, A57 (2015)
41. Milli, J., Mouillet, D., Lagrange, A.M., Boccaletti, A., Mawet, D., Chauvin, G., Bonnefoy, M.: Impact of angular differential imaging on circumstellar disk images. A&A **545**, A111 (2012)
42. Milli, J., Mouillet, D., Mawet, D., Schmid, H.M., Bazzon, A., Girard, J.H., Dohlen, K., Roelfsema, R.: Prospects of detecting the polarimetric signature of the Earth-mass planet α Centauri B b with SPHERE/ZIMPOL. A&A **556**, A64 (2013)
43. Mouillet, D., Lagrange, A.M., Beuzit, J.L., Renaud, N.: A stellar coronograph for the COME-ON-PLUS adaptive optics system. II. First astronomical results.. A&A **324**, 1083 (1997)

44. Mugnier, L.M., Cornia, A., Sauvage, J.F., Védrenne, N., Fusco, T., Rousset, G.: Maximum likelihood-based method for angular differential imaging. In: Hubin, N., Max, C.E., Wizinowich, P.L. (eds.) Adaptive Optics Systems, Marseille, 23 June 2008. Society of Photo-Optical Instrumentation Engineers (SPIE) Conference Series, vol. 7015, p. 2 (2008)
45. Norris, B., Schworer, G., Tuthill, P., Jovanovic, N., Guyon, O., Stewart, P., Martinache, F.: The VAMPIRES instrument: imaging the innermost regions of protoplanetary discs with polarimetric interferometry. MNRAS **447**, 2894 (2015)
46. Oppenheimer, B.R., Baranec, C., Beichman, C., et al.: Reconnaissance of the HR 8799 exosolar system. I. Near-infrared spectroscopy. ApJ **768**, 24 (2013)
47. Oppenheimer, B.R., Beichman, C., Brenner, D., et al.: Project 1640: the world's first ExAO coronagraphic hyperspectral imager for comparative planetary science. In: Ellerbroek, B.L., Marchetti, E., Véran, J.-P. (eds.) Adaptive Optics Systems III, Amsterdam, 1 July 2012. Society of Photo-Optical Instrumentation Engineers (SPIE) Conference Series, vol. 8447, p. 20 (2012)
48. Paul, B., Mugnier, L.M., Sauvage, J.F., Ferrari, M., Dohlen, K.: High-order myopic coronagraphic phase diversity (COFFEE) for wave-front control in high-contrast imaging systems. Opt. Express **21**, 31 751 (2013)
49. Pepe, F., Ehrenreich, D., Meyer, M.R.: Instrumentation for the detection and characterization of exoplanets. Nature **513**, 358 (2014)
50. Perrin, M.D., Duchene, G., Millar-Blanchaer, M., et al.: Polarimetry with the Gemini Planet imager: methods, performance at first light, and the Circumstellar Ring around HR 4796A. ApJ **799**, 182 (2015)
51. Petit, C., Sauvage, J.F., Fusco, T., et al.: SPHERE eXtreme AO control scheme: final performance assessment and on sky validation of the first auto-tuned LQG based operational system. In: Marchetti, E., Close, L.M., Véran, J.-P. (eds.) Adaptive Optics Systems IV, Montréal, 22 June 2014. Society of Photo-Optical Instrumentation Engineers (SPIE) Conference Series, vol. 9148 (2014)
52. Racine, R., Walker, G.A.H., Nadeau, D., Doyon, R., Marois, C.: Speckle noise and the detection of faint companions. PASP **111**, 587 (1999)
53. Ragazzoni, R.: Pupil plane wavefront sensing with an oscillating prism. J. Mod. Opt. **2**, 43 (1996)
54. Ragazzoni, R., Farinato, J.: Sensitivity of a pyramidic wave front sensor in closed loop adaptive optics. A&A **350**, L23 (1999)
55. Roddier, F.: The Effects of Atmospheric Turbulence in Optical Astronomy. Progress in Optics, vol. 19, p. 281. North-Holland, Amsterdam (1981)
56. Roddier, F.: Curvature sensing and compensation: a new concept in adaptive optics. Appl. Opt. **27**, 1223 (1988)
57. Roddier, F.: Adaptive Optics in Astronomy. Cambridge University Press, Cambridge (1999)
58. Rouan, D., Riaud, P., Boccaletti, A., Clénet, Y., Labeyrie, A.: The four-quadrant phase-mask coronagraph. I. Principle. PASP **112**, 1479 (2000)
59. Rousset, G.: Wave-Front Sensors. Cambridge University Press, Cambridge (1999)
60. Rousset, G., Fontanella, J.C., Kern, P., Gigan, P., Rigaut, F.: First diffraction-limited astronomical images with adaptive optics. A&A **230**, L29 (1990)
61. Sauvage, J.F., Fusco, T., LeMignant, D., Petit, C., Sevin, A., Dohlen, K., Robert, C., Mugnier, L.: SPHERE non-common path aberrations measurement and pre-compensation with optimized phase diversity processes: experimental results. In: Second International Conference on Adaptive Optics for Extremely Large Telescopes, Victoria, p. 48 (2011)
62. Schmid, H.M., Beuzit, J.L., Mouillet, D., Waters, R., Buenzli, E., Boccaletti, A., Dohlen, K., Feldt, M., SPHERE Consortium: Polarimetry of extra-solar planets and circumstellar disks with ZIMPOL/SPHERE. In: Boccaletti, A. (ed.) Proceedings of the Conference in the Spirit of Lyot 2010: Direct Detection of Exoplanets and Circumstellar Disks, Paris, 25–29 Oct 2010
63. Schneider, G.: A quick study of science return from direct imaging exoplanet missions: detection and characterization of circumstellar material with an AFTA or EXO-C/S CGI (2014). arXiv:1412.8421

64. Serabyn, E., Wallace, K., Troy, M., Mennesson, B., Haguenauer, P., Gappinger, R., Burruss, R.: Extreme adaptive optics imaging with a clear and well-corrected off-axis telescope subaperture. ApJ **658**, 1386 (2007)
65. Soummer, R., Aime, C., Falloon, P.E.: Stellar coronagraphy with prolate apodized circular apertures. A&A **397**, 1161 (2003)
66. Soummer, R., Perrin, M.D., Pueyo, L., et al.: Five Debris disks newly revealed in scattered light from the Hubble space telescope NICMOS archive. ApJ **786**, L23 (2014)
67. Soummer, R., Pueyo, L., Larkin, J.: Detection and characterization of exoplanets and disks using projections on Karhunen-Loève Eigenimages. ApJ **755**, L28 (2012)
68. Vigan, A., Langlois, M., Moutou, C., Dohlen, K.: Long slit spectroscopy for exoplanet characterization in SPHERE. In: McLean, I.S., Casali, M.M. (eds.) Ground-Based and Airborne Instrumentation for Astronomy II, Marseille, 23 June 2008. Society of Photo-Optical Instrumentation Engineers (SPIE) Conference Series, vol. 7014, p. 4 (2008)
69. Wahhaj, Z., Liu, M.C., Biller, B.A., et al.: The Gemini NICI planet-finding campaign: the offset ring of HR 4796 A. A&A **567**, A34 (2014)

Chapter 3
Aperture Masking Imaging

Michael J. Ireland

3.1 Introduction

The problem of achieving the full diffraction limit of a telescope through the use of aperture masking has a long history. Hippolyte Fizeau suggested in 1868 that this could be achieved through the use of a non-redundant (2-hole) aperture mask well before a modern understanding of the effects of astronomical seeing had been developed [5]. When combined with the human eye as a fast detector, such aperture masks, or later, an interferometer [15], could be used to both resolve close binary stars and measure the sizes of stars.

In early eyepiece observing, the human brain was also an effective computer in recovering moderate contrast binary stars from speckle patterns at separations well below the seeing limit [2]. In this case, a single speckle pattern is easily distinguished from two identical speckle patterns separated by more than the diffraction-limit but less than a seeing disc size if there is sufficient diversity in the observed speckle patterns, or if occasional "lucky" images show reduced Full-Width-Half-Maximum (FWHM) along the projected axis of the binary star. The dominance of the photographic plate in seeing-limited observations meant that after the decommissioning of the 20 ft interferometer for the Mt Wilson 100 in telescope, diffraction limited observing had a period of hiatus – only to be re-born with video-rate detection and optical Fourier transforms [6].

Since the 1980s, electronic detectors and fast computers have enabled both a movement from 2-hole aperture masks to many-hole masks, and the exploitation of bispectral phase when speckle imaging with an unmasked pupil [12]. These techniques have been limited both by detector noise at fast readout rates as well

M.J. Ireland (✉)
Research School of Astronomy and Astrophysics, Australian National University, Canberra, ACT 2611, Australia
e-mail: michael.ireland@anu.edu.au

H.M.J. Boffin et al. (eds.), *Astronomy at High Angular Resolution*, Astrophysics and Space Science Library 439, DOI 10.1007/978-3-319-39739-9_3

as low signal-to-noise in the many speckle regime. Some of the greatest imaging achievements from bispectrum speckle interferometry prior to regular adaptive optics observing remain unsurpassed. Full pupil speckle masking achieved dynamic ranges of 10^4 [19] while non-redundant masking achieved precision calibration and reliable super-resolved imaging [17].

The decade from 2005 to 2015 has been characterised by adaptive optics imaging with low to moderate Strehl ratios, and infrared detectors with relatively high noise when short exposures are used, combined with limited data rates. This has meant that image analysis techniques for typical AO imaging have been often borrowed from space-based or seeing-limited observing. Aperture mask interferometry has played a key role in precision calibration of interferometric observables – visibility amplitudes and closure phases – that enable imaging and model fitting and image reconstruction at or beyond the formal diffraction limit.

The coming years will offer further revolutions in imaging. Despite great advances in adaptive optics performance, limitations such as the cone effect in laser guide star adaptive optics and the need for very bright stars for extreme adaptive optics means that imaging will generally be done in the low to moderate Strehl ratio regime. In this regime, deconvolution or model fitting methods borrowed from spaced-based observing are flawed, and obtaining point-spread function references from wide field observing is only possible in crowded fields. However, low to moderate Strehl ratio imaging need not remain the same – for example, new low noise infrared detectors mean that speckle imaging behind adaptive optics will again be possible, providing a clear role for advanced diffraction-limited analysis techniques. Indeed, the move towards infrared (rather than visible) adaptive optics corrected laser guide star tip/tilt references means that infrared data will begin to be routinely collected on stars bright enough for speckle techniques. However, as we will see in this chapter, optimising pupil geometry using a non-redundant aperture mask has clear calibration advantages in the low to moderate Strehl ratio regime, and it will continue to have a clear role 150 years after its birth.

In this chapter, I will attempt to put aperture masking imaging in a proper context by first outlining the general problem of narrow-field imaging in Sect. 3.2, focusing on the difficulties in precision calibration. In Sect. 3.3 I will describe the principles of aperture masking, focusing on the key observables of visibility amplitude and closure-phase, and will describe how closure-phase differs from kernel-phase and bispectral phase in Sect. 3.4. Finally, I will show some examples of aperture mask imaging in Sect. 3.5 and then conclude.

3.2 Narrow Field Imaging

In the case of generic narrow-field imaging, neglecting effects of polarisation, we consider the observed image $i(\boldsymbol{\alpha})$ of an object intensity distribution $o(\boldsymbol{\alpha})$ in the presence of a space-invariant Point-Spread Function (PSF) $m(\boldsymbol{\alpha})$, and can write:

$$i(\boldsymbol{\alpha}) = m(\boldsymbol{\alpha}) * o(\boldsymbol{\alpha}) \tag{3.1}$$

Here, $\boldsymbol{\alpha}$ the 2-dimensional angular image vector in the image plane and $*$ represents convolution. In the Fourier domain, this is:

$$I(\mathbf{u}) = M(\mathbf{u}) \cdot O(\mathbf{u}) \tag{3.2}$$

Here we have taken I to be the 2-dimensional Fourier transform of i (same for m and o), and written the image spatial frequency vector as u. Multiplication in the Fourier domain is computationally simpler than convolution in the image domain, but is otherwise equivalent. As long as the recorded image i is Nyquist-sampled, the *modulation transfer function (MTF)* M and therefore the image Fourier transform I is bounded in the Fourier domain by the auto-correlation of the pupil.

At this point, in speckle imaging and aperture-mask interferometry, it is conventional to split into modulus and phase components:

$$|O(\mathbf{u})| = |I(\mathbf{u})|/|M(\mathbf{u})| \tag{3.3}$$

$$\mathrm{Arg}[O(\mathbf{u})] = \mathrm{Arg}[I(\mathbf{u})] - \mathrm{Arg}[M(\mathbf{u})], \tag{3.4}$$

with $|M|$ and $\mathrm{Arg}[M]$ e.g. estimated from observations of unresolved reference sources or reconstructed from a large number of observations of the bispectrum [12]. Rather than going down this path, let us consider first the general problem of image reconstruction with both uncertainties in the data and with uncertain point-spread functions.

When presented with an image i, it is tempting to want to derive object properties directly from the image. For example, the use of aperture-photometry on an adaptive optics image where sources are well-resolved uses the approximation that each aperture contains flux from only one object. This approximation breaks down in the presence of large PSF wings, or when objects are barely resolved. Where an image of a bright source dominates an image, it can be tempting to subtract the a model point-spread function from the image, and make inferences from the residual image. This kind of approach is not always possible or necessary, and rarely optimal. Instead, all we need to be able to do is to forward-model an object o to see if it fits an image i. The inverse process can then be viewed as a detail – simply minimising chi-squared or maximising likelihood in the presence of a regulariser and prior knowledge using Bayesian techniques. Explicitly:

$$L(o|i) = L(i|o)P(o), \tag{3.5}$$

where $P(o)$ is the prior probability of a given object o. For readability, we will not write the prior probability $P(o)$ on the right hand side of the following equations, but note that it is often important to include, e.g. in the use of a maximum entropy regulariser.

The fundamental problem in determining if an object o is consistent with an image i is that the point-spread function m is variable. Imagine being able to perfectly characterise m experimentally by taking images of unresolved sources in all possible conditions. Then we could write:

$$L(o|i) = \int_m \exp(-\Sigma_j \frac{((m*o)_j - i_j)^2}{2\sigma_j^2}) f(m), \tag{3.6}$$

where i_j is now the image observed at pixel j, and the function $f(m)$ is the assumed known probability density function of PSFs m. Of course, in a more general case, we have a set of observed images $\{i_k\}$, in which case the likelihood function becomes a product over likelihoods for each image k:

$$L(o|\{i_k\}) = \Pi_k \int_m \exp(-\Sigma_j \frac{((m*o)_j - i_{k,j})^2}{2\sigma_{k,j}^2}) f(m) \tag{3.7}$$

The only reason that more complex approaches than this (e.g. Fourier domain imaging) are used is that this is computationally difficult. For e.g. a 10×10 pixel arbitrary point-spread function, the integrals are integrals over 100 dimensions, almost certainly computed with Monte-Carlo techniques. The space of PSFs m is also difficult to characterise – for the Hubble Space Telescope, all breathing modes could be parameterised by a few numbers, but e.g. speckle imaging in the presence of both random and static aberrations is more complex. It is important to realise, however, that moving beyond this equation is only done for reasons of computational simplicity, and not anything fundamentally *better* about Fourier, bispectral or kernel-phase techniques.

One simplification of the problem is to only consider the *high-Strehl* regime where the image is *dominated by diffraction of quasi-static speckles*. A *quasi-static speckle* is in our context seen as resulting from slowly varying pupil-plane phase aberrations, which cause slowly varying image artefacts (speckles) in standard adaptive optics imaging. In this case, the diversity of images can be characterised by a mean $\bar{i}$ and a covariance matrix of the mean, taking into account the uncertainties due to PSF variation, which are added to the pixel-based uncertainties $\sigma_{j,k}$. This removes the product over all images k, leaving us with:

$$L(o|\{i\}) = \int_{\bar{m}} \exp(-\frac{1}{2}((\bar{m}*o)^k - \bar{i}^k)(C^{-1})_k^j((\bar{m}*o)_j - \bar{i}_j)) f(\bar{m}) \tag{3.8}$$

Here C_k^j is the covariance matrix derived from the set of images and the knowledge of pixel uncertainties, with an Einstein summation convention assumed. Although it is not in general possible to create an invertible sample covariance matrix directly from a small set of images, it is certainly possible once pixel variances are explicitly taken into account (which gives a diagonal approximation to C_k^j). As only the mean image $\bar{i}$ is explicitly considered, the PSF must also be considered as a mean over the number of exposures $\bar{m}$, and not a set of instantaneous PSFs. A move from here to the Fourier domain is not trivial, but is simplified by assuming that Fourier amplitude and phase are independent. We will write the amplitude as $\mathbf{O}_A$, $\mathbf{M}_A$ and $\mathbf{I}_A$ – vectors of length equal to the number of sampled Fourier points, and will write the phase as $\mathbf{O}_\phi$, $\mathbf{M}_\phi$ and $\mathbf{I}_\phi$. For the amplitude, we have:

$$L(\mathbf{O}_A|\{\mathbf{I}_A\}) = \int_{\bar{\mathbf{M}}_A} \exp(-\frac{1}{2}((\bar{\mathbf{M}}_A \circ \mathbf{O}_A) - \bar{\mathbf{I}})^T \underline{C_{I,A}}^{-1}((\bar{\mathbf{M}}_A \circ \mathbf{O}_A) - \bar{\mathbf{I}}_A)) f(\bar{\mathbf{M}}_A). \quad (3.9)$$

It is possible to change to this vector and matrix notation following e.g. [13] from the subscripted notation of equation 3.8 because the restriction of the object Fourier transform $\mathbf{O}$ to a pixel grid in Fourier space does not lose information so long as the image i goes to zero at the edge of the observed field of view. The symbol $\circ$ represents element-wise multiplication. If we can finally approximate the components of $\mathbf{M}$ to have a large amplitude relative to their dispersion (again, as part of the high-Strehl regime), then marginalising over all possible values of $\mathbf{M}$ is equivalent to adding a "calibration error" to the data covariance:

$$L(\mathbf{O}|\{\mathbf{I}\}) = \exp(-\chi_A^2/2) \times \exp(-\chi_\phi^2/2) \quad (3.10)$$

$$\chi_A^2 = ((\bar{\mathbf{M}}_A \circ \mathbf{O}_A) - \bar{\mathbf{I}}_A)^T (\underline{C_{A,I}} + \underline{C_{A,M}})^{-1}((\bar{\mathbf{M}}_A \circ \mathbf{O}_A) - \bar{\mathbf{I}}_A) \quad (3.11)$$

$$\chi_\phi^2 = (\bar{\mathbf{M}}_\phi - \bar{\mathbf{I}}_\phi - \mathbf{O}_\phi)^T (\underline{C_{I,\phi}} + \underline{C_{M,\phi}})^{-1}(\bar{\mathbf{M}}_\phi - \bar{\mathbf{I}}_\phi - \mathbf{O}_\phi) \quad (3.12)$$

Even in the high-Strehl regime, this is not an especially simple process, as the large covariance matrices are in general highly non-diagonal in the presence of quasi-static speckle errors. The approach of the kernel-phase technique [8, 13] is to ignore Fourier amplitudes and to project the Fourier phase uncertainties onto a subspace that is independent of pupil-plane phase errors to first order:

$$\underline{C_{I,\phi}} + \underline{C_{M,\phi}} = \underline{P}^T \mathrm{Diag}(\{\sigma^2\}) \underline{P}. \quad (3.13)$$

Here $\underline{P}$ is a projection matrix, and $\{\sigma^2\}$ are uncertainties on linear combinations of Fourier phases taken to be independent. We will discuss kernel-phase more in Sect. 3.4.

3.3 Non-redundant Aperture Masking

In non-redundant aperture masking, a mask is placed in the pupil-plane of a telescope to enable more robust measurements of the object Fourier transform O. This robustness comes through two key avenues: higher amplitude of the modulation transfer function M_A, especially at low Strehls, and a smaller dispersion in Modulation Transfer Function (MTF) amplitude and phase when influenced by phase errors [8] (Figure 3.1). The complexity to aperture-masking imaging comes in forming quantities that enable direct χ^2 minimisation over possible object brightness

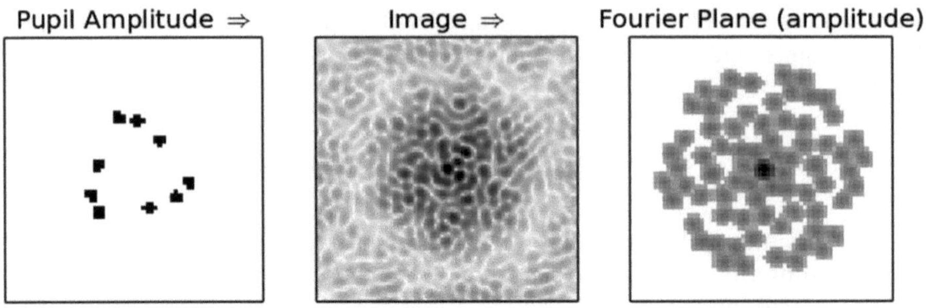

Fig. 3.1 A 9-hole non-redundant aperture mask (*left*), the image formed by this in the case of moderate aberrations (*centre*), showing some asymmetry, and the Fourier amplitude of the image (*right*)

distributions, independently of some distribution of PSFs (see Fig. 3.2). One of these quantities is Fourier amplitude (as in Eq. 3.3), also called visibility amplitude, especially when it applies to the object $|O|$ rather than the image. In order to make an unbiased estimator for amplitude at low signal-to-noise, the squared visibility is typically used for fitting [7]. The second key quantity is the closure-phase, also called bispectral phase.

The closure phase is a quantity derived from the phase of three baselines that form a closing triangle in the telescope pupil. Figure 3.3 shows a single closing triangle. The bispectral point formed by these three complex Fourier amplitudes is given by their triple product, and the phase of this triple product is the closure phase. Once two baselines are defined, the third is given by the closure condition (two defined vectors give the reason for the *bi* in bispectrum). In two dimensional imaging, the bispectrum is therefore a 4-dimensional quantity. Where apertures are considered discrete, there can be a much larger number of bispectrum vectors (closing triangles) than there are either baselines or apertures. The utility of the closure-phase is that a phase piston applied to any subaperture (caused by e.g. turbulence or thermal variations in an instrument) cancels in the closure phase, *so long as each Fourier point maps to a unique baseline in the pupil plane*. This is the property of *non-redundancy*.

For the simple non-redundant aperture mask shown in Fig. 3.3, there are 4 closure-phases, of which only 3 are linearly independent:

$$\beta_1 = \phi_{AB} + \phi_{BC} + \phi_{CA} \tag{3.14}$$

$$\beta_2 = \phi_{AB} + \phi_{BD} + \phi_{DA} \tag{3.15}$$

$$\beta_3 = \phi_{AC} + \phi_{CD} + \phi_{DC} \tag{3.16}$$

$$\beta_4 = \phi_{BC} + \phi_{CD} + \phi_{DA} \tag{3.17}$$

$$= \beta_1 - \beta_2 + \beta_3 \tag{3.18}$$

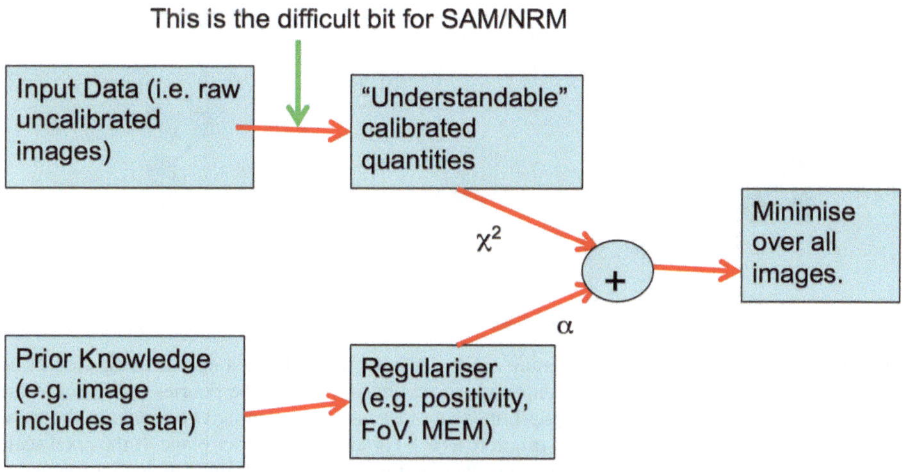

Fig. 3.2 Generalised regularised imaging, showing schematically where the shift from raw pixel data to visibility amplitude and closure-phases occurs. The key difference to imaging with e.g. equation 3.6 is that there is no marginalisation over all possible PSFs – the process of "calibration" is to form observables that relate directly to the target brightness distribution

The final line follows once the simple relationship $\phi_{AB} = -\phi_{BA}$ is appreciated. This follows directly from the well-known property of Fourier transforms of real functions (i.e. the image) – the Fourier transform is the complex conjugate of itself, rotated by 180 degrees. This means that for a 4-hole aperture mask, there are 3 linearly independent closure-phases, 4 closure-phases and 6 baselines. In general, for an N_h hole *non-redundant* aperture mask, there are $N_b = N_h(N_h - 1)/2$ baselines, $N_{\rm cp} = N_h(N_h - 1)(N_h - 2)/6$ closure-phases and $N_{\rm ind} = (N_h - 1)(N_h - 2)/2 = N_b - N_h + 1$ linearly independent closure-phases.

Once squared Fourier amplitudes and closure-phases are formed for an image, the next step is to find appropriate calibrated quantities that refer to the true object brightness distribution. The averaged square amplitude of the MTF (i.e. the power spectrum) and the MTF bispectral phases are estimated from one or several point-spread function references. Following Eqs. 3.3 and 3.4, and neglecting the effects of photon bias which is discussed elsewhere [7], we have the following estimators for amplitude and bispectral phase:

$$\widehat{\mathbf{O}_A^2} = \langle \mathbf{I}_A^2 \rangle / \langle \mathbf{I}_{A,\rm Cal}^2 \rangle \tag{3.19}$$

$$\widehat{\mathbf{O}_\beta} = \langle \mathbf{I}_\beta \rangle - \langle \mathbf{I}_{\beta,\rm Cal} \rangle \tag{3.20}$$

Here the subscript "Cal" represents observations of a point-source calibrator, used to estimate the modulation transfer function. Note that this calibration of phase by subtraction is generally applicable to any point-spread function (Eq. 3.4), but the statement $\langle \mathbf{I}_{\beta,\rm Cal} \rangle = 0$ is only approximately true in general in the presence

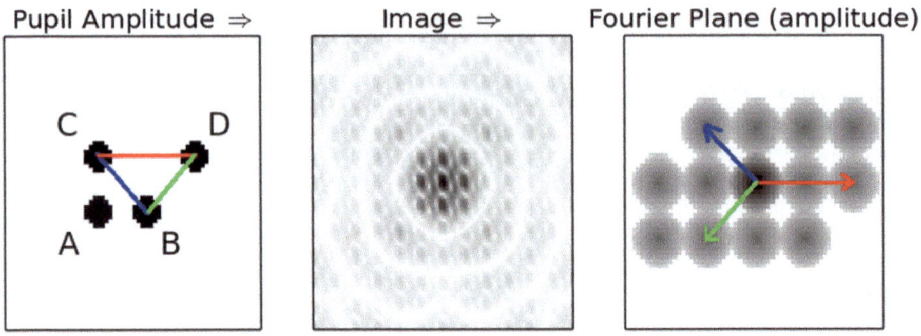

Fig. 3.3 A 4-hole non-redundant aperture mask (*left*), the image formed by this in the case of no aberrations (*centre*) and the Fourier amplitude of the image (*right*). The Fourier transform of the image is the autocorrelation of the, pupil, meaning that baselines joining two points in the pupil plane correspond to vectors joining to origin to a (u, v) point in the Fourier plane. If the pixel scale on the left is in Δx metres, then the Fourier plane on the right has pixel units of spatial frequency in (dimensionless) units of $\Delta x/\lambda$, with λ the observing wavelength

of phase aberrations [8]. In order for easy model-fitting or image reconstruction of the object brightness distribution o (i.e. Fig. 3.2), for example using the first version of the OIFITS data format, these estimators are stored together with only the diagonal elements of their covariance matrices. Even neglecting photon bias, Eq. 3.19 is in general a biased estimator for object square Fourier amplitude, with a relative uncertainty of order $\mathrm{SNR}_{\mathrm{Cal}}^{-2}$, where $\mathrm{SNR}_{\mathrm{Cal}}$ is the signal-to-noise ratio of the averaged calibrator square Fourier amplitudes. This is generally not taken into account, because typical observations have very high values of $\mathrm{SNR}_{\mathrm{Cal}}$, and because effects of non-Gaussian errors and incorrectly estimated covariance matrices for $\widehat{\mathbf{O}_A^2}$ are more significant effects. For aperture-masking observations behind adaptive optics, the effect of photon bias on the bispectrum is typically negligible, but not for the power spectrum.

Now that we have considered the process of aperture-masking data calibration, let us consider and compare with the simpler process of imaging with an unobstructed aperture. Considering the data in the Fourier plane and neglecting PSF variations, this process amounts to measuring $\mathbf{O}_A$ and $\mathbf{O}_\phi$ for all pixels in the Fourier plane, and calibrating the Fourier data ready for image reconstruction or model fitting:

$$\widehat{\mathbf{O}_A} = \langle \mathbf{I}_A \rangle / \langle \mathbf{I}_{A,\mathrm{Cal}} \rangle \tag{3.21}$$

$$\widehat{\mathbf{O}_\phi} = \langle \mathbf{I}_\phi \rangle - \langle \mathbf{I}_{\phi,\mathrm{Cal}} \rangle \tag{3.22}$$

For a given incident flux rate, we can then ask how the photon shot noise limited signal to noise compares between non-redundant aperture masking and full pupil imaging. Figure 3.4 shows the result for the 9-hole aperture mask illustrated in Fig. 3.3. Both techniques have comparable signal-to-noise for the longest baselines,

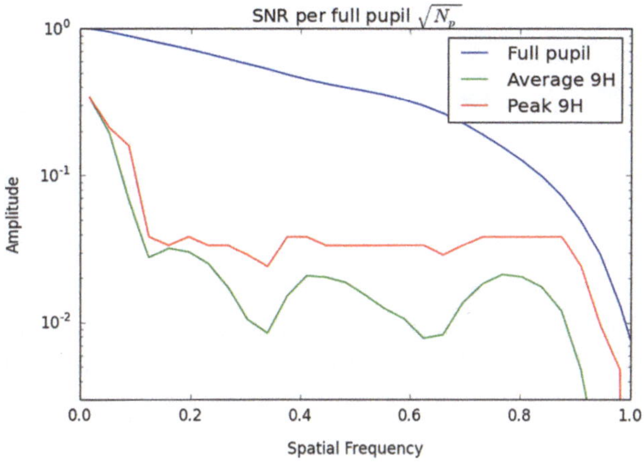

Fig. 3.4 Fourier amplitude as measured in units of photons as a function of spatial frequency, scaled to a photon rate of 1 detected photon per unobstructed aperture. Three cases are shown: an unobstructed pupil, the azimuthally averaged amplitude for observations of an unresolved object with a 9-hole non-redundant aperture mask, and the peak azimuthal amplitude for a 9-hole non-redundant aperture mask. As the 9-hole aperture-mask blocks most of the light, the amplitude is significantly lower at zero spatial frequency (a measure of total object flux within the field of view), but is comparable to the unobstructed pupil for the highest spatial frequencies

but aperture masking loses significant signal to noise at short and intermediate baselines. The comparison is much less favourable for aperture masking if readout noise (or thermal noise from a warm mask) are significant. This demonstrates, as expected, that aperture masking should only be used when full pupil imaging is dominated by calibration (e.g. quasi-static phase) errors, and if information contained in the longest baselines is essential for the science goal.

3.4 Kernel and Bispectral Phase

In recent years, the Fourier imaging techniques have been applied to redundant pupils geometries (such as an unobstructed pupil) in the case of adaptive optics imaging at moderately high Strehl ratios [13]. In kernel-phase imaging, the calibrated Fourier phase is viewed as a vector of phases at discrete locations (e.g. Eq. 3.22), but all phase information is not used. The Fourier phase information is split into 2 subspaces, one that is independent to first order of pupil-plane aberrations, the kernel-phase, and one that can be used to determine the pupil-plane phase aberrations, which we will call the eigenphase [14].

In aperture masking when considering sub apertures as point apertures, bispectral phases are closure-phases, and all closure-phases are independent of pupil-plane phase. Linear combinations of Fourier phases that are independent of pupil-plane phase are kernel-phases, so the terms *kernel-phase*, *closure phase* and *bispectral*

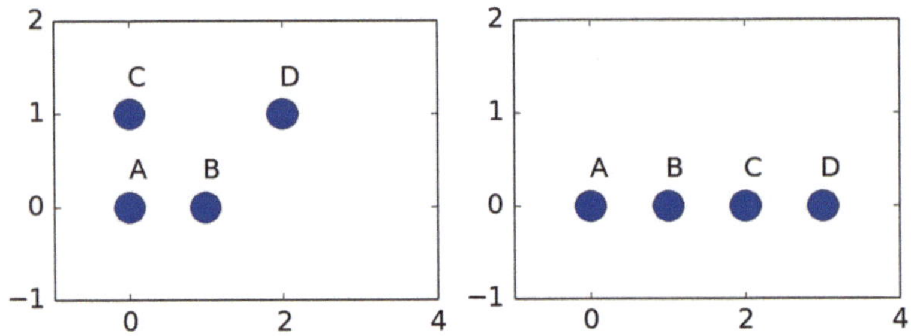

Fig. 3.5 *Left*: A four-element non-redundant array with 3 kernel-phases and 3 independent bispectral phases. *Right*: A four-element redundant array with 1 kernel-phase and 2 independent bispectral phases

phase can be used almost interchangeably. This is not the case for redundant pupil geometries – we will briefly consider the differences here.

In the example on the left of Fig. 3.5, there are 4 sub-apertures, 6 baselines, 4 closure-phases and 3 statistically independent closure-phases or kernel-phases (discussed in Sect. 3.3). Consider the redundant example on the right in Fig. 3.5, where we label phase for the baseline joining apertures A and B as ϕ_{AB}. Due to redundancy, we have the identities:

$$\phi_{AB} = \phi_{BC} = \phi_{CD} \tag{3.23}$$

$$\phi_{AC} = \phi_{BD}, \tag{3.24}$$

Instead of 6 baselines, we only have 3, formed by the baselines between sub apertures A-B, A-C, and A-D. There are two bispectral phases, which are linearly independent of each other:

$$\beta_1 = \phi_{AB} + \phi_{BC} + \phi_{CA} \tag{3.25}$$

$$= 2\phi_{AB} - \phi_{AC} \tag{3.26}$$

$$\beta_2 = \phi_{AC} + \phi_{CD} + \phi_{DA} \tag{3.27}$$

$$= \phi_{AC} + \phi_{AB} - \phi_{AD} \tag{3.28}$$

However, neither of these are kernel-phases. As an example, imagine a 0.6 radian piston of aperture C. To compute the observed phase on baseline *AB*, we need to sum the complex fringe visibilities on the contributing baselines *AB*, *BC*, and *CD*. We can use the small angle approximation, with $\mathrm{Arg}(\exp(i\phi_1) + \exp(i\phi_2)) \approx \phi_1 + \phi_2$, or simply sum complex visibilities. Either approach gives $\beta_1 = -0.3$ and $\beta_2 = +0.3$. The single kernel-phase in this example is formed from a linear combination of

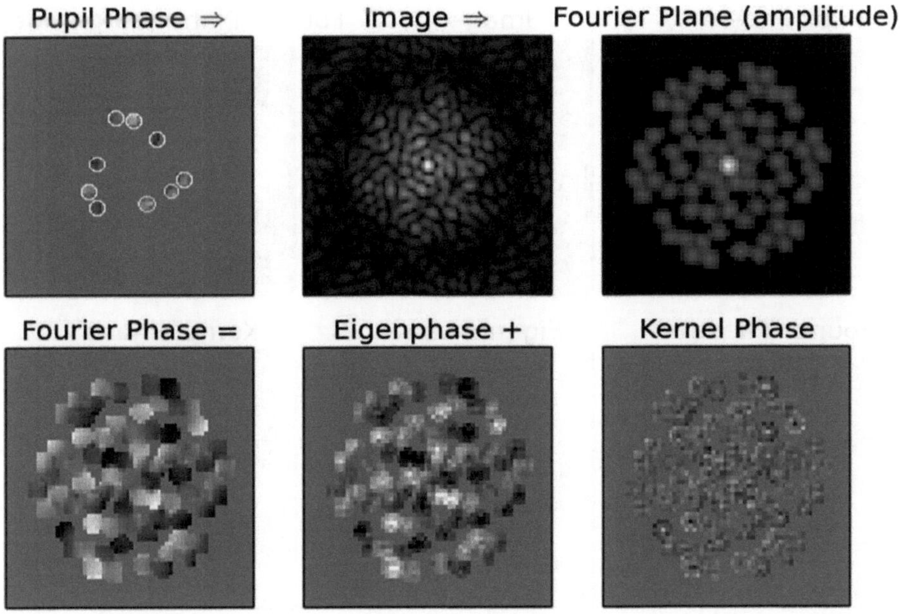

Fig. 3.6 An illustration of splitting the information in the Fourier plane into amplitude, eigen-phase [14] and kernel-phase [13] for a 9-hole non-redundant aperture mask. Phase aberrations are dominated by pistons on each hole, which are not seen in the kernel-phase. The phase colour map corresponds to ± 1.25 radians of phase, while the amplitude is shown in a cube-root stretch

these bispectral phases:

$$\theta_1 = \beta_1 + \beta_2 \tag{3.29}$$

$$= 3\phi_{AB} - \phi_{AD} \tag{3.30}$$

This observable is robust to small phase errors – indeed, in a comparable way to closure-phases in the non-redundant array example. For larger phase errors, the redundant array has a clear disadvantage. For example, imagine a $2\pi/3$ piston on sub-aperture C. This would produce fringes on baselines AB, BC and CD that would cancel when they are summed, producing no fringes and an indeterminate phase.

We can also apply the kernel-phase technique to highly redundant apertures, including for example non-redundant aperture masks if baselines *within* each sub aperture are included. The application of kernel-phase to a non-redundant mask and the Keck telescope pupil is shown in Figs. 3.6 and 3.7. In this case, the pupil is represented by a square pixel grid. Although kernel-phases and eigen-phases are linear combinations of Fourier phases so can not be represented in the Fourier plane, we can project the Fourier phases onto the subspaces spanned by the set of eigen-phases and kernel-phases. It is this representation of phase that is illustrated in Figs. 3.6 and 3.7. The peak to valley phase aberrations in this example are ~2

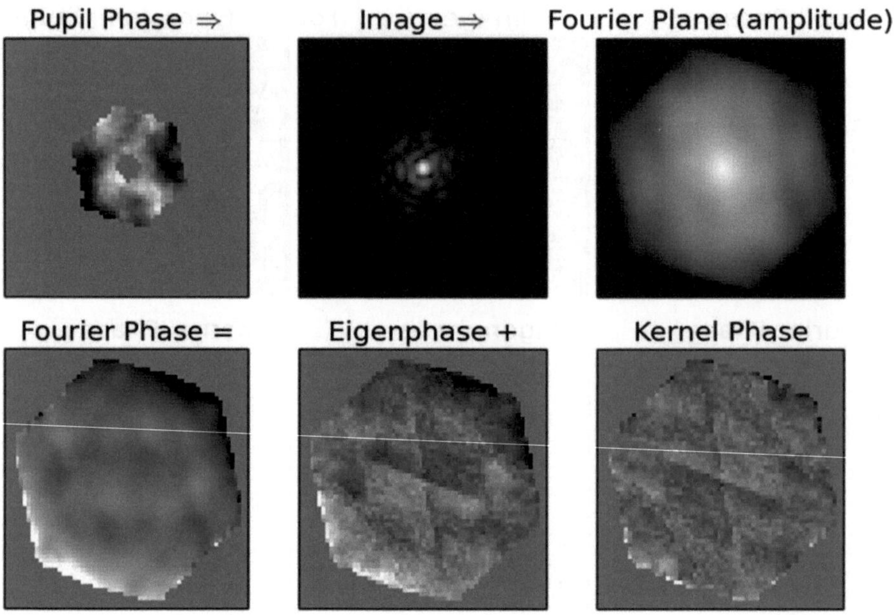

Fig. 3.7 The same as Fig. 3.6 but for an unobstructed pupil. The redundancy in the pupil means that aberrations can be clearly seen in the Fourier amplitude map, and Fourier phase does not separate as well into kernel-phase and eigen-phase

radians, which produces Strehl ratios of order 0.5 – the *moderate* Strehl ratio regime. In this regime, the redundant pupil geometry means that high Fourier amplitudes are retained for the case of the non-redundant mask, and kernel-phase remains robust to sub-aperture pistons.

3.5 Applications of Aperture-Masking Imaging

Given the improved calibration to quasi-static speckle errors for aperture-masking imaging over imaging with an unobstructed pupil, there are a small number of significant uses for the technique. It is clearly not the technique of choice when key observables are at moderate spatial frequency (e.g. objects resolved by several λ/D that are faint enough to require long integrations). There are two key uses for aperture mask imaging that have emerged recently. Each of them applies only to observations where structures to be resolved are near the diffraction limit (e.g. a field of view less than about $4\lambda/D$, with D the telescope diameter).

3.5.1 Precision Binary Astrometry

It is possible to measure accurate positions of stars in a partially resolved binary star system with aperture-masking, because the calibrated observables (Eqs. 3.19 and 3.20) relate directly to the object brightness distribution. There are numerous examples of this in the literature, for example [4], where binary astrometry is possible down to separations of $\sim 0.5\lambda/D$. This is by no means a fundamental limitation, although contrast and separation become degenerate at the smallest separations, and require signal-to-noise inversely proportional to the cube of the separation. Unpublished orbits (e.g. [16]) include astrometry down to $\sim 0.3\lambda/D$.

3.5.2 Faint, Low-Strehl Imaging

Perhaps surprisingly, laser guide star observations of relatively faint object (down to K$\sim$15 with Keck) can be observed with higher fidelity using an aperture mask than with full pupil imaging. A recent example of this is [3], where SDSS J105213.51+ 442255.7AB, a binary brown dwarf with *K* magnitudes of 15 and 15.5, was observed as part of an orbit monitoring programme down to a separation of $\sim$30 milli-arcsec, with 1 milli-arcsec uncertainties. Due to the highly variable nature of laser guide star PSFs, these astrometric measurements would not have been possible without the use of the aperture mask. By searching over the space of all possible binary stars, data such as these are also useful for surveying for companions [11], or imaging structures at the diffraction limit. Imaging complex, barely resolved structures is more difficult, because visibility amplitude calibrates very poorly, so only point antisymmetric image components can be seen. Nonetheless, this is an exciting future prospect for aperture masking.

3.5.3 High-Contrast Imaging (e.g. LkCa 15)

Aperture masking imaging has also been successfully used to detect faint structures and objects next to bright stars. Perhaps the most well known example of this is LkCa 15, a young star with a significant gas and dust depletion in the inner 20 AU of its disc [18], likely due to planetary formation. Aperture-masking imaging showed faint resolved structures, argued to be associated with planet formation [10]. Subsequent observations of this system showed evidence for orbital motion [9] and also variability (e.g. compare Fig. 3.8 to [8]). One difficulty in interpreting these observations in general is that imaging based on closure-phases is not sensitive to point-symmetric structure, and both point-symmetric structure and barely-resolved

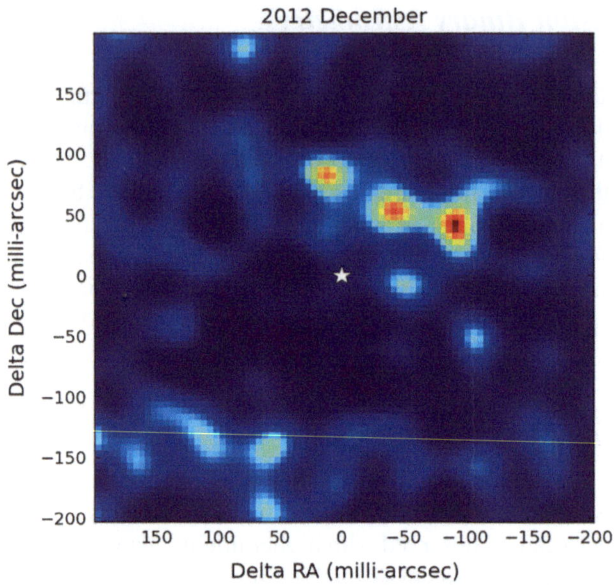

Fig. 3.8 An example maximum entropy image reconstruction of LkCa 15, using the same data as presented in [9], and the algorithms as presented in [8]. As this is a fit to linear combinations of closure-phases only (i.e. kernel-phases), it is not sensitive to symmetric structures. The central point source is modelled separately from the image, an the brightest structures seen have a contrast of ∼250:1 from the central source

flux close to the star is removed in an image regularisation process. As long as these issues are known, they can be robustly taken into account when interpreting aperture mask imaging [1].

3.6 Conclusions

The general problem of narrow-field imaging in astronomy is really a problem of reconstructing a model (or image) of an object brightness distribution from a noisy image in the presence of an only partially characterised PSF. This problem has a long history, and placing a mask with non-redundant sub aperture spacings over the pupil of a telescope is 150 years old. Through the use of an aperture-mask, the primary observables of closure-phase (or kernel-phase) and visibility amplitude can be accurately calibrated, enabling direct model fitting or image reconstruction based on these observables. Where Strehl ratios are high or targets are very faint, aperture masking may not be the best technique, but it excels at resolving and robustly calibrating structures right at the diffraction limit of a telescope. Given that adaptive optics is unlikely to deliver very high Strehl ratio images with high sky coverage in at least the next decade, aperture mask imaging will retain its role as the technique of choice for single-telescope diffraction limited imaging for some time to come.

References

1. Cheetham, A., Huélamo, N., Lacour, S., de Gregorio-Monsalvo, I., Tuthill, P.: Near-IR imaging of T Cha: evidence for scattered-light disc structures at Solar system scales. Mon. Not. R. Astron. Soc. **450**, 1 (2015)
2. Doolittle, E., Hough, G.W.: Catalogue and re-measurement of the 648 double stars discovered by Professor G.W. Hough. Publ. Univ. Pa. Flower Astron. Observ. **3**, 1 (1907)
3. Dupuy, T.J., Liu, M.C., Leggett, S.K., Ireland, M.J., Chiu, K., Golimowski, D.A.: The mass-luminosity relation in the L/T transition: individual dynamical masses for the new J-band flux reversal binary SDSSJ105213.51+442255.7AB. Astrophys. J. **805**, 56 (2015)
4. Evans, T.M., Ireland, M.J., Kraus, A.L., et al.: Mapping the shores of the Brown Dwarf desert. III. Young moving groups. Astrophys. J. **744**, 120 (2012)
5. Fizeau, H.: Prix bordin: Rapport sur le concours de l'anneé 1867. C. R. Acad. Sci. **66**, 932 (1868)
6. Gezari, D.Y., Labeyrie, A., Stachnik, R.V.: Speckle interferometry: diffraction-limited measurements of nine stars with the 200-INCH telescope. Astrophys. J. **173**, 1 (1972)
7. Gordon, J.A., Buscher, D.F.: Detection noise bias and variance in the power spectrum and bispectrum in optical interferometry. Astron. Astrophys. **541**, 46 (2012)
8. Ireland, M.J.: Phase errors in diffraction-limited imaging: contrast limits for sparse aperture masking. Mon. Not. R. Astron. Soc. **433**, 1718 (2013)
9. Ireland, M.J., Kraus, A.L.: Orbital motion and multi-wavelength monitoring of LkCa15 b. IAU Symp. **299**, 199–203 (2014)
10. Kraus, A.L., Ireland, M.J.: LkCa 15: a young exoplanet caught at formation? Astrophys. J. **745**, 5 (2012)
11. Kraus, A.L., et al. 2015. A survey of Kepler objects of interest. Accepted for the Astronomical Journal on 17 Apr 2016. http://adsabs.harvard.edu/abs/2016arXiv160405744K
12. Lohmann, A.W., Weigelt, G., Wirnitzer, B.: Speckle masking in astronomy – triple correlation theory and applications. Appl. Opt. **22**, 4028 (1983)
13. Martinache, F.: Kernel phase in Fizeau interferometry. Astrophys. J. **724**, 464 (2010)
14. Martinache, F.: The asymmetric pupil fourier wavefront sensor. Publ. Astron. Soc. Pac. **125**, 422 (2013)
15. Michelson, A.A., Pease, F.G.: Measurement of the diameter of Alpha-Orionis with the interferometer. Astrophys. J. **53**, 249 (1921)
16. Rizzuto, A.C., Ireland, M.J., Dupuy, T.J., Kraus, A.L.: Dynamical masses of young stars. I. Discordant model ages of upper scorpius. ApJ **817**, 164 (2016)
17. Tuthill, P.G., Monnier, J.D., Danchi, W.C.: A dusty pinwheel nebula around the massive star WR104. Nature **398**, 487 (1999)
18. van der Marel, N., van Dishoeck, E., Bruderer, S., Perez, L.M., Isella, A.: Gas density drops inside dust cavities of transitional disks around young stars observed with ALMA. Astron. Astrophys. **579**, A106 (2015)
19. Weigelt, G., Balega, Y.Y., Preibisch, T., Schertl, D., Smith, M.D.: Bispectrum speckle interferometry of the massive protostellar object S140 IRS 1: evidence for multiple outflows. Astron. Astrophys. **381**, 905 (2002)

Chapter 4
Optical Long Baseline Interferometry

Jean-Baptiste Le Bouquin

4.1 Linking the *Object* to the *Interference Fringes*

The purpose of Optical Long Baseline Interferometry (OLBI) is to overcome the diffraction limit of classical single-aperture telescopes by combining coherently the light coming from several small apertures separated by large distances. The link between these interference fringes (what an interferometer measures) and the object image (what the astronomer is interested in) can be expressed mathematically by modeling three simple progressive steps:

- interference fringe from a single emitter,
- linear relation between a small displacement on the sky and of the fringe,
- integration over many non-coherent emitters to build an extended source.

At the end of these three steps, we obtain the Van Cittert – Zernike theorem.

4.1.1 Interference of a Single Emitter

The light travels through the two arms of the interferometer (Fig. 4.1). The wave from one telescope is delayed by the projection of the baseline **B** in the direction of observation **S**. The beam of the second telescope is delayed artificially by the so-called delay lines by a variable quantity δ.

$$E_1 = A\, e^{\frac{2i\pi}{\lambda}\, \mathbf{S}\cdot\mathbf{B}} \tag{4.1}$$

J.-B. Le Bouquin (✉)
Université Grenoble Alpes and CNRS, IPAG, F-38000 Grenoble, France
e-mail: jean-baptiste.lebouquin@obs.ujf-grenoble.fr

H.M.J. Boffin et al. (eds.), *Astronomy at High Angular Resolution*, Astrophysics and Space Science Library 439, DOI 10.1007/978-3-319-39739-9_4

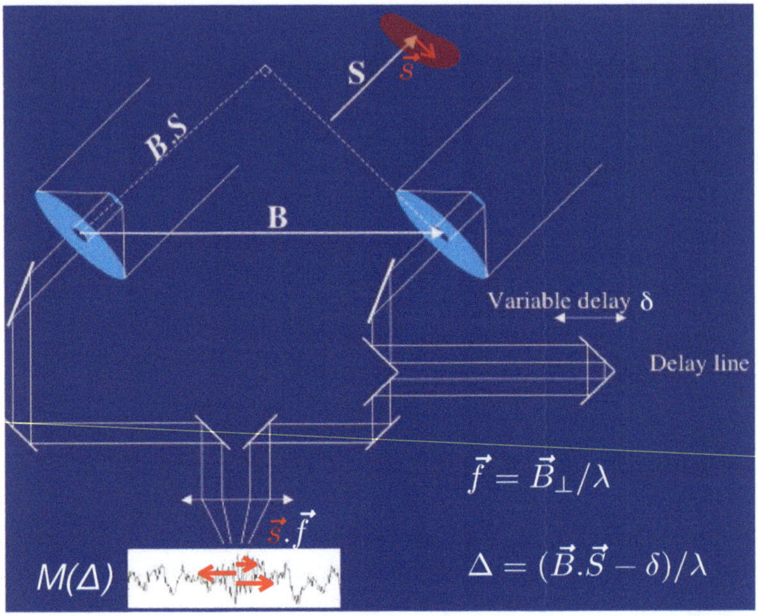

Fig. 4.1 A schematic representation of a two telescope optical interferometer observing an extended source. At the output of the interferometer, the fringe patterns created by each individual emitters are slightly displaced with respect to each other. This results in a reduced fringe contrast when integrating over all the emitters that compose the source. The mathematical relationship between the source angular brightness distribution and the resulting fringe contrast is given by the so-called Van Cittert – Zernike theorem

$$E_2 = A\, e^{\frac{2i\pi}{\lambda}\,\delta} \tag{4.2}$$

The intensity at the recombined output of the interferometers is given by the superposition of these fields:

$$M =< (E_1 + E_2)(E_1 + E_2)^* > \tag{4.3}$$

The two waves interfere because they originate from the same emitter. We note here that we suppose that the temporal coherence of light is infinite, which corresponds to an hypothetical *monochromatic wave* (first hypothesis). It is beyond the scope of this paper to explore the behaviour of polychromatic light. The flux measured at the output of the interferometer therefore writes:

$$M =< A.A^* > + < A.A^* > \cos(2\pi \frac{\mathbf{S} \cdot \mathbf{B} - \delta}{\lambda}) \tag{4.4}$$

The measured intensity depends on the optical path difference (OPD), hereafter $\Delta = \mathbf{B} \cdot \mathbf{S} - \delta$. Depending on OPD, the interference is constructive ($\Delta = 0$) or destructive ($\Delta = \lambda/2$). Equation 4.4 describes the so-called fringe pattern. Here I rewrite this cosine pattern as the real part of the OPD modulation and introduce the total photometry $I =< A.A^* >$, so that the following notation will be easier:

$$M/I = 1 + \Re\{\exp(2i\pi\Delta/\lambda)\} \tag{4.5}$$

4.1.2 *Linearity Between the Emitter and the Fringes Displacements*

The next step is to understand how the fringe pattern changes when the emitter is slightly displaced in the plane of the sky. We consider the emitter displacement **s** is a *small angle* such as $|\mathbf{s}| << 1$rad (source is compact on sky, second hypothesis). We consider the internal delay δ is not changed. The fringe pattern is displaced at the recombination point such as:

$$M/I = 1 + \Re\{\exp(2i\pi\ \Delta/\lambda + 2i\pi\ \mathbf{B}_{\perp} \cdot \mathbf{s}/\lambda)\} \tag{4.6}$$

The fringe displacement is proportional to the emitter angular displacement and to the baseline projected onto the plane of the sky. Here you can think of the whole interferometer as a big Young experiment, which is another way to introduce this mathematical formalism. The key concept is that the angular displacement of the emitter is proportional to the fringe displacement:

$$M/I = 1 + \Re\{\exp(2i\pi\ \Delta/\lambda + 2i\pi\ \mathbf{f} \cdot \mathbf{s})\} \tag{4.7}$$

where the proportionality factor $\mathbf{f} = \mathbf{B}_{\perp}/\lambda$ is often called the spatial frequency. We will see now why.

4.1.3 *Integration Over Many Emitters*

The final step is to consider an extended target as a collection of emitters. So we simply integrate the previous equation but for all angles over the target image. To do so, we suppose that the individual emitters have *zero mutual coherence* (third hypothesis). That is correct if the source has no spatial coherence, which is always true for astronomical objects (no spatially resolved optical sources equivalent to masers in the radio have been observed so far).[1] The integration writes:

$$M_0 = \int I(\mathbf{s}) + \Re\{\, I(\mathbf{s})\ \exp(2i\pi\ \Delta/\lambda + 2i\pi\,\mathbf{f}\cdot\mathbf{s})\}\ \mathrm{d}\mathbf{s} \tag{4.8}$$

$$M_0 = \int I(\mathbf{s})\mathrm{d}\mathbf{s}\ +\ \Re\{\exp(2i\pi\ \Delta/\lambda) \int I(\mathbf{s})\exp(2i\pi\,\mathbf{f}\cdot\mathbf{s})\mathrm{d}\mathbf{s}\ \} \tag{4.9}$$

$$M_0/I_0 = 1 +\ \Re\{\ V\ \exp(2i\pi\ \Delta/\lambda)\ \} \tag{4.10}$$

with $I_0 = \int I(\mathbf{s})\mathrm{d}\mathbf{s}$ the total flux of the source and $V(\mathbf{f}) = \mathrm{FT}\{I(\mathbf{s})\}/I_0$ its normalized Fourier transform at the spatial frequency **f**. Thus, once integrated over all the emitters of the source, the new contrast of the fringe pattern is the Fourier component of the angular brightness distribution of the source, at the spatial frequency **f** defined by the projected baseline. This is the so-called Van Cittert – Zernike theorem.

4.2 Interpreting Interferometric Observations

To summarise the foundation of the technique: by modulating the OPD, we form fringes whose contrast and phases define the measured complex visibility $V(\mathbf{f})$, the latter being the 2D Fourier Transform of the angular brightness distribution $I(\mathbf{s})$ of the source at the spatial frequency $\mathbf{f} = \mathbf{B}_\perp/\lambda$. How does this work in practice to measure something of an astrophysical target?

As illustrated in Fig. 4.2, one can define three regimes: the source is partially resolved, the source is resolved with a sparse uv-sampling, and the source is resolved with a good uv-sampling (many **f** observed).

4.2.1 Partially Resolved: Diameter Measurements

Measuring diameters has been the first use of astronomical optical interferometry, by Michelson and Pease in 1921 [1]. The authors measured the apparent diameter of Betelgeuse, which is arguably the largest star in the sky. Measuring diameters is still one of the major applications for the technique. Its most important contributions to astrophysics are closely related to the distance scale ladder:

- The calibration of surface brightness relations for stars (link between color and intrinsic size, so intrinsic luminosity).

[1] An illustrative example of coherent source is an hypothetical object made of two fibres fed by the same laser source. The two individual emitters of this "object" do have strong mutual coherence.

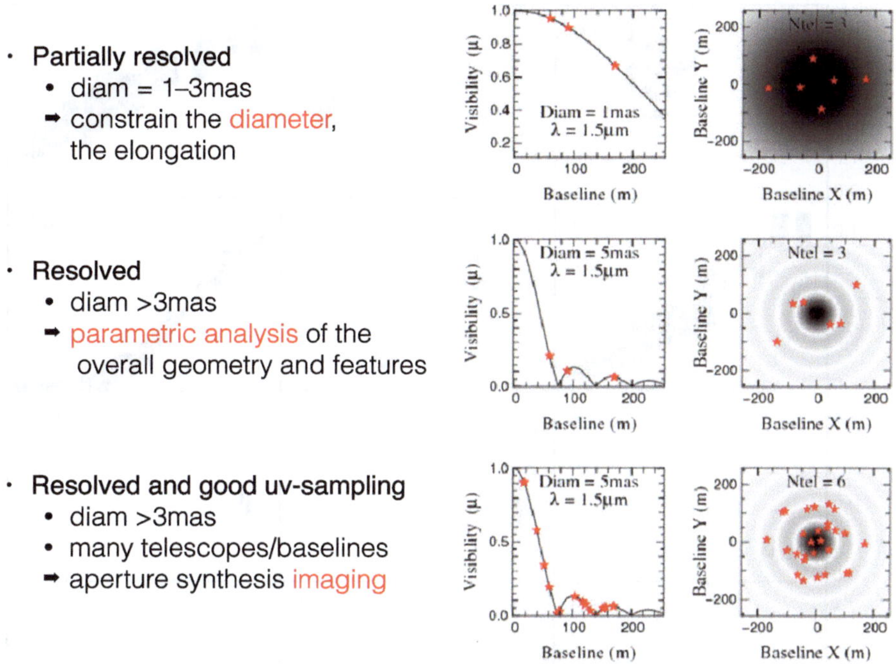

Fig. 4.2 The three observing regimes of optical interferometry with a modern facility. The *middle panels* show an illustrative target visibility versus the radial spatial frequency (*black line*), overlaid with some observations (*red stars*). The *right panels* show the corresponding uv-sampling of the spatial frequencies

- The calibration of the Cepheid distance scale via a direct measurement of the size of their pulsation, with the Baade-Wesselink method.

The last 15 years have seen a renewed interest for stellar physics, in particular stellar surfaces processes characterisation, to better interpret the discovery and characterization of exo-planets. Of particular interest is the prospect of synergy between asteroseismology and OLBI for precise diameter measurements.

But let's illustrate this section with a completely different, and more dynamic, example. Figure 4.3 shows the visibility versus spatial frequencies, for different dates after the Nova Delphini exploded in 2013. All the visibility measurements are represented on the same scale. With time, the visibility drops more and more rapidly versus spatial frequencies, indicating the Nova is growing in size. The basic ingredient is that the Fourier Transform of a disc with uniform brightness is the well known Airy function. If this simple model is applicable to the observed object, it becomes possible to estimate the diameter even with only one visibility measurement. Therefore one can provide an independent guess of the diameter every night, with maybe less than 1 h of 2-telescope observing time.

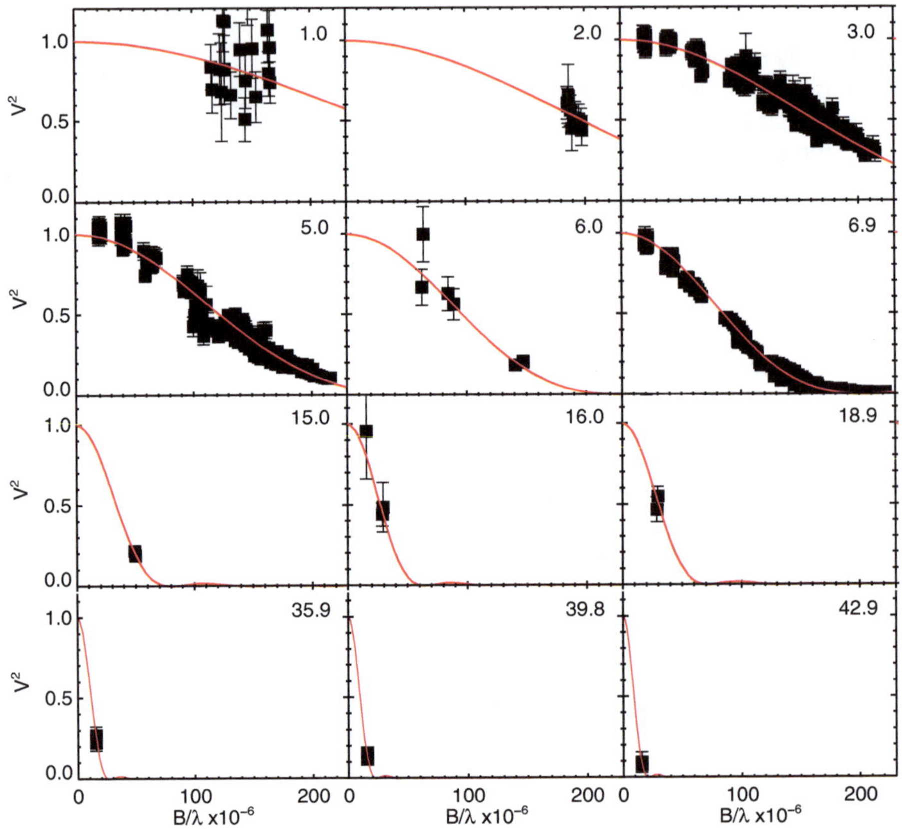

Fig. 4.3 Sub-sample of the visibility curves versus spatial frequencies observed on the Nova Delphini 2013 with the CHARA array. The time in days since the explosion is reported in the *upper-right corners*. The *red lines* are the best fit model with a uniform disc model (Airy function). The best fit diameters range from 0.5 mas at day 1 up to 10 mas at day 40

The authors monitored the measured size as a function of time after the explosion. They revealed that the dimension increased by more than a decade in 40 days. This very nice observational result was published in Nature recently [2]. Combining the expansion in apparent size and radial velocities, the authors could determine the distance of the explosion at 4.5 kpc.

Of course we don't want to measure only diameters. A technique dedicated to High Angular Resolution should say something about more complex geometries. To do so, one needs to resolve sufficiently the source and to gather a sufficient number of measurements. It is still predominantly done through a parametric analysis.

4.2.2 Parametric Analysis

Binaries are typical objects that are best studied with parametric modelling. The study of multiple systems is the second major application for optical interferometry. Recent surveys for binary search include more than 100 targets observed within 15 nights (see [3] and [4]). Figure 4.4 is an example of a difficult but successful parametric analysis: the interacting binary SS Lep. Clearly, the complex visibility curves cannot be reproduced with a single uniform disc (Airy function).

Past photometry and spectroscopy already revealed the components of this system: a main-sequence AV, a MIII evolved giant, and a cool shell dominating at infrared wavelengths. A simple model of orbiting MIII+AV and a surrounding shell was able to reproduce the full complexity of the interferometric observations at any time. This model provides: (a) the stellar diameters, (b) the orbital motion and thus the individual masses, and (c) the individual Spectral Energy Distributions (SED). Altogether the authors found that the system is *not* filling its Roche lobe,

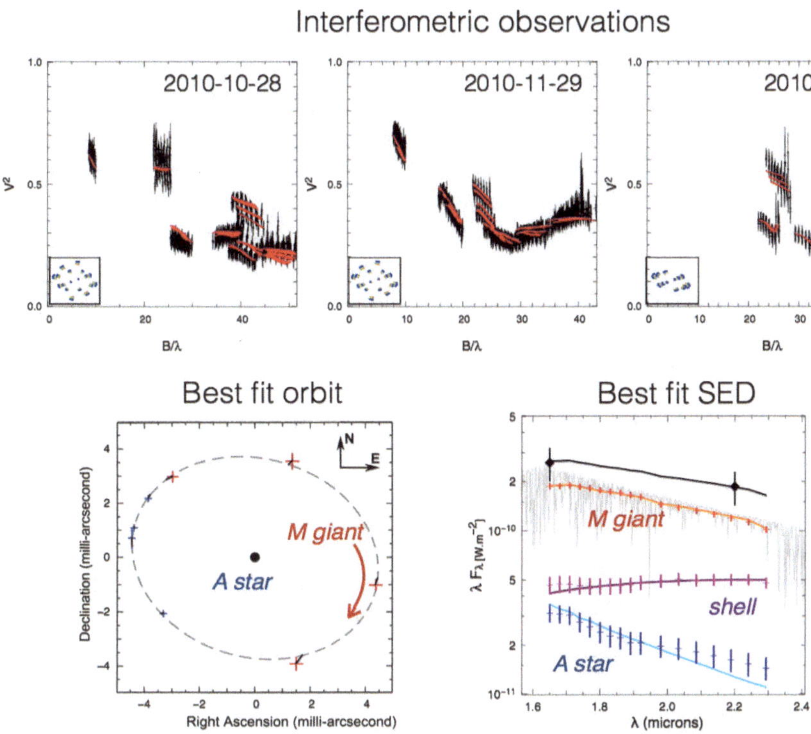

Fig. 4.4 Sample of observations (*top*) of the interacting binary SS Lep with the VLTI. Each subplot represent typically few hours of observation with four telescopes. The *red lines* is the best-fit binary model whose primary component is marginally resolved (M giant), plus a surrounding shell. The recovered orbital motion and the SED decomposition are shown in the lower panels

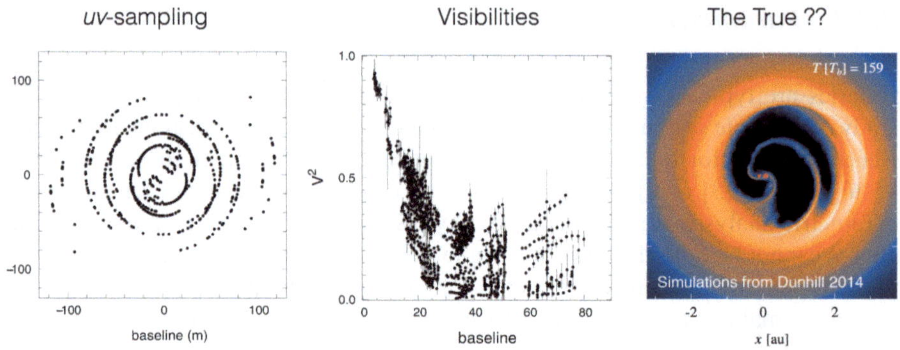

Fig. 4.5 Observations on a young binary whose parameters are similar to SS Lep (Fig. 4.4). The uv sampling is very good, but it is still impossible to build a geometrical model to capture the spatial complexity. Thus the fitted parameters are model-dependent and cannot be used quantitatively

and they had to propose an alternative scenario to explain the mass-transfer. These results have been published by [5].

It is critical to understand that optical interferometry could deliver all these new informations only because the *underlying model was adequate*. This is well illustrated in Fig. 4.5 which shows observations of a binary with PIONIER at VLTI. This young binary has similar parameters than SS Lep, but with a very complex circumbinary disc with possibly accretion streamers. This target was observed a lot, as seen in the well sampled uv-plane. But even with these high quality observations the binary orbit or its circumstellar environment cannot be constrained. The geometry is too complex and rapidly changing.

In the data we obviously see the binary motion, we see the contribution of the shell, but so far no proper geometric model able to reproduce the observation quantitatively has been found. Therefore there is no scenario that can reconcile all the data. This examples leads to point out that, for the parametric analysis to work, the model needs to represent *all* the (true) complexity of the target up to the dynamic range corresponding to the accuracy of the data (typically $\approx$15 in current instruments). The consequence is that in order to match the model complexity one needs to gather enough observations in the uv plane.

The availability of moderate ($\approx$100) or high spectral resolution ($\approx$1000) opens completely new observables for the parametric analysis technique. The most common is the *differential visibility* between the continuum and a spectral line. The main interest of differential quantities is that they are often much more robust against calibration issues. The differential quantities are generally linked to difference of size or subtle astrometric shifts. For interferometers with the highest spectral resolution, this technique provides interesting clues on the dynamic of the emitting/absorbing material at the highest angular resolution. Many results of modern interferometry have been obtain with such an analysis [9–11].

4.2.3 Aperture Synthesis Imaging

Of course, the ultimate goal of any high angular resolution technique is to unveil unpredicted morphologies in a model-independent way. This requires image reconstruction, or aperture synthesis imaging from the interferometry data. The technique boils down to reconstruct the image that would have been observed by the gigantic telescope of similar diameter but with a sparse uv-sampling . In order to inverse the sparsely populated Fourier Transform, it is necessary to add some regularisation in order to prescribe how the missing spatial frequencies should be interpolated. Therefore the result may depend on the regularisation choice. Accordingly, to perform unambiguous image reconstruction, one needs a good sampling of the baselines in term of size and orientation. The reader is referred to the chapter by Baron (this book) for a thorough description of the image reconstruction techniques.

Figure 4.6 shows a dense uv-coverage obtained at VLTI, with several nights of observations of a Mira-type star R Car that involve relocating the telescopes between each nights. These observations were taken during technical time in order to demonstrate the imaging power of VLTI on true data. The observed visibilities follow the overall trend for a disc (Airy function), but with major discrepancies.

Six image reconstructions from this dataset are shown in the bottom panels, from different softwares using different regularisations. Clearly, the basic structures are consistently recovered on all reconstructions: a star with spots and an envelope. This is a superb result. But the subtle details are not fully consistent between reconstructions. Again this refers to the fact that the object complexity at the dynamic range of the observation is still larger than the number of measurements.

Figure 4.7 presents another iconic example: the image reconstruction of Altair by Monnier et al. [6]. This result is in apparent contradiction with my previous statement. The authors gather only a limited uv-sampling, and still they were able to reconstruct an impressive model-independent image. But the key is that the object intrinsic complexity is manageable. Altair has no circumstellar envelope as R Car. Thus the authors forced the algorithm, through the use of a carefully defined regularisation and prior, to put all the reconstructed flux inside an ellipse corresponding to the stellar photosphere. Its shape was determined simultaneously than the reconstruction. This considerably reduces the degrees of freedom and allows a better convergence.

Lastly, the availability of spectral resolution opens completely new ways to implement the aperture synthesis imaging approach. A large variety of algorithms have been proposed and used to exploit true observations [12–14].

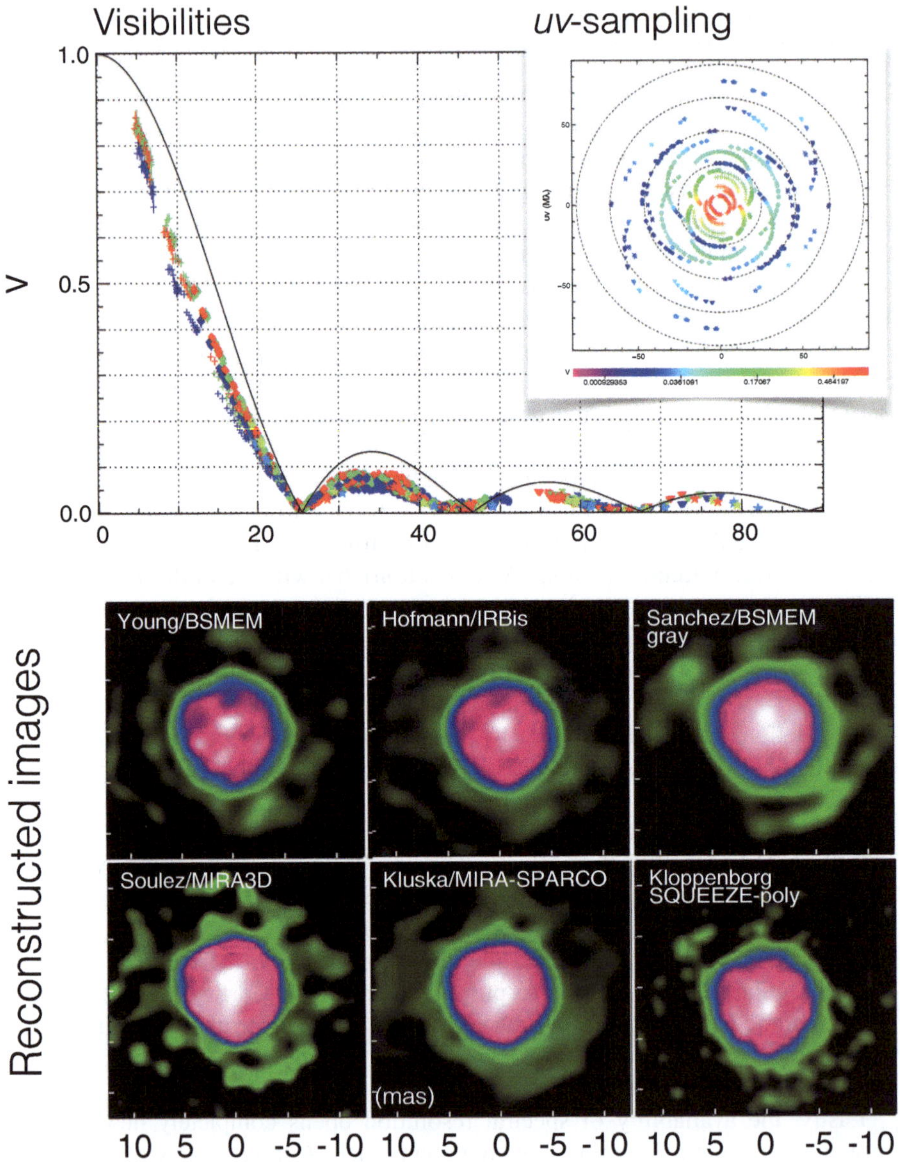

Fig. 4.6 Image reconstruction of the Mira-type star R Car

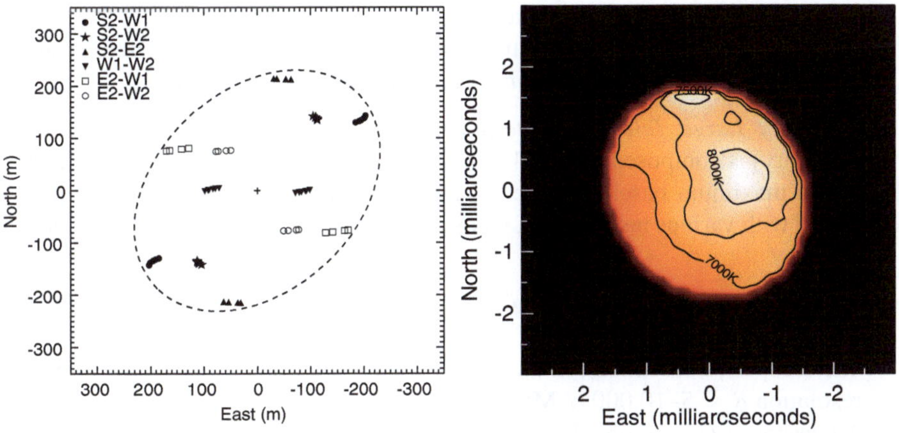

Fig. 4.7 *Left*: coverage of the spatial frequencies (uv-plane) gathered in Altair by Monnier et al. [6] with the MIRC/CHARA interferometer. *Right*: corresponding image reconstruction. This truly remarkable image compares directly with the prediction of the von-Zeipel theory for fast rotating stars

4.3 Instrumentation Suite

Several facilities exist, but the two most productives and the most open are CHARA and the VLTI (Fig. 4.8). Interestingly, CHARA is located in the Northern hemisphere while VLTI is located in the Southern hemisphere. Both of them are supported by modern tools to prepare, calibrate, reduce and analyse observations.

4.3.1 Observing Facilities

The *Center for High Angular Resolution Astronomy* (CHARA,[2] from Georgia State University) is composed of six 1 m telescopes located in a fixed array of 330 m maximum baseline. See [7] for recent results. This facility is equipped with various instruments, whose overall performances can be summarised as follow:

- MIRC: combines up to 6 telescopes (15 baselines), $Hmag \approx 5$, spatial resolution $\approx$0.7 mas, with spectral resolution $R \approx$ 40–400. There is an on-going effort to increase sensitivity and expand it to the K band.
- CLASSIC, CLIMB, FLUOR: combine 2 to 3 telescopes (1 to 3 baseline), $Kmag \approx 8$, spatial resolution $\approx$ 1 mas, with spectral resolution $R \approx 5$. There is an on-going effort to increase the number of telescopes.

[2] http://www.chara.gsu.edu

- VEGA, PAVO: combine 2 to 3 telescopes (1 to 3 baseline), *Vmag* $\approx$ 7.5, spatial resolution $\approx$ 0.3mas, with spectral resolution $R \approx$ 100–10,000. There is an ongoing effort to increase the number of telescopes and sensitivity.

The *Very Large Telescope Interferometer* (VLTI,[3] from European Southern Observatory) is composed of four 1.8 m Auxiliary Telescopes in a relocatable array of 150 m maximum baselines. See [8] for a review of operations. This facility is equipped with various instruments, whose overall performances can be summarised as follow:

- 4 telescopes (6 baselines), *Hmag* $\approx$ 8, spatial resolution $\approx$ 1.5 mas, with spectral resolution $R \approx$ 40 (PIONIER).
- 4 telescopes (6 baselines), *Kmag* $\approx$ 7, spatial resolution $\approx$2.5 mas, with spectral resolution $R \approx$ 5–10,000 (AMBER, GRAVITY).
- 4 telescopes (6 baselines), $L,M,N \approx 6$, spatial resolution $\approx$5–10 mas, with spectral resolution $R \approx$ 50–10,000 (MATISSE in 2018).

VLTI also combines the four 8 m Unit Telescopes of the Paranal observatory but only few nights per months. Using these telescopes, the limiting magnitudes are increased by about +2 mag.

Other facilities are currently in operation (NPOI,[4] SUSI,[5] LBTI[6]) or in development (MROI[7]). The interferometric mode of the Keck telescopes, the so called Keck Interferometer is now stopped, as well as many other facilities or prototypes (PTI,[8] IOTA,[9] Gi2T, COAST[10]...) that produced early science in the past decades and pioneered the technologies used today. More information is available in the OLBI publication database at http://jmmc.fr/bibdb hosted by the Jean-Marie Mariotti Centre. The community is now exploring the design of a next generation optical array dedicated to planet formation studies: the Planet Formation Imager (PFI). Alternatives approaches, such as specific smaller arrays, may emerge in the future with the goal to address well-defined science cases.

[3] https://www.eso.org/sci/facilities/paranal/telescopes/vlti.html

[4] http://lowell.edu/research/research-facilities/npoi/

[5] http://www.physics.usyd.edu.au/sifa/Main/SUSI

[6] http://lbti.as.arizona.edu/LBTI/index.html

[7] http://www.mro.nmt.edu/about-mro/interferometer-mroi/

[8] http://nexsci.caltech.edu/missions/Palomar/

[9] http://tdc-www.harvard.edu/IOTA/

[10] http://www.mrao.cam.ac.uk/outreach/radio-telescopes/coast/

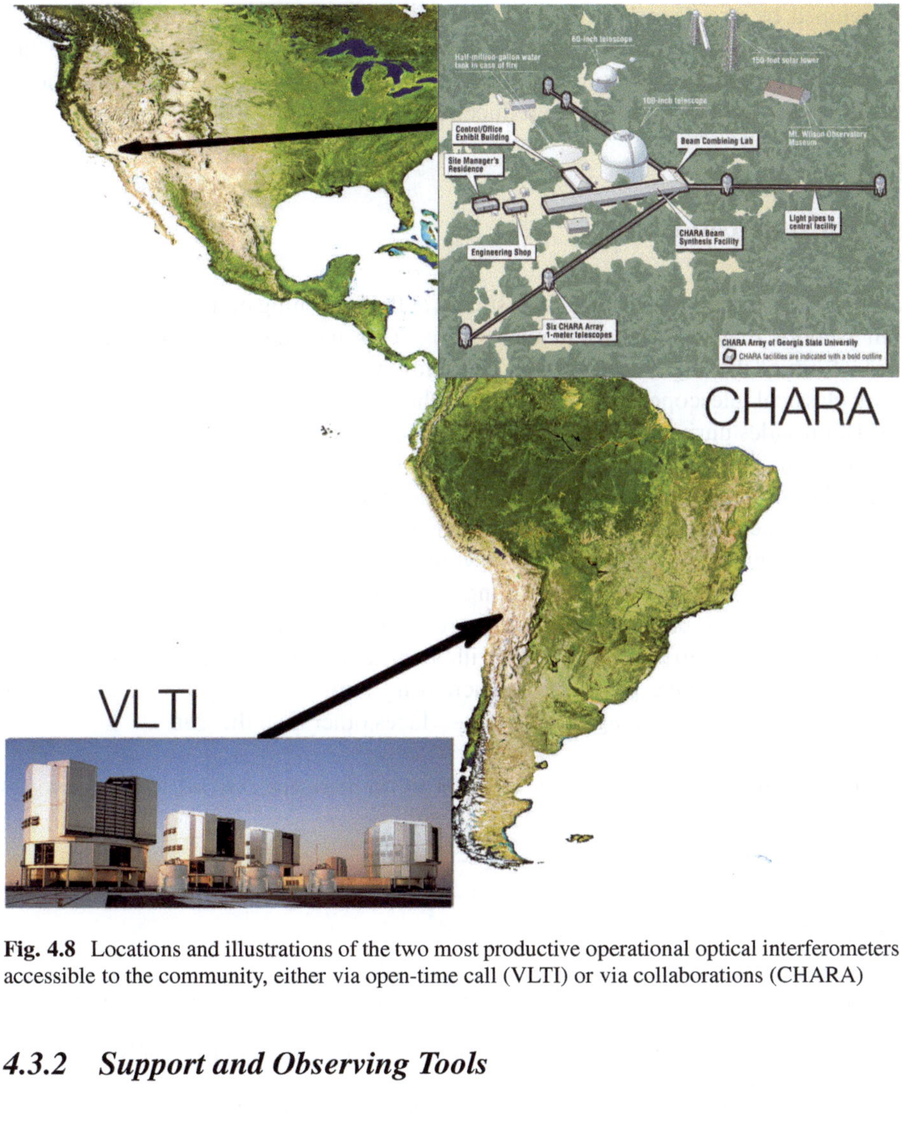

Fig. 4.8 Locations and illustrations of the two most productive operational optical interferometers accessible to the community, either via open-time call (VLTI) or via collaborations (CHARA)

4.3.2 *Support and Observing Tools*

Considering the level of support, the OLBI community has soon developed a strong sense of international collaboration and sharing. The best example is the OIFITS format used to store reduced interferometric measurement. Its first version was defined in 2005[15] and an evolution is being proposed by the community. This format is now widely used worldwide by all the above-mentioned facilities. This allows data to be easily shared and analysed. Another example are the tools developed by the Jean-Marie Mariotti Centre to prepare,[11] calibrate,[12] reduce[13] and analyse[14] observations with efficient and user-friendly interfaces. The last effort from JMMC is to setup a global database of existing optical interferometric observations, with easy access to reduced products.[15]

4.4 Conclusions

Optical long baseline interferometry started at the beginning of the twentieth century but is only now entering a mature phase. This technique samples the Fourier Transform of the angular brightness distribution at spatial frequencies inaccessible with classical telescopes, even considering the Extremely Large Telescope. As such, OLBI provides unrivalled spatial resolution.

For years, measuring the stellar diameters or detecting and monitoring binaries were the main applications of the technique. Large surveys of hundreds of stars have now been achieved. Recently, parametric analysis of sources of manageable complexity brought a completely new insight on interesting systems, such as interacting binaries, discs around young or evolved stars, stellar surface structures, AGNs... However, unknown complexity remains very hard to handle. To go further one has to do true aperture synthesis with an array of a large number of telescopes. The recent years have provided an increasing number of impressive and iconic results such as the first images of stellar surfaces other than the Sun.

Acknowledgements The participation of JBLB to the workshop that led to this book was supported by OSUG@2020 ANR-10-LABX-56.

References

1. Michelson, A.A., Pease, F.G.: Measurement of the diameter of alpha Orionis with the interferometer. ApJ **53**, 249 (1921)
2. Schaefer, G.H., Brummelaar, T.T., Gies, D.R., et al.: The expanding fireball of Nova Delphini 2013. Nature **515**, 234 (2014)
3. Sana, H., Le Bouquin, J.-B., Lacour, S., et al.: Southern massive stars at high angular resolution: observational campaign and companion detection. ApJS **215**, 15 (2014)
4. Marion, L., Absil, O., Ertel, S., et al.: Searching for faint companions with VLTI/PIONIER. II. 92 main sequence stars from the Exozodi survey. A&A **570** A127 (2014)

[11] Aspro2: http://www.jmmc.fr/aspro

[12] SearchCal: http://www.jmmc.fr/searchcal

[13] PIONIER data reduction: http://www.jmmc.fr/pndrs

[14] LITpro http://www.jmmc.fr/

[15] OiDb: http://www.jmmc.fr/oidb

5. Blind, N., Boffin, H.M.J., Berger, J.-P., et al.: An incisive look at the symbiotic star SS Leporis. Milli-arcsecond imaging with PIONIER/VLTI. A&A **536**, A55 (2011)
6. Monnier, J. D., Zhao, M., Pedretti, E., et al.: Imaging the surface of Altair. Science **317**, 342 (2007)
7. ten Brummelaar, T. A., Huber, D., von Braun, K., et al.: Some recent results from the CHARA array. ASPC **487**, 389 (2014)
8. Merand, A., Abuter, R., Aller-Carpentier, E., et al.: VLTI status update: a decade of operations and beyond. SPIE **9146** (2014)
9. Weigelt, G., Kraus, S., Driebe, T., et al.: Near-infrared interferometry of η Carinae with spectral resolutions of 1500 and 12000 using AMBER/VLTI. A&A **464**, 87 (2007)
10. Malbet, F., Benisty, M., de Wit, W.-J., et al.: Disk and wind interaction in the young stellar object MWC 297 spatially resolved with AMBER/VLTI. A&A **464**, 43 (2007)
11. Le Bouquin, J.-B., Absil, O., Benisty, M., et al.: The spin-orbit alignment of the Fomalhaut planetary system probed by optical long baseline interferometry. A&A **498**, L41 (2009)
12. Mourard, D., Monnier, J.D., Meilland, A., et al.: Spectral and spatial imaging of the Be+sdO binary Φ Persei. A&A **577**, A51 (2015)
13. Millour, F., Meilland, A., Chesneau, O., et al.: Imaging the spinning gas and dust in the disc around the supergiant A[e] star HD 62623. A&A **526**, A107 (2011)
14. Kluska, J., Malbet, F., Berger, J.-P., et al.: SPARCO: a semi-parametric approach for image reconstruction of chromatic objects. Application to young stellar objects. A&A **564**, A80 (2014)
15. Pauls, T.A., Young, J.S., Cotton, W.D., Monnier, J.D.: A data exchange standard for optical (visible/IR) interferometry. PASP **117**, 1255 (2005)

Chapter 5
Image Reconstruction in Optical Interferometry: An Up-to-Date Overview

Fabien Baron

5.1 Introduction

Optical astronomical interferometry started in the 1970s [29, 39] as a way to overcome the angular resolution limit imposed to monolithic telescopes by the laws of diffraction. By combining light from an array of several small telescopes, the imaging power is commensurate with the extent of the array rather than with the size of a single telescope. However, because interferometric measurements only deliver partial information about the brightness distribution of observed sources, the technique strongly relied on parametric models to interpret the observed data. As such, it was mostly confined to stellar diameter measurements for several decades.

Two major advances have since completely transformed the field, gradually putting the focus on model-independent imaging. The first one was the development of instruments that could combine the light from three telescopes or more simultaneously. This was initially demonstrated with aperture masking [31], and later then with diluted aperture interferometers that delivered the first interferometric optical images of a binary system [1] and of a supergiant [79]. The second major advance was the development of modern interferometers much longer baselines than the first generation, for example the VLTI (Very Large Telescope Interferometer) or the CHARA (Centre for High Angular Resolution Astronomy) Array. This opened the field to science cases requiring milliarcsecond angular resolution in the visible or infrared, leading to many "firsts": the first resolved image of a main sequence star beyond the Sun [54], of interacting binaries [5, 12, 80], eclipsing discs [36], rapid-rotators [21], young stellar objects [8, 38, 59], supergiant hot spots [4, 32] and magnetic cool spots [55]. In many cases, interferometric imaging challenged conventional models established with other observing techniques: good examples

F. Baron (✉)
Georgia State University, 25 Park Place NE, Atlanta, GA 30303-2911, USA
e-mail: baron@chara.gsu.edu

H.M.J. Boffin et al. (eds.), *Astronomy at High Angular Resolution*, Astrophysics and Space Science Library 439, DOI 10.1007/978-3-319-39739-9_5

include the gravity-darkening law on the aforementioned rapid rotators, as well as the nature of Young Stellar object discs [69].

Reconstructing an image from interferometric data constitutes a challenging signal processing problem. In particular, image reconstruction algorithms have yet to meet astronomers' expectations of imaging fidelity on the most complex objects (Young Stellar Objects, AGB stars, spotted stars) that require high-contrast or polychromatic imaging. In this chapter we give an up-to-date summary of the state of the field in terms of reconstruction algorithms, yet without going into the mathematical details. For a more formal overview of the field, the reader is referred to the excellent reviews in [74] and [73].

5.2 Principles of Optical Interferometry

In this section we give a brief overview of data acquisition in order to understand the process of image reconstruction. We model an interferometer as an array of N telescopes, with $\mathbf{r}_n(t)$ the position of the telescope n at time t. The baseline formed between two telescopes n and m is the vector $\mathbf{b}_{nm}(t) = \mathbf{r}_m(t) - \mathbf{r}_n(t)$. As the telescope beams interfere in the combiner instrument, each baseline forms a fringe pattern. From these fringes, quantities known as complex visibilities are extracted. For the baseline formed by the telescopes n and m, the sampled spatial frequency is $\nu_{nm}(t) = \mathbf{b}_{nm}(t)/\lambda$. If we call a_{nm} the measured contrast of the fringe, and ϕ_{nm} the phase of the fringes relative to a fixed reference, then we can define the so-called complex visibility on this baseline as $V_{nm} = a_{nm}\exp(j\phi_{nm})$. The reason for forming this quantity is the van Cittert-Zernike theorem, which indicates that the complex visibility is the normalized Fourier Transform of the source brightness at the fringe frequency $\nu_{nm}(t)$. The brightness distribution of the object on the sky (its image) will be modeled as a discrete matrix $\mathbf{i}$ of $N \times N$ pixels, with the discrete-continuous Fourier matrix $\mathbf{H}(t)$ ensuring the image brightness to visibility transformation. We note here that it may also be described by the coefficients $\mathbf{x}$ of its projection onto another spatial basis, so that $\mathbf{i} = \mathbf{W}\mathbf{x}$. In the first part of this paper we will assume the image is described in the "canonical" impulsion basis, i.e. the image is modeled as a conventional rectangular pixel grid. $\mathbf{W}$ is then simply the identity (meaning $\mathbf{x} = \mathbf{i}$). The van Cittert-Zernike theorem is:

$$\mathbf{V} = \mathbf{H}(t)\mathbf{i} = \mathbf{H}(t)\mathbf{W}\mathbf{x} \tag{5.1}$$

In radio interferometry (except mm-VLBI), correlation and calibration techniques give direct access to the complex visibilities $\mathbf{V}$. In optical interferometry however, the atmospheric turbulence perturbs the visibility phases on timescales of the order of one tenth of a second. As a result, part of the phase information is definitively lost, and reliable observables can only be obtained by averaging fringe data. The most commonly used observables are the power spectra, the closure phases, and the differential phases. The power spectrum (or "squared visibility") on the baseline nm

is defined as $P_{nm} = V_{nm}V^*_{nm}$. A bispectrum is formed from a triangle of baselines: for telescopes n, m, p, the corresponding bispectrum is $B_{nmp} = V_{nm}V_{mp}V^*_{np}$. The phases of the bispectra, the closure phases, contain some of the remaining phase information not destroyed by atmospheric perturbations [50, 52], and thus is critical for imaging. Each interferometric snapshot captures $N(N-1)/2$ powerspectra, but only $(N-1)(N-2)/2$ independent closure phases. Consequently, optical arrays with more telescopes recover a higher percentage of the object phase. In addition to closure phases and power spectra in each wavelength channel, the differential phases between channels can be computed [49], and are shown to remain impervious, to a first order, to atmospheric perturbations. Finally, we note that other observables such as differential closure phases [81] or closure amplitudes [50] are currently rarely used.

5.3 Bayesian Framework of Image Reconstruction

The image reconstruction procedure is tasked with the reconstruction of the image $\mathbf{i}$ given the data $\mathbf{D}$ and a "model of image reconstruction" $\mathbf{M}$. In the following we will express the image reconstruction process in a Bayesian framework.

5.3.1 Bayes Equation for Image Reconstruction

One searches the image that maximizes the posterior probability of the reconstruction $\Pr(\mathbf{i}|\mathbf{D},\mathbf{M})$. The Bayes theorem expressed as:

$$\Pr(\mathbf{i}|\mathbf{D},\mathbf{M}) = \frac{\Pr(\mathbf{i}|\mathbf{M})\Pr(\mathbf{D}|\mathbf{i},\mathbf{M})}{\Pr(\mathbf{D}|\mathbf{M})}, \tag{5.2}$$

where $\Pr(\mathbf{D}|\mathbf{M})$ is a factor called the evidence, which normalises the denominator, and principally depends on $\mathbf{M}$. The imaging model $\mathbf{M}$ encompasses our choices of priors, the choice of image representation (e.g. as pixels on a rectangular grid with a given pixellation) and the choice of initial conditions (starting image). This ensemble defines the minimisation problem to be tackled by the reconstruction algorithm. Thus for a given reconstruction, the evidence factor is a constant. The imaging equation simplifies into:

$$\Pr(\mathbf{i}|\mathbf{D},\mathbf{M}) \propto \Pr(\mathbf{i}|\mathbf{M})\Pr(\mathbf{D}|\mathbf{i},\mathbf{M}). \tag{5.3}$$

The most probable image is the one that maximises the right hand of Eq. 5.3, or equivalently minimises its negative logarithm criterion $J(\mathbf{i})$:

$$J(\mathbf{i}) = -\log\left[\Pr(\mathbf{i}|\mathbf{M})\Pr(\mathbf{D}|\mathbf{i},\mathbf{M})\right] \tag{5.4}$$

5.4 Likelihood

The likelihood encompasses our knowledge of the statistical distribution of the data.

For simplicity, in the following, the errors on the power spectrum, triple amplitudes and closure phases will be assumed to be independent and normally distributed. This assumption forms the basis of the OIFITS data exchange standard in optical interferometry [56].

In general the observables are assumed to follow a normal distribution, unless they are phase quantities (closure phases, differential phases), in which case they follow a wrapped normal distribution [30, 72]. In addition they are often supposed to be roughly independent/uncorrelated, in agreement with the OIFITS standard. The likelihood of the data can then be expressed as:

$$\Pr(\mathbf{D}|\mathbf{i}, \mathbf{M}) \propto \exp\left\{-\chi^2_{\mathbf{D}}(\mathbf{i})/2\right\}. \quad (5.5)$$

The reader should note however that there is evidence for minor departures from Gaussianity for some instruments [64], and that correlations between spectral channels are to be taken into account when working with real data – see e.g. [53]. In fact, the addition of covariance tables constitute a key addition to the current draft of the OIFITS v.2 format.

5.4.1 Non-convexity and Multi-modality of the Likelihood

While the resulting metrics are by themselves statistically well-behaved, the overall likelihood is in general non-convex, and thus cannot readily be maximised with convex optimisation tools. This non-convexity is due to the non-linearity of most of the observables with respect to the image flux, though some observables are worse offenders than others. In particular, the bispectrum has both its modulus and phase normally distributed, so that the resulting probability density in the complex plane is "banana-shaped" and thus clearly non-convex. A convex approximation of this probability distribution can be easily obtained by using best-fitting ellipsoidal distributions [46]. The cost of this approximation, the likelihood can be made globally convex. However it is known in practice that this approximation is not precise enough, and it will negatively affect the most difficult (high-contrast) reconstructions (spotted stars, exoplanet detection).

Another possibility to convexify the likelihood is to approximate the wrapped normal distribution of the phases (non-convex, non-differentiable) by the convex and differentiable von Mises distribution [62]. It is currently uncertain whether it suffers from the same drawbacks as the previous approach.

Regardless of its convexity, by construction the likelihood is still missing some of the phase information from the object, i.e. the phase corrupted by the atmosphere. Thus, it is fundamentally multimodal. The missing phase can be represented by $N-1$

unknown parameters per snapshot, and while the overall image centering accounts for part of these parameters, other degeneracies with non-obvious effects exist [40]. As a result, encountering local minima of the likelihood cannot be avoided during optimisation.

5.5 Regularisation

The reconstruction problem is an ill-posed problem (in the sense of Hadamard): the information contained in the data is not sufficient to constrain all the unknowns to recover. Typically, the data is composed of a few hundreds or thousands points, and we try to reconstruct an image containing several thousands pixels. The discrepancy may not seem too great, but the data points are likely to be nearby uv points and thus present great redundancy of information.

Attempting to reconstruct an image solely by maximising the likelihood leads only to over-fitting; which means that artefact-ridden images are obtained as the noise in the data is interpreted as true signal. To circumvent this issue, one has to regularise the problem [76], i.e. introduce reasonable prior expectations on the object such as e.g. its positivity, a support constraint, the smoothness or sharpness of its brightness distribution. In this approach, called regularised maximum likelihood, the reconstruction becomes the process of examining the probability of images that both obey the regularisation constraints and are compatible with the data.

Regularisation plays a fundamental role in interferometric image reconstruction. In addition to prevention of overfitting, its main purpose is to mitigate the effects of the lack of Fourier coverage on the reconstruction by imposing prior information on the solution. Ideally, it should also make the combination of likelihood and regulariser less sensitive to local minima. To prevent biasing the solution, though, the regularisers should be as generic and non-committal as possible. In optical interferometry, the positivity of the image fluxes is one of the simplest regularisers, yet one of the most effective. It is de facto presupposed in all reconstructions. Classic regularisers are presented and discussed in the review by Renard et al. (2011) [60]. Newer regularisers are generally based on Compressed Sensing, introduced in Sect. 5.5.3.

By choosing an adapted regulariser for the target, reconstructions routinely achieve super-resolution, i.e. an effective image resolution greater than the Rayleigh interferometer resolution [27, 60].

The general form of a regulariser expression is often chosen to be:

$$\Pr(\mathbf{i}|\mathbf{M}) \propto \exp\{\mu R(\mathbf{i})\} \tag{5.6}$$

where R the regularisation function and μ is called the regularisation hyperparameter [65, 74].

Accordingly, the most probable image $\tilde{\mathbf{i}}$ jointly maximises the likelihood and the regularisers, and can be found through the minimisation of the criterion $J(\mathbf{i})$ from Eq. 5.4, under the constraints of normalisation and positivity:

$$\tilde{\mathbf{i}} = arg\, min_{\mathbf{i}\in\mathbb{R}^N} \left\{\chi^2_{\mathbf{D}}(\mathbf{i}) + \mu R(\mathbf{i})\right\} \quad \text{s.t.} \quad \forall n,\, \tilde{i}_n \geq 0 \quad \text{and} \quad \sum_{n=1}^{N} \tilde{i}_n = 1, \qquad (5.7)$$

5.5.1 Separable Regularisation Functions

The majority of regularisers R in use in optical interferometry work by constraining the location and/or the intensity of the image flux. Amongst them, some effective flux regularisers will penalise excursions from a default wanted state, or sometimes a prior image $\mathbf{m}$. This default state reflects our expectations for the image in the absence of data. Early in the history of optical interferometry, the most common regularisers were often separable functions: i.e. the regularisation for a given pixel depends solely on the flux in this pixel, and not on the neighbouring pixels. Historically, the first image regulariser in optical interferometry was Maximum Entropy [28, 31], $R_{\text{MEM}}(\mathbf{x}) = \sum_n x_n - m_n - x_n \log(x_n/m_n)$, which directly compares the current image to the model with the Kullback-Leibler divergence metric. Most expressions of separable functions nowadays are using ℓ_p norms, defined by $R_{\ell_p}(\mathbf{x}) = \left(\sum_n |x_n|^p\right)^{\frac{1}{p}}$. When applied directly onto the image pixels, these norms penalise departure from the default image. The penalty will increase more or less than linearly with the departure, depending on the power coefficient p (often chosen between 0 and 2). Typical examples are R_{ℓ_2} and R_{ℓ_0}, which favour respectively smooth images and sharp features. Under our assumptions of image positivity and normalisation to unity, R_{ℓ_1} is always unity and thus a useless regulariser. Other separable regularisers can be constructed to behave $\ell_2 - \ell_1$ norm [43]: small excursions from the default image are penalised quadratically (ℓ_2 behaviour), while larger excursions are penalised linearly (ℓ_1 behaviour). In all these cases, the role of the prior image is key to imaging fidelity: without it the regularisation happens pixel by pixel, and does not take into account local pixel correlations.

5.5.2 Example of Regulariser: Prior Images

Prior images are often used in conjunction with regularisers belonging to maximum entropy type. Typically, they are either chosen to be the zero image (also abusively referred to as using "no prior"), or simpler (or blurred) versions of the object to image derived from model-fitting. Prior images will act as "masks" during the reconstruction, preventing spurious noise outside a well-defined zone where the flux is known to reside. A typical example is reconstructing a spotted stellar surface.

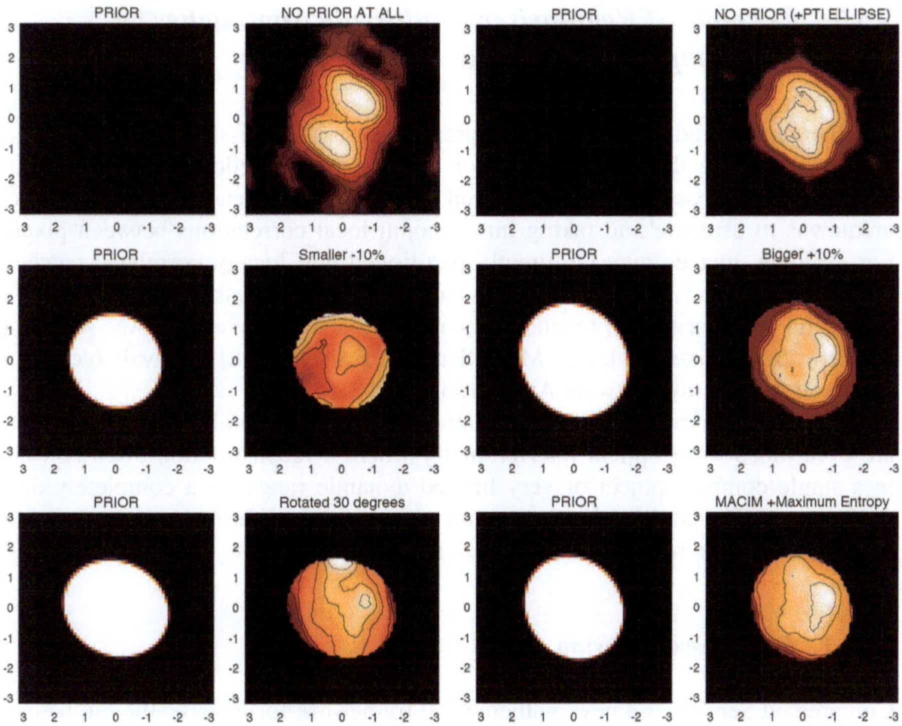

Fig. 5.1 The "Goldilocks" sensitivity to the image prior of the reconstruction of Altair (from CHARA-MIRC 4-telescope data). Using a flat image ("no prior") as a prior barely allows the shape of the star to be recovered (*top row*, *left*), unless the short frequencies are constrained by additional data (*top row*, *right*, data from Palomar Testbed Interferometer). Using an entropic prior regulariser of the form $R(\mathbf{i}) = -\sum_n \log(\mathbf{i}_n - \mathbf{m}_n)$ constrains the shape of the star better, provided the size and orientation of the prior are correct within a few percents (*center* and *bottom rows*). In this example, an undersized prior leads to severe artifacts on the disc edges (*center row*, *left*). An oversized prior will produce extraneous flux in an attempt to fill the whole prior region (*center row*, *right*); though the central flux distribution will remain roughly untouched. As would be expected, an over-inclined disc behaves both as an undersized and oversized prior (*bottom row*, *left*). A good image be recovered only if the prior is just right [54] (All images courtesy of John Monnier (private comm.))

An effective procedure is to first fit the data with a limb-darkened disc. Then this disc is blurred with a Gaussian width of a few pixels, and used as prior for image reconstruction. In practice this procedure is effective, as it fills the short frequencies with likely data. The reason for blurring the prior is twofold: to avoid aliasing and sharp edges, and to prevent under-sizing the prior which strongly damages the reconstruction fidelity. As illustrated in Fig. 5.1, one may encounter a Goldilocks dilemma: the prior has to be "just right" to recover the exact brightness distribution on the stellar surface. A fair criticism of this process is therefore that its dependency on good prior model makes it stray too far from the ideal model-independent/non-committal approach.

5.5.3 *Example of Regulariser: Multiscale Approaches and Compressed Sensing*

The literature in radio interferometry presents several multi-scale approaches that aim at decreasing the effective number of degrees of freedom by replacing the canonical pixel basis by more adapted bases, both decreasing the number of unknowns to estimate and taking into account local correlations between pixels. For example, images may be directly partitioned into locally correlated patches, using image plane partition schemes such as Voronoi tessellation [15]. Most successful multi-scale approaches rely on wavelets: wavelet-CLEAN [66], the Adaptive Scale Pixel [11], and Multi-Scale CLEAN [23, 58] are available in the Common Astronomy Software Applications (CASA).

Most targets in radio interferometry present very compressible images, and they are even more so in optical interferometry where a reconstruction often consists in a single compact object of very limited dynamic range on a completely dark background. The multiscale approaches in optical interferometry are therefore strongly tied to the recent Compressed Sensing theory.

5.5.3.1 Compressed Sensing

Compressed Sensing is a new mathematical framework formalising the multi-scale ideas. If an astronomical signal can be expressed as a small number of coefficients $\mathbf{c}$ in an adequately chosen basis $\mathbf{W}$ (or "dictionary"), it is said sparse in this basis: its information content may be much smaller than its effective bandwidth. Compressed sensing theory then supersedes the Shannon sampling theory and allows optimal reconstruction of the signal even if it is sampled at less than its Nyquist frequency [17, 26]. Compressed sensing literature discusses the linear measurement problem (rewritten with our former interferometric notations):

$$\mathbf{v} = \mathbf{H}\mathbf{i} + \mathbf{n} = \mathbf{H}\mathbf{W}\mathbf{c} + \mathbf{n} \tag{5.8}$$

where $\mathbf{v}$ is the measured data, $\mathbf{n}$ its associated noise, and $\mathbf{H}$ describes the measurement process. Compressed sensing then states that the compressible signal $\mathbf{i}$ can be recovered as $\mathbf{W}\tilde{\mathbf{c}}$, where $\tilde{\mathbf{c}}$ is the solution to the sparsity-regularised optimisation problem:

$$\underset{\mathbf{c}\in\mathbb{R}^P}{arg\,min}\,||\mathbf{c}||_0 \quad \text{s.t.} \quad ||\mathbf{v} - \mathbf{H}\mathbf{W}\mathbf{c}||_2 \leq \epsilon. \tag{5.9}$$

where ϵ is the desired residual level for the optimisation. [9] demonstrated the existence of one-to-one correspondence between ϵ in Eq. 5.9 and μ in Eq. 5.7. Compressed Sensing can therefore be understood as regularised maximum likelihood, with the regulariser taken to be the ℓ_0 norm of the coefficients of the image in the sparsity basis. As the ℓ_0 norm intervening in Eq. 5.9 is non-convex and non-differentiable, its minimisation is numerically intractable using conventional convex optimisation algorithms (the majority of image reconstruction codes). For a restricted set of problems, [17] demonstrated the convex ℓ_1 norm to possess similar properties, but as previously mentioned, it remains a constant (unity) and thus useless for optical reconstructions. Non-convex but differentiable ℓ_p norms $(0 < p < 1)$ may constitute a possible alternative to the ℓ_0 and ℓ_p norm [20]: Carrillo et al. [18] reports successfully approximating the effect of ℓ_0 using a reweighed ℓ_1.

5.5.3.2 Applying Compressed Sensing to Optical Interferometry

To improve the recovery of the signal, Compressed Sensing states that the measurement process **H** should ideally have a dense representation in **W**, i.e. these bases must be as incoherent as possible [16, 17]. The signal processing literature generally prescribes random sampling, and proceeds to assess the probability of exact reconstruction in such a case. In a practical imaging scenario though, the UV sampling is non-random and pre-determined, due to the fixed telescope configuration or to the observing schedule. An observer will ideally attempt to sample all the available UV points. Consequently, given **H**, the user has only the choice of dictionary **W** to improve the reconstruction. Both this choice and that of the minimisation engine will ultimately determine the quality of the reconstruction.

Image reconstruction in interferometry seems nearly a textbook application of Compressed Sensing theory, at least in the radio domain. Wiaux et al. [78] first demonstrated the application of Compressed Sensing to the reconstruction of simulated radio interferometry data, and all the classic algorithms from the Compressed Sensing literature such as Matching Pursuit or Iterative Soft Thresholding Algorithm have corresponding radio packages [44, 77].

For Compressed Sensing recovery via ℓ_0 to work as intended, Eq. 5.8 requires a linear transformation from the image basis to the measurements. As underlined earlier, the observables in optical interferometry are non-linear functions of the complex visibilities, leading to a multimodal and non-convex likelihood. The underlying measurement process however, remains essentially the same as in radio: the problem to solve is similar to the Basis Pursuit Denoise problem, only less numerically tractable. As the theoretical benefits of a Compressed Sensing approach are not clear in the non-linear regime, the practical benefits are to be assessed by simulations.

5.5.3.3 Dictionaries for Optical Interferometry

Sparsity dictionaries may be used alone or coupled, which is equivalent to using several regularisers at once [77].

The most immediate dictionary that may come to mind is the image pixel basis, also known as the impulsion basis. Imposing sparsity on the impulsion basis during reconstruction leads to reconstructions with the least amount of lit-up pixels. This idea was first exploited by the original Building Block Method [34], inspired by the CLEAN method, in which pixels were populated one after another as the likelihood progressively increased. In practice image plane sparsity is only useful when imaging barely-resolved or unresolved objects, e.g. companions around a star or very compact objects.

Another useful dictionary consists in the basis formed by the spatial gradient of the image, or by extension any higher order spatial derivative of the image. Imposing sparsity in the gradient plane will favour patches of uniform brightness with sharp edges in the reconstructions. Total variation [61], as the ℓ_1 norm of the spatial gradient, measures the amount of variation in an image. This regulariser has been used in medical imaging and deconvolution as an edge-preserving regulariser [22, 68], and shown to be an effective all-purpose regulariser in optical interferometry [60]. Through dimensional analysis, [4] recently argued that $\ell_{0.5}$ norm is more effective to reconstruct spotted surfaces. The combination of spatial derivative dictionaries, known as total generalised variation [13], is also particularly effective.

As the strength of these dictionaries is due to the actual correlation between neighbouring pixels introduced by the regularisation, using wavelets is also a possibility. Wavelets tend to offer good sparsity bases for astronomical images [67]. During the regularisation the image is turned into wavelet coefficients, a natural multi-scale representation on which to apply Compressed Sensing. At the time of writing this paper, the dictionary choices are restricted in optical interferometry: in SQUEEZE [3, 5] the Isotropic Undecimated Wavelets (IUW) from [67] and the Cohen-Daubechies-Feauveau-9/7 from the JPEG2000 format [70]. In the PAINTER [62, 63] package, the IUW or a dictionary made of the concatenation of a Haar wavelet basis with the first eight orthonormal Daubechies wavelet bases (Db1-Db8). This latter basis was demonstrated to be superior to IUW on typical radio images [18]. In the optical domain, stellar objects are often isotropic and the IUW seems to fare well on test images as shown on Fig. 5.2. One could foresee the use of more specialised wavelets in the future, in order to enhance specific image features (e.g. arcs/discs for Young Stellar Objects).

Finally, we note that when doing spectro-interferometry, other sparsity bases may be imposed as trans-spectral regularisers. Some possible choices were reviewed in [77] and are now of common use: total variation, a structured ℓ_2 norm [74], or the Discrete Cosine Transform [62].

5.5.4 Regularisation Weight

The relative weight of the likelihood and of the regularisation prior is set by the hyperparameter μ in Eq. 5.7. From a strict Bayesian analysis viewpoint, $P(\mu)$ is likely significant over a large range of values, and the proper procedure would be to marginalise the problem over this nuisance parameter. However, for historical and ease of use reasons and because the marginalisation is a very non-trivial process, it is

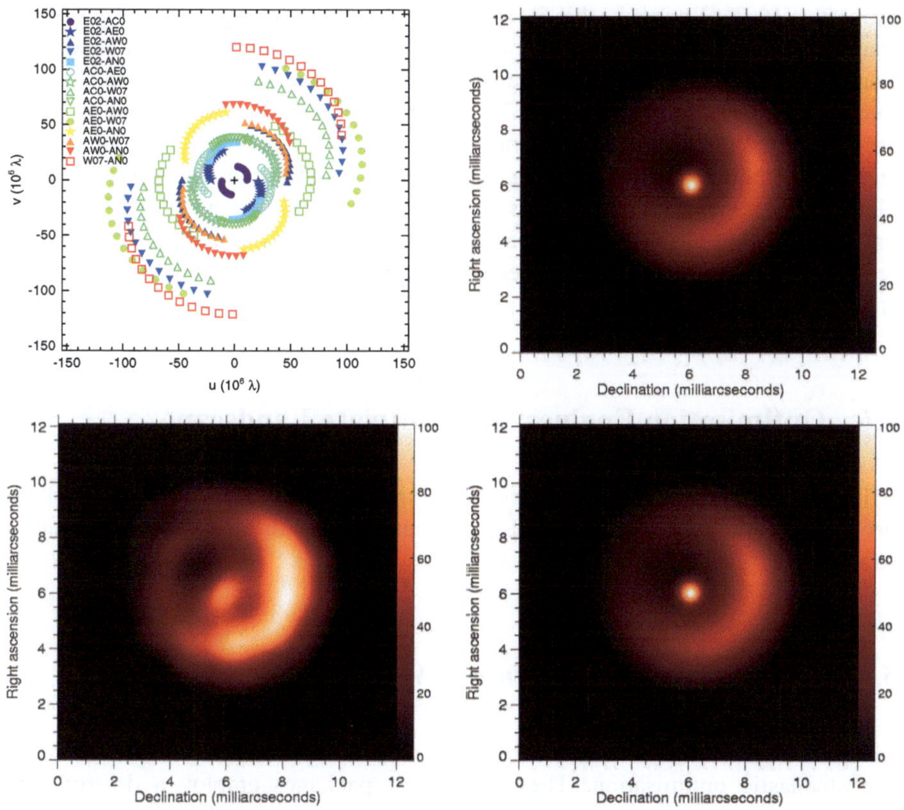

Fig. 5.2 Reconstruction of the Beauty Contest 2004 dataset simulating the observation of the Young Stellar Object LkHα 101. *Top left*: the UV coverage for the data set. *Top right*: the truth image. *Bottom left*: the BSMEM maximum entropy reconstruction, and a SQUEEZE Compressed Sensing image using the Undecimated Isotropic wavelets. Two flaws are apparent on the BSMEM regularisation: the reconstructed central star is spread out and therefore dimmer; and the flux distribution on the shell is different. The peak fluxes of the reconstructed image differ sensibly from the ground truth. Though still not perfect, the Compressed Sensing reconstruct is a significant improvement

often considered as a Dirac function $P(\mu) = \delta(\mu - \mu_0)$, and the minimisation is set to use the single value μ_0. Very low values of μ favours the likelihood over the prior and will tend to over-fit of the data, resulting in spurious high frequency artefacts on the reconstruction. Conversely, very high values of μ favour the regulariser and the reconstruction may not depart significantly from the prior image. Consequently, an optimal range of μ_0 values exists.

As predicted theoretically and demonstrated in simulations [60], this range only depends on the type of regularisation and object type, but not on the data coverage or quality. Provided one knows the nature of the object to reconstruct, this range can be determined empirically by simulating the observation of an object with similar properties [60]. Obviously this is not applicable when observing a target whose shape is mostly unknown (a good example of these would be enshrouded AGB stars). The best solution in this case (at the moment) would be to rely on L-curves to determine the optimum range [37, 60]: unfortunately on real data the L-curves are not very smooth due to local minima, the inflexion point hard to determine, and the method requires launching a large number of reconstructions within the plausible range of μ_0. A more Bayesian approach is attempted in BSMEM [6] based on the method of [28] where the marginal probability of μ is maximised along the reconstruction; however it has not been completely convincing in terms of numerical stability.

5.6 Optimisation Engines: The Software Landscape

All image reconstruction packages solve the regularised maximum likelihood problem by finding the solution of Eq. 5.7, but they still differ in their optimisation strategy.

5.6.1 Stochastic vs Deterministic Approaches

The first distinction we can make between reconstruction engines is whether they use stochastic optimisation. The majority of packages employ a deterministic criterion minimisation scheme based on constrained gradient descent (constrained meaning positivity is imposed during minimisation), where the gradient of $J(\mathbf{i})$ with respect to the pixel fluxes is computed analytically.

5.6.1.1 Classic Deterministic Algorithms

The classic deterministic algorithms for optical interferometry are based on convex optimisation methods, i.e. line search or trust region methods. Arguably, nowadays both these approaches have been superseded by the superior Alternating Direction Method of Multipliers (ADMM), discussed further in Sect. 5.6.1.4.

In the line search methods, a descent direction is computed from this gradient, then the image is changed by this descent direction adequately scaled by a line search. The most classic example is the steepest descent method used in the Building Block Method [34] and the conjugate gradients in its successor IRBis [33]. The constrained semi-Newton method used in MiRA [72] is more powerful: based on the Limited-memory BFGS with gradient projection for positivity, it approximates the inverse Hessian to improve convergence [71]. MiRA is often the go-to reconstruction software thanks to its vast choice of classic regularisers, but it cannot employ non-convex or non-differentiable regularisations such as ℓ_0 and relies on a convex approximation of the likelihood.

In the trust region method employed by BSMEM [6, 14], the criterion $J(\mathbf{i})$ from Eq. 5.7 is approximated by its local Taylor expansion. This requires computing the gradient and diagonal Hessian of $J(\mathbf{i})$ with respect to the pixel fluxes. The region where this approximation remains valid is tested, and called a trust region. By only moving within this trust region, the method can tackle the non-convex likelihood. Positivity is imposed quite artificially through the single available regulariser: Maximum Entropy.

5.6.1.2 Stochastic Algorithms

The two stochastic packages available are MACIM [35] and its successor SQUEEZE [7]. Both engines use Monte Carlo Markov Chains (MCMC). In a single chain, a set of elements of same unit flux are moving randomly within the image plane, forming different image configurations as iterations go. MACIM uses simulated annealing: the acceptance or rejection of a new configuration is based on a "temperature", a parameter that follows a decreasing schedule and directly influence the relative strength of the likelihood. The initial configuration is "hot", and it is free to evolve within an image space dominated by the regularisation. As the temperature drops, the likelihood is progressively introduced, and the configurations start to favour regions of higher and higher likelihood. Ideally the configurations should then partially "freeze" (i.e. they should reach an equilibrium close to the global criterion optimum, but deviate from the perfect solution at each iteration due to the stochastic motion of elements). To obtain the final image of a chain, the Monte Carlo ensemble average of all frozen configurations is computed after the final iteration. SQUEEZE [7] employs two slightly different methods: parallel simulated annealing and parallel tempering. In parallel simulated annealing, independent simulated annealing chains with different starting points are run in parallel, and the resulting images are co-aligned and averaged together to create a final image independent of the minimisation path. In parallel tempering, several chains are also run in parallel, but a given chain always remains at a constant temperature during the whole optimisation. Chains with close temperatures can exchange images, so that from proxy to proxy, most possible configurations can be tested. Parallel tempering is consequently more robust than simulated annealing. The main issue with MCMC lies in its slow speed. The main advantage of the MCMC engines resides in the ease with which their optimise non-convex, non-differentiable, or non-smooth likelihood and regularisation (including ℓ_0). They can tackle multi-wavelength and multi-epoch imaging, as well as model fitting in addition to image reconstruction with minimal coding [37]. Parallel tempering in particular can compute the marginal likelihood to do Bayesian model selection [7].

5.6.1.3 Explicit vs Implicit Approach and Self-Calibration

A second distinction that can be made between the optimisation approaches is whether or not they attempt to recast the data into complex visibilities. Minimising a likelihood made only of complex visibilities is (under some assumptions) a convex problem, thus easily solvable by convex optimisation. The underlying scheme is somewhat similar to the Expectation-Maximisation algorithm [25]. The algorithm relies on the explicit introduction of latent variables, which here correspond to the missing phase parameters.

The self-calibration algorithm [57] used in radio is an iterative algorithm that alternatively optimises the image then the latent variables until convergence. While it generally brings a definite improvement over the traditional CLEAN method, the self-calibration procedure is not robust at low signal-to-noise data and will simply not converge. WISARD [47] is an attempt to apply self-calibration to optical interferometry, under the assumption of a uniform distribution of phase noise amongst the baselines (unfortunately probably unjustified in the optical domain). Because the interferometric data contains no information on the missing phases, the corresponding parameters are, in effect, optimised with respect to the regularisation functions. Unfortunately, compared to implicit methods that do not attempt to express these parameters, WISARD has not demonstrated any improvement in imaging fidelity, and it suffers from the drawbacks of self-calibration at low signal-to-noise. ADMM provides a better numerical approach that can also recast the problem into complex visibilities (see Sect. 5.6.1.4). The only area where self-calibration has proved useful is multi-wavelength imaging from differential visibilities [48].

5.6.1.4 Multiwavelength: Proximal Methods

A new set of methods based on proximal methods has recently appeared. Instead of directly optimising Eq. 5.7, the image parameters on which the likelihood and the regularisation act are decoupled:

$$\tilde{\mathbf{i}} = arg\, min_{\mathbf{i}\in\mathbb{R}^N}\, \chi^2(\mathbf{i}) + \mu R(\mathbf{k}) \quad \text{with} \quad \mathbf{i} = \mathbf{k} \tag{5.10}$$

At first, this may seem like an awful way of rewriting the initial regularised maximum likelihood, but this new form allows the global problem to be split into several sub-problems, including minimising with respect to **i** and **k**, imposing their equality, and global positivity. Skipping here lots of mathematical steps, the end result is an iterative algorithm where all the sub-problems present closed-form solutions, except one. The only non-closed sub-problem is the unconstrained minimisation of the likelihood plus a ℓ_2 norm, i.e. an easily solvable problem.

The main numerical advantage of ADMM over classic convex optimisation techniques is the possibility to use non-differentiable regularisers, including the ℓ_1 norm for Compressed Sensing. Non-smooth norms such as ℓ_0 are still not doable, and the likelihood still has to be convexified, but this alone is sufficient to justify ADMM. In addition, the sub-problems can be trivially parallelised for multi-wavelength or multi-epoch reconstructions. MiRA-3D [75] was the first demonstration of the potential of ADMM for optical interferometry, which proved that multi-wavelength imaging with ADMM was significantly more efficient than with classic MiRA. MiRA-3D remained limited to complex visibility likelihoods, but its successor PAINTER [62, 64] is able to handle actual optical interferometry observables and do ℓ_1 Compressed Sensing.

The only drawback of the proximal methods is currently the number of tweaking parameters required for convergence. As the field of proximal methods is maturing, further improvements over the standard ADMM method are to be expected and in fact are already implemented in radio [19].

5.6.2 *Fidelity of Current Reconstructions*

Nowadays, all the previously named algorithms will in practice produce images which are very close to one another. Yet these may still differ considerably from the truth images. This lack of fidelity on key test cases is well-known [10], and can only be solved by more research on better regularisers or dictionaries.

To objectively assess the imaging fidelity of algorithms, the best solution is to simulate realistic data sets from known "truth" images, then measure the distance between the reconstructions and the reference truth images. There are two pitfalls when adopting this procedure. First, because the conventional imaging observables (power spectra and bispectra) are invariant to translation, several reconstructions from the same datasets may present different offsets. Hence the reconstructions and the reference truth images needs to be registered (co-aligned). Second, defining a proper metric for registration and image comparison is non-trivial. The Mean Square Error is a conventional default choice, but not particularly satisfying [60], and interferometrists are now exploring other options.

The IAU Interferometry Beauty Contest [2, 24, 41, 42, 45, 51] is a competition organised every two years during the SPIE Astronomical Instrumentation Conferences, which aims at showcasing the performance of image reconstruction software in optical interferometry. Science cases are selected by the organisers, then synthetic data sets in the OIFITS format are generated from model images, and the truth images remain secret during the competition. The data sets are then given to the competitors for reconstruction, and the algorithm producing the reconstructions closest to the truth images (using the MSE) is declared the winner. Overall these contests have not only confirmed the reliability of image reconstruction in terms of detection of the main features of an image, but also helped maintaining the dynamism of the field.

5.7 Conclusion

The field of optical interferometry is now actively developing algorithms to solve increasingly complex imaging problems. In this chapter we mostly explored solutions to the snapshot model-independent imaging problem: a non-moving object to image in a two-dimensional plane.

But as the instruments are getting more powerful, a new set of challenges appear. Interferometrists are gearing toward 5D imaging: imaging the surfaces of moving and rotating stars at multiple epochs and wavelengths. Novel regularisers for spectral and temporal regularisation are to be developed and tested, with useful ideas to exchange with the fields of computer vision and machine learning. Staying in astronomy, the synergy between interferometric imaging and complementary inverse methods is already being explored (light-curve inversion, Doppler imaging), and the applicability of optical algorithms to mm-VLBI is being actively studied.

Acknowledgements My work on image reconstruction is partially supported by NSF-AST 1445935.

References

1. Baldwin, J., Beckett, M., Boysen, R.C., et al.: The first images from an optical aperture synthesis array: mapping of Capella with COAST at two epochs. A&A **306**, L13 (1996)
2. Baron, F., Cotton, W.D., Lawson, P.R., et al.: The 2012 interferometric imaging beauty contest. Proc. SPIE **8445**, 1 (2012)
3. Baron, F., Monnier, J., Young, J., Buscher, D.: New theoretical frameworks for interferometric imaging. In: Creech-Eakman, M.J., Guzik, J.A., Stencel, R.E. (eds.) Resolving the Future of Astronomy with Long-Baseline Interferometry. ASP Conference Series, vol. 487, p. 229. Astronomical Society of the Pacific, San Francisco (2014)
4. Baron, F., Monnier, J.D., Kiss, L.L., et al.: CHARA/MIRC observations of Two M supergiants in perseus OB1: temperature, bayesian modeling, and compressed sensing imaging. ApJ **785**, 46 (2014)
5. Baron, F., Monnier, J.D., Pedretti, E., et al.: Imaging the algol triple system in the H band with the CHARA interferometer. ApJ **752**, 20 (2012)
6. Baron, F., Young, J.S.: Image reconstruction at Cambridge University. Proc. SPIE **7013**, 70133X (2008)
7. Baron, F., Monnier, J., Kloppenborg, B.: A novel image reconstruction software for optical/infrared interferometry. Proc. SPIE **7734**, 77342I (2010)
8. Benisty, M., Renard, S., Natta, A., et al.: A low optical depth region in the inner disk of the Herbig Ae star HR 5999. A&A **531**, A84 (2011)
9. van den Berg, E., Friedlander, M.P.: Probing the Pareto frontier for basis pursuit solutions. SIAM J. Sci. Comput. **31**, 890 (2009)
10. Berger, J., Malbet, F., Baron, F., et al.: Imaging the heart of astrophysical objects with infrared interferometry. A&ARA **20**, 53 (2012)
11. Bhatnagar, S., Cornwell, T.J.: Scale sensitive deconvolution of interferometric images I. Adaptive Scale Pixel (ASP) decomposition. A&A **426**, 747 (2004)
12. Blind, N., Boffin, H.M.J., Berger, J.P., et al.: An incisive look at the symbiotic star SS Leporis. A&A **536**, A55 (2011)

13. Bredies, K., Kunisch, K., Pock, T.: Total generalized variation. SIAM J. Imaging Sci. **3**, 492 (2010)
14. Buscher, D.F.: Direct maximum-entropy image reconstruction from the bispectrum. In: Robertson, J.G., Tango, W.J. (eds.) IAU Symposium on Very High Angular Resolution Imaging, Sydney vol. 158, p. 91 (1994)
15. Cabrera, G.F., Casassus, S., Hitschfeld, N.: Bayesian image reconstruction based on Voronoi diagrams. ApJ **672**, 1272 (2008)
16. Candès, E., Romberg, J.: Sparsity and incoherence in compressive sampling. Inverse Probl. **23**, 969 (2007)
17. Candès, E., Romberg, J., Tao, T.: Robust uncertainty principles: exact signal reconstruction from highly incomplete frequency information. IEEE Trans. Inf. Theory **52**, 489 (2006)
18. Carrillo, R.E., McEwen, J.D., Wiaux, Y.: Sparsity Averaging Reweighted Analysis (SARA): a novel algorithm for radio-interferometric imaging. MNRAS **426**, 1223 (2012)
19. Carrillo, R.E., McEwen, J.D., Wiaux, Y.: PURIFY: a new approach to radio-interferometric imaging. MNRAS **439**, 3591 (2014)
20. Chartrand, R., Staneva, V.: Restricted isometry properties and nonconvex compressive sensing. Inverse Probl. **24**, 1 (2008)
21. Che, X., Monnier, J.D., Zhao, M., Pedretti, E., Thureau, N., Mérand, A., ten Brummelaar, T., McAlister, H., Ridgway, S.T., Turner, N., Sturmann, J., Sturmann, L.: Colder and Hotter: Interferometric imaging of β Cassiopeiae and α Leonis. ApJ **732**, 68 (2011)
22. Chen, S., Donoho, D., Saunders, M.: Atomic decomposition by basis pursuit. SIAM J. Sci. Comput. **20**, 33 (1999)
23. Cornwell, T.: Multiscale CLEAN deconvolution of radio synthesis images. IEEE Sel. Topics Signal Process. **2**, 793 (2008)
24. Cotton, W., Monnier, J., Baron, F., et al.: 2008 imaging beauty contest. Proc. SPIE **7013**, 70131N (2008)
25. Dempster, A.P., Laird, N.M., Rubin, D.B.: Maximum likelihood from incomplete data via the EM algorithm. J. R. Stat. Soc.: Ser. B **39**, 1 (1977)
26. Donoho, D., Elad, M., Temlyakov, V.: Stable recovery of sparse overcomplete representations in the presence of noise. IEEE Trans. Inf. Theory **52**, 6 (2006)
27. Donoho, D., Johnstone, I., Hoch, J., Stern, A.: Maximum entropy and the nearly black object. J. R. Stat. Soc. Ser. B (Methodol.) **54**, 41 (1992)
28. Gull, S.F.: Development in Maximum Entropy data analysis. In: Fougére, P.F. (ed.) Maximum Entropy and Bayesian Methods. vol. 5491, pp. 53–71. Kluwer Academic, Dordrecht (1990)
29. Hanbury Brown, R., Davis, J., Allen, L.R.: The angular diameters of 32 stars. MNRAS **167**, 121 (1974)
30. Haniff, C.A.: Least-squares Fourier phase estimation from the modulo 2pi bispectrum phase. J. Opt. Soc. Am. A **8**, 134 (1991)
31. Haniff, C.A., Mackay, C.D., Titterington, D.J., Sivia, D., Baldwin, J.E.: The first images from optical aperture synthesis. Nature **328**, 694 (1987)
32. Haubois, X., Perrin, G., Lacour, S., et al.: Imaging the spotty surface of Betelgeuse in the H band. A&A **508**, 923 (2009)
33. Hofmann, K.H., Weigelt, G., Schertl, D.: An image reconstruction method (IRBis) for optical/infrared interferometry. A&A **565**, A48 (2014)
34. Hofmann, K., Weigelt, G.: Iterative image reconstruction from the bispectrum. A&A **278**, 328 (1993)
35. Ireland, M.J., Monnier, J.D., Thureau, N.: Monte-Carlo imaging for optical interferometry. Proc. SPIE **6268**, 62681T (2006)
36. Kloppenborg, B., Stencel, R., Monnier, J.D., et al.: Infrared images of the transiting disk in the ϵ Aurigae system. Nature **464**, 870 (2010)
37. Kluska, J., Malbet, F., Berger, J.P., et al.: SPARCO: a semi-parametric approach for image reconstruction of chromatic objects. Application to young stellar objects. A&A **564**, A80 (2014)

38. Kraus, S., Hofmann, K.H., Menten, K.M., et al.: A hot compact dust disk around a massive young stellar object. Nature **466**, 339 (2010)
39. Labeyrie, A.: Interference fringes obtained on VEGA with two optical telescopes. ApJ **196**, L71 (1975)
40. Lannes, A.: Integer ambiguity resolution in phase closure imaging. J. Opt. Soc. Am. A **18**, 1046 (2001)
41. Lawson, P.R., Cotton, W.D., Hummel, C.A., et al.: 2006 interferometry imaging beauty contest. Proc. SPIE **6268** (2006)
42. Lawson, P.R., Cotton, W.D., Hummel, C.A., Monnier, J.D., Zhao, M., Young, J.S., Thorsteinsson, H., Meimon, S.C., Mugnier, L., Le Besnerais, G., Thiebaut, E., Tuthill, P.G.: The 2004 Optical/IR Interferometry Imaging Beauty Contest. BAAS **36**, 1605 (2004)
43. Le Besnerais, G., Lacour, S., Mugnier, L.M., Thiebaut, E., Perrin, G., Meimon, S.: Advanced imaging methods for long-baseline optical interferometry. IEEE J. Sel. Top. Signal Process. **2**, 767 (2008)
44. Li, F., Cornwell, T.J., de Hoog, F.: The application of compressive sampling to radio astronomy. A&A **528**, A31 (2011)
45. Malbet, F., Cotton, W., Duvert, G., Lawson, P.: The 2010 interferometric imaging beauty contest. Proc. SPIE **7734**, 12 (2010)
46. Meimon, S., Mugnier, L.M., Le Besnerais, G.: Convex approximation to the likelihood criterion for aperture synthesis imaging. J. Opt. Soc. Am. A **22**, 2348 (2005)
47. Meimon, S., Mugnier, L.M., Le Besnerais, G.: Self-calibration approach for optical long-baseline interferometry imaging. J. Opt. Soc. Am. A **26**, 108 (2008)
48. Millour, F., Meilland, A., Chesneau, O., Stee, P., Kanaan, S., Petrov, R., Mourard, D., Kraus, S.: Imaging the spinning gas and dust in the disc around the supergiant A[e] star HD 62623. A&A **526**, A107 (2011)
49. Millour, F., Petrov, R.G., Vannier, M., Kraus, S.: AMBER closure and differential phases: accuracy and calibration with a beam commutation. Proc. SPIE **7013**, 1 (2008)
50. Monnier, J.D.: Astrophysics with closure phases. In: Perrin, G., Malbet, F. (eds.) EAS Publications Series, vol. 6, p. 213. EDP Sciences, Les Ulis (2003)
51. Monnier, J.D., Berger, J.P., Le Bouquin, J.B., et al.: The 2014 interferometric imaging beauty contest. Proc. SPIE **9146**, 1 (2014)
52. Monnier, J.D., Berger, J., Millan-Gabet, R., et al.: Few skewed disks found in first closure-phase survey of Herbig Ae/Be stars. ApJ **647**, 444 (2006)
53. Monnier, J.D., Che, X., Zhao, M., Ekström, S., Maestro, V., Aufdenberg, J., Baron, F., Georgy, C., Kraus, S., McAlister, H., Pedretti, E., Ridgway, S., Sturmann, J., Sturmann, L., ten Brummelaar, T., Thureau, N., Turner, N., Tuthill, P.G.: Resolving Vega and the inclination controversy with CHARA/MIRC. ApJL **761**, L3 (2012)
54. Monnier, J.D., Zhao, M., Pedretti, E., Thureau, N., Ireland, M., Muirhead, P., Berger, J.P., Millan-Gabet, R., Van Belle, G., ten Brummelaar, T., McAlister, H., Ridgway, S., Turner, N., Sturmann, L., Sturmann, J., Berger, D.: Imaging the surface of Altair. Science **317**, 342 (2007)
55. Parks, J.R., White, R.J., Schaefer, G.H., Monnier, J.D., Henry, G.W., McAlister, H.A., ten Brummelaar, T.A.: Starspot Imaging with the CHARA Array. In: Creech-Eakman, M.J., Guzik, J.A., Stencel, R.E. (eds.) Resolving The Future Of Astronomy With Long-Baseline Interferometry. ASP Conference Series, vol. 487, p. 345. Astronomical Society of the Pacific, San Francisco (2014)
56. Pauls, T.A., Young, J.S., Cotton, W.D., Monnier, J.D.: A data exchange standard for optical (Visible/IR) Interferometry. PASP **117**, 1255 (2005)
57. Pearson, T.J., Readhead, A.C.S.: Image formation by self-calibration in radio astronomy. ARAA **22**, 97 (1984)
58. Rau, U., Cornwell, T.J.: A multi-scale multi-frequency deconvolution algorithm for synthesis imaging in radio interferometry. A&A **532**, A71 (2011)
59. Renard, S., Malbet, F., Benisty, M., Thiébaut, E., Berger, J.P.: Milli-arcsecond images of the Herbig Ae star HD 163296. A&A **519**, A26 (2010)

60. Renard, S., Thiébaut, E., Malbet, F.: Image reconstruction in optical interferometry: benchmarking the regularisation. A&A **533**, A64 (2011)
61. Rudin, L., Osher, S., Fatemi, E.: Nonlinear total variation based noise removal algorithms. Phys. D: Nonlinear Phenom. **60**, 259 (1992)
62. Schutz, A., Ferrari, A., Mary, D., Thiébaut, E., Soulez, F.: Large scale 3D image reconstruction in optical interferometry. ArXiv e-prints (2015)
63. Schutz, A., Ferrari, A., Mary, D., Soulez, F., Thiébaut, E., Vannier, M.: Painter: a spatiospectral image reconstruction algorithm for optical interferometry. J. Opt. Soc. Am. A **31**, 2334 (2014)
64. Schutz, A., Vannier, M., Mary, D., Ferrari, A., Millour, F., Romain, P.: Statistical characterization of polychromatic absolute and differential squared visibilities obtained from amber/vlti instrument. A&A **565**, A88 (2014)
65. Skilling, J.: Quantified maximum entropy. In: Fougère, P.F. (ed.) Maximum Entropy and Bayesian Methods, vol. 5491, pp. 341–350. Kluwer Academic, Dordrecht (1990)
66. Starck, J.L., Bijaoui, A., Lopez, B., Perrier, C.: Image reconstruction by the wavelet transform applied to aperture synthesis. A&A **283**, 349 (1994)
67. Starck, J.L., Fadili, J., Murtagh, F.: The undecimated wavelet decomposition and its reconstruction. IEEE Trans. Image Process. **16**, 297 (2007)
68. Strong, D., Chan, T.: Edge-preserving and scale-dependent properties of total variation regularisation. Inverse Probl. **19**, S165 (2003)
69. Tannirkulam, A., Monnier, J.D., Harries, T.J., Millan-Gabet, R., Zhu, Z., Pedretti, E., Ireland, M., Tuthill, P., ten Brummelaar, T., McAlister, H., Farrington, C., Goldfinger, P.J., Sturmann, J., Sturmann, L., Turner, N.: A Tale of Two Herbig Ae Stars, MWC 275 and AB Aurigae: comprehensive models for spectral energy distribution and interferometry. ApJ **689**, 513 (2008)
70. Taubman, D.S., Marcellin, M.W.: JPEG2000: image compression fundamentals, standards, and practice. Kluwer Academic, Boston (2002)
71. Thiébaut, E.: Optimisation issues in blind deconvolution algorithms. Proc. SPIE **4847**, 174 (2002)
72. Thiébaut, E.: MIRA: an effective imaging algorithm for optical interferometry. Proc. SPIE **7013**, 70131I (2008)
73. Thiébaut, É.: Principles of Image Reconstruction in Interferometry. In: Mary, D., Theys, C., Aime, C. (eds.) EAS Publications Series, vol. 59, pp. 157–187. EDP Sciences, Les Ulis (2013)
74. Thiébaut, E., Giovannelli, J.F.: Image reconstruction in optical interferometry. IEEE Signal Proc. Mag. **27**, 97 (2010)
75. Thiébaut, É., Soulez, F., Denis, L.: Exploiting spatial sparsity for multiwavelength imaging in optical interferometry. J. Opt. Soc. Am. A **30**, 160 (2013)
76. Titterington, D.M.: General structure of regularisation procedures in image reconstruction. A&A **144**, 381 (1985)
77. Vannier, M., Mary, D., Millour, F., Petrov, R.G., Bourguignon, S., Theys, C.: Spectral regularisation and sparse representation bases for interferometric imaging. Proc. SPIE **7734**, 2 (2010)
78. Wiaux, Y., Jacques, L., Puy, G., Scaife, A.M.M., Vandergheynst, P.: Compressed sensing imaging techniques for radio interferometry. MNRAS **395**, 1733 (2009)
79. Young, J.S., Baldwin, J.E., Boysen, R.C., Haniff, C.a., Lawson, P.R., Mackay, C.D., Pearson, D., Rogers, J., St.-Jacques, D., Warner, P.J., Wilson, D.M.A., Wilson, R.W.: New views of Betelgeuse: multi-wavelength surface imaging and implications for models of hotspot generation. MNRAS **315**, 635 (2000)
80. Zhao, M., Gies, D., Monnier, J., Thureau, N., Pedretti, E., Baron, F., Merand, A., ten Brummelaar, T., McAlister, H., Ridgway, S., Others: First resolved images of the eclipsing and interacting binary β Lyrae. ApJL **684**, L95 (2008)
81. Zhao, M., Monnier, J.D., ten Brummelaar, T., Pedretti, E., Thureau, N.D.: Exoplanet studies with CHARA-MIRC. Proc. SPIE **7013**, 1 (2008)

Chapter 6
Tori, Discs, and Winds: The First Ten Years of AGN Interferometry

Sebastian F. Hönig

After languishing for a decade largely through lack of data, this field should now see a revival, as it is refreshed by detailed infrared imaging. The dynamical problems guessed at years ago can be brought into clearer focus.

Julian Krolik, Nature News & Views, 2004

6.1 Active Galactic Nuclei 101

Every large galaxy in the universe hosts a supermassive black hole in its centre. Despite its enormous mass of several millions to billions of solar masses, it contributes typically less than 1% to the mass of the central bulges. Yet, the black hole mass is tightly correlated with at least the mass and velocity dispersion of the stars in the bulges. The question is: Why would the bulge care about the black hole? The common answer is that the growth phases and evolution of bulges and galaxies are interconnected via feeding and feedback mechanisms. These processes are particularly important in the active growing phases of the black holes, making active galactic nuclei (AGN) a central piece of the puzzle of how galaxies evolved over cosmic times.

The AGN phase is characterised as an episode of significant mass accretion, and number counts suggest that a supermassive black hole spends up to 10 % of its lifetime as an AGN. The accreting mass forms a disc of hot gas that emits radiation from the X-rays to optical wavelengths (see Fig. 6.1). Fast (>1000 km/s) gas clouds orbit above and below the disc, giving rise to broad ultraviolet (UV) and optical emission lines ("Broad-line region"; BLR). At high elevation from the disc, highly ionised gas clouds are seen receding from the AGN, emitting narrow forbidden and permitted emission line ("Narrow-line region"; NLR). In the plane of the disc and at distances of >0.1 pc away from the AGN, the temperature in the gas is cool enough for dust to survive. It is commonly pictured that this dusty, molecular gas

S.F. Hönig (✉)
Department of Physics & Astronomy, University of Southampton, Southampton, SO17 1BJ, UK
e-mail: S.Hoenig@soton.ac.uk

H.M.J. Boffin et al. (eds.), *Astronomy at High Angular Resolution*, Astrophysics and Space Science Library 439, DOI 10.1007/978-3-319-39739-9_6

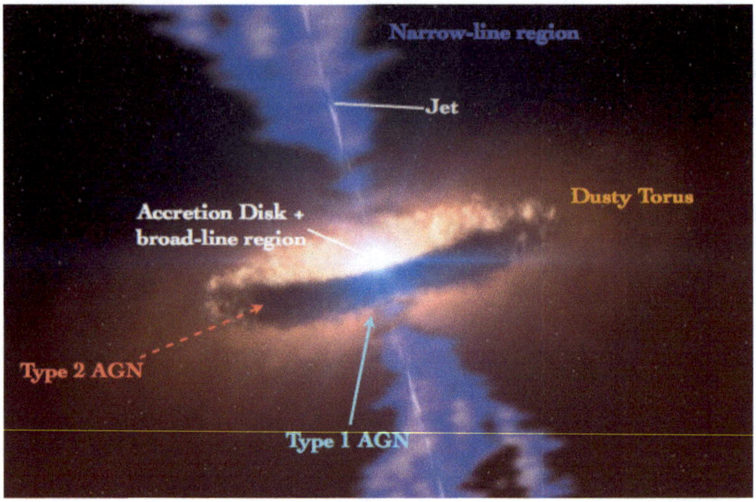

Fig. 6.1 Illustration of the mass distribution of the AGN phase (Image: ESO, with modifications by the author)

forms a geometrically and optically thick torus, which obscures intersecting sight lines. At the same time, the absorbed UV and optical emission is re-radiated in the infrared. Such obscured AGN can only be identified by their IR emission and the narrow lines from high-ionised species from the NLR while the disc and broad lines are not visible. If we see the AGN unobscured, it is referred to as a "type 1" AGN; in the obscured case, we call it "type 2" [1]. It should be noted that this picture represents an important simplification. Since the proposal of this geometric unification scheme in the 1980s, it has been realised that the degree of obscuration and the strength, broadness, and equivalent width of the various emission lines also depends on factors like specific accretion rate, environment, or evolutionary state. However, for the purpose of this review on AGN interferometry, we focus on galaxies that host "Seyfert" AGN – the garden variety class of AGN and poster child for the simple unification picture (barring controversial exceptions, of course).

6.2 The Dusty Environment

The dusty region around the black hole plays an important role beyond obscuration. It is the most extended region that contributes to accretion, arguably connecting the black hole environment on sub-parsec scales to the host galaxy on scales of several tens to hundreds of parsec. Given its size, it harbours a huge reservoir of gas that eventually feeds the black hole, making it a prime target of interest when trying to uncover interactions between black hole and galaxy that lead to coevolution. In addition, the infrared (IR) emission from the dust contributes a significant fraction to the overall radiative power output of the AGN.

As with the other structures of the AGN, the dusty region has typical angular sizes of few to several 10 milliarcseconds, which makes it impossible to directly resolve it with single telescopes. Therefore, most of what we know about the "torus" has been inferred indirectly. The basic framework we have is that the region must be dense enough to provide the obscuration seen in type 2 AGN. The spectral energy distribution (SED) in the IR suggests that dust temperatures range from about 200 K up to the sublimation temperature of $\sim$1500 K, which leads to the expectation of the size range from sub-parsec to tens of parsec scales under the assumption of radiative equilibrium [2].

One important point has been inferred observationally and theoretically from a wide range of arguments [23, 24, 35]: The mass in the torus cannot be smoothly distributed but is rather arranged in clouds. These clouds have been described as cores or fragments of galactic molecular clouds, and must be very dense and compact. In order to withstand the shear of the gravitational potential of the black hole and be stable to its internal pressure, the clouds can only be as large as a small fraction of their distance from the AGN – typically less than 1 %. In addition, their densities must be greater than about 10^8 cm^3 to be resistant against gas pressure and external forces (gravity and radiation pressure).

6.3 Infrared Long-Baseline Interferometry of AGN: Pushing the Limits

The biggest hurdle to test our picture of how mass accretion works for the supermassive black holes in the AGN phase are the small spatial scales involved. The accretion disc has a typical size of few light-days or less while the BLR reaches out to several light days in the Seyfert AGN. However, for objects with distances generally of the order of tens of Mpc, the angular sizes correspond to the sub-milliarcsecond scale. These scales are impossible to resolve with even the largest optical telescopes available or planned. The dusty region is our best shot since it extends out the furthest.

Within the last 10 years, infrared (IR) long-baseline interferometry has matured to a degree that we are now able to observe AGN. The major challenge in IR interferometry concerns the need to coherently combine the light "on the fly" in the presence of strong atmospheric disturbances of the signal, which influence the coherence strength and phase. The infrastructure required by IR interferometry involves dozens of optical elements, reducing the overall transmission and increasing the thermal background. This means that observations of AGN are usually sensitivity-limited. However, even under these circumstances, the persistent improvements on the two major large-telescope interferometers – the Very Large Telescope Interferometer (VLTI) on Paranal, and the Keck Interferometer (KI) on Mauna Kea – have led to AGN becoming routine targets, and the sample of objects is continuing to grow. While in the early 2000s, only the brightest sources

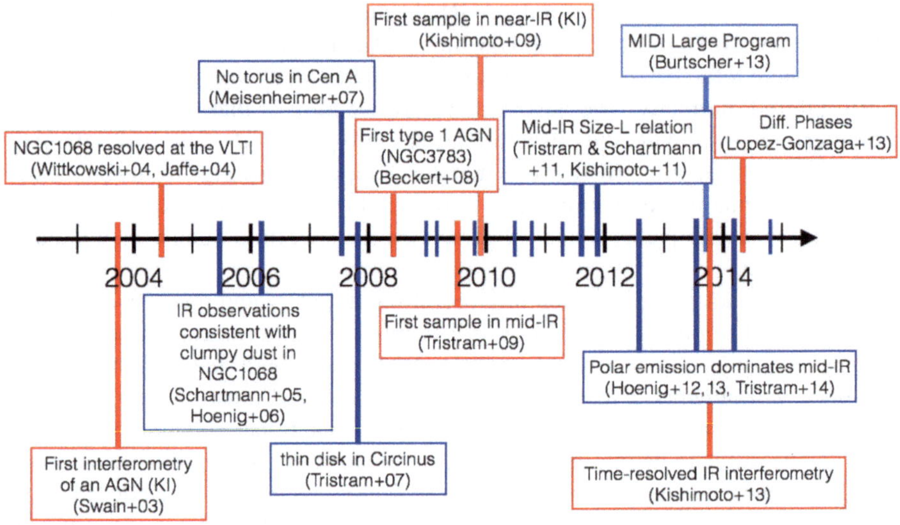

Fig. 6.2 Timeline showing the evolution of AGN publications from IR interferometry and its related interpretation/modelling. Several publications are highlighted based on a subjective selection by the author

NGC 4151 and NGC 1068 could be targeted [16, 34, 41], we are now nearing a sample of 50 AGN that have been successfully observed in the near- or mid-IR [3–5, 12, 13, 18–21, 26–29, 36–38], with some of them having both visibility (= centro-symmetric spatial information) and phase data (=non-symmetric spatial information). Furthermore, a few objects are being monitored to study how their sizes evolve in the presence of variable central illumination [22, 28].

6.4 Science Results

Figure 6.2 shows a timeline of interferometry papers on AGN. Some key results[1] are highlighted that will be discussed in more detail in the following.

6.4.1 Sizes and What They Mean

At the current stage of long-baseline interferometry, the main observable that can be used for scientific exploitation is visibility. Technically, visibility is the absolute

[1]The label "key results" is a fully subjective judgment by the author of this chapter. They include the most cited papers in AGN IR interferometry.

value of the Fourier transform of the intensity distribution of an object, but for the purpose of AGN[2] we can consider it as the fraction of unresolved emission (or correlated flux) at a given baseline and position angle as compared to the total flux within the single-telescope aperture. Most of the time, we observe few combinations of baseline and position angle, which makes is impossible to properly reconstruct a full 2-dimensional brightness distribution. Therefore, we have to invoke models for the brightness distribution based on which we can infer a size. To summarize; interferometry does not measure a size; but it is rather the size that is inferred from visibility.

Since 2004, four different models have been used to turn AGN visibilities into sizes, which are more accessible to non-interferometrists.

- **Gaussian brightness distribution and FWHM size.** The beauty of the Gaussian function is its behaviour under Fourier transformation: it remains a Gaussian. Therefore early results used this function and reported the FWHM as inferred size of the respective object. The major problem with this model is that with better and more data available it became clear that the IR intensity distribution is not a Gaussian. As illustrated in [21], this leads to the inferred FWHM strongly dependent on the visibility level (see Fig. 6.3), so that the sizes are essentially representing λ/baseline rather than the object. Moreover, the better data required more than just one Gaussian to be employed for reproducing the brightness distribution.
- **Power-law brightness distribution and half-light radii.** Using the constraints from multi-baseline data, [18] suggested a power law brightness distribution as the better representation. The corresponding size of such a distribution is the half-light radius (see Fig. 6.3). As shown in [21], such a model results is almost no dependence of size on baseline in the mid-IR data of six AGN. This should come as no surprise. The radiative transfer and dilution of light with distance from the AGN essentially set up a power-law envelope. Even when convolved with a non-power-law mass distribution, power-law modes will remain, so that, in some sense, power laws may be considered a "natural brightness distribution" for the AGN IR emission. The major drawback of this model is that it requires sets of interferometric measurements at least at two different baseline lengths for any given position angle, which comes at the expense of significant observing time. However, angular and radial interpolation methods have been created to make use of measurements with offset position angles.
- **Point source plus Gaussian.** The AGN Large Programme (LP) performed with the mid-IR MIDI instrument at the VLTI shows many objects for which the visibility drops with baseline initially but then seemingly flattens. Thus, [5] took the approach to model the sources with a two-component model of a Gaussian plus a point source, the latter being responsible for the flattening. While one may argue that the Gaussian part introduces similar problems than with a single Gaussian (see above), the LP took the advantage of the large dataset for each

[2]Where in all objects except of Circinus and maybe NGC 1068 we are observing in the first lobe of the Fourier transform.

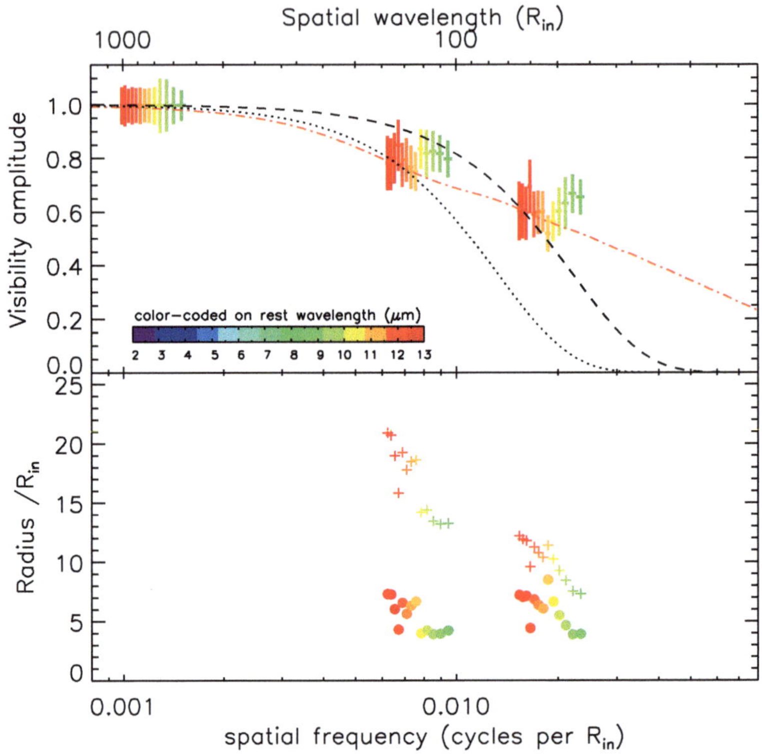

Fig. 6.3 Example for size extraction from visibilities (*upper label*) of three different baseline lengths using Gaussian models (*black dotted* and *dashed lines*) and a power-law (*red dash-dotted line*). While the power law is able to reproduce all data sets simultaneously, providing a unique size (see *lower panel*), the Gaussian model sizes depend on baseline length (From [21])

individual object, modelled the distribution in 2D, and inferred half-light radii from these models. However, these half-light radii are different from the ones inferred from power laws and a proper comparison as to how they compare and how a power law brightness distribution fares for the LP sources is still pending.

- **Ring (plus point source).** Near-IR interferometry probes only the hottest dust near the sublimation front. No dust emission emerges from closer than the sublimation radius. Therefore, it became a standard to use a thin ring model to infer sizes from near-IR interferometry [17, 19, 20]. Actually, the accretion disc or "big blue bump" contributes significantly to the near-IR emission – of the order of a few to 25 %, depending on object. This skews the visibilities of the dust emission to higher values since the big-blue bump emission is much smaller and unresolved with current interferometers. Therefore, it is practical to combine a ring model with a point source, for which the contribution has been determined from spectral decomposition using optical to near-IR photometry.

For non-interferometrists it is important to remember that these sizes are model specific. They refer to a characteristic level to which the flux has been resolved, which could be 50 % or any other "isophote". As such, significant flux is present at larger radii than these size estimates, which has to be taken into account when making comparisons to models. Moreover, these sizes are wavelength-dependent and will typically increase with wavelength since cooler dust at larger distances from the AGN will contribute to the emission.

6.4.2 The Dust Is Clumpy, Indeed!

The first long-baseline IR interferometry was recorded with the KI in 2003 in the near-IR K-band of the brightest type 1 AGN NGC 4151 [34]. The near-IR is supposed to trace the hottest dust close to being sublimated. However, the size of the inferred sublimation radius was smaller than expected, so the authors remained cautious on interpreting their result either as the ring of hot dust at the inner edge of the torus or gas emission from the allegedly much smaller accretion disc.

In 2004, the VLTI achieved its first fringes of an AGN in both the near- and mid-IR using the VINCI and MIDI instruments, respectively [16, 41]. The target for both observations was NGC 1068, which is widely considered a prototype of local AGN. It is important to note that the MIDI instrument does not only provide single wavelength data, but covers the full 8 to 13 μm range – an interferometric IFU in some sense.

With the combination of these data, it could be shown that the wavelength-dependent size is consistent with dust in local thermal equilibrium with the AGN and confirmed that the IR emission is really from dust, something that has sometimes been questioned in the literature given the comparably flat IR SEDs. Furthermore, with the aid of two sets of models [9, 30], it was possible to demonstrate that the fluxes at different spatial resolution elements is not consistent with a smooth distribution of dust, but requires the dust to be clumpy. This result was subsequently confirmed by observation of another nearby AGN, where the interferometric fluxes for a given baseline wiggled with position angle of the source (see Fig. 6.4) that can be only explained if the source is inhomogeneous [36]. These results were a first direct confirmation of substructure within the dusty gas and confirmed theoretical predictions from more than a decade earlier.

6.4.3 The Inner Radius Scales with Luminosity, But What Kind of Dust Are We Seeing?

In parallel to the mid-IR work performed at the VLTI, observations at the KI probed the innermost region of the dust distribution in the near-IR [19, 20, 22, 28].

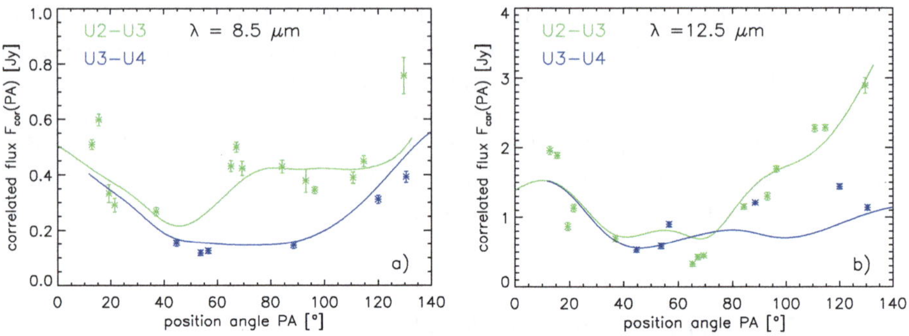

Fig. 6.4 "Wiggling" of the visibility at 8.5 μm (*left*) and 12.5 μm (*right*) for position-angle scans at two different fixed baseline lengths in the Circinus galaxy (Figure from [36])

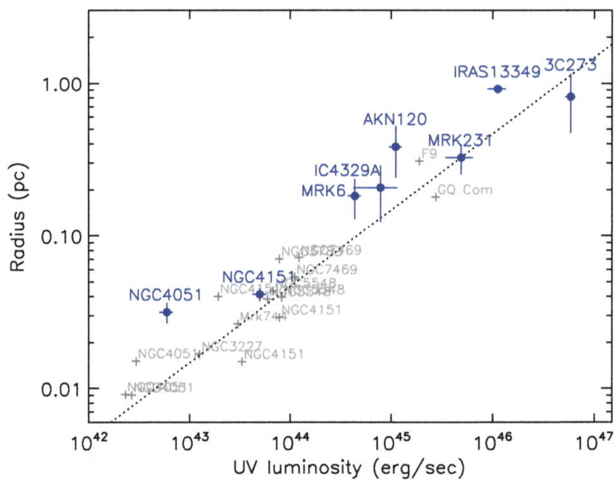

Fig. 6.5 Sublimation radius inferred from near-IR interferometry with the KI (*blue data points*) as well as near-IR reverberation mapping (*grey crosses*). The *dotted line* shows a dependence of radius on the *square root* of the luminosity as expected from the Stefan-Boltzmann law. The normalisation is consistent with large graphite grain opacities (Figure from [20])

This region is defined by the hottest temperature of about ∼1500 K where dust can still survive. From simple radiative equilibrium considerations and the Stefan-Boltzmann law, it can be estimated that the sublimation radius should depend on the square-root of the AGN luminosity, $r_{\rm sub} \propto L^{1/2}$, with the absolute scaling depending on dust properties. This dependence was previously indicated using optical-to-IR time lag measurements [33] and could be confirmed with KI interferometry of 8 AGN [19].

While the size-luminosity relation held, its absolute scaling came as a surprise (see Fig. 6.5). From opacities of standard astronomical dust, it was expected that the sublimation radius would be about 1.3 pc for an AGN luminosity of 10^{46} erg/s. The scaling that was found from interferometry and time-lag measurements is about a factor of 3 smaller [17, 19]. This led to the suggestion that the sublimation region is dominated by large graphite grains. Indeed, we may expect that the hottest dust is composed of those species and sizes that can survive the highest temperatures, consistent with the idea that at the inner radius we are left over with only the tip of the original ISM dust distribution. Moreover, since silicate grains are supposed to sublimate at lower temperatures than graphite, it is perfectly reasonable that the hottest dust is dominated by the latter type of grains.

These results should be put in light of the bigger picture. It has been noted that silicate emission features are lacking strength in observed mid-IR spectra of AGN while they feature prominently in any kind of dust torus model. Indeed, at interferometric resolution, silicate emission features are notably absent in spatially-resolved spectra of the inner few parsec of type 1 AGN [3–5, 21, 37]. On the other hand, silicate absorption features, which indicate rather cool dust, are present in obscured type 2 AGN, albeit generally being rather weak. In addition, the strongest absorption features are probably not associated with the obscuring material in the nuclear environment, but originate from galactic scale extinction, e.g. from host dust lanes [8, 14]. Putting these pieces together, the interferometry results are a strong indication of significant evolution of the dust composition with distance from the AGN.

6.4.4 Constraints on the Volume Filling Factor

The dust size measurements lead to another interesting constraint; by relating the sizes to the flux emitted from this same region, we can calculate surface emissivities. [21] assume that the typical hot dust temperature is 1400 K as inferred from spectral fits [17]. A black body emission model for the inferred interferometric half-light radius is then compared to the correlated flux from this radius (see Fig. 6.6).

For the hot dust emission near the sublimation front, the inferred surface emissivity turns out to be $\sim$0.3 in the six objects investigated by [21]. The interesting constraint we get from this comes from the fact that surface emissivity is the product of dust emissivity and surface filling factor. If this product is of the order 0.1, and considering that both individual factors must be smaller than or equal to unity,[3] both factors are quite strictly limited to > 0.1. For the dust emissivity, this is further evidence for the domination of large grains. The surface filling factor, on the other

[3]Strictly speaking, the colour temperature inferred from spectral fits is a lower limit to the physical temperature, given the uncertainty from the dust distribution, which may make the emission appear redder than the temperature in this region. However, this effect should be minor at the inner rim of the dust distribution. If we assume that the "true" temperature is 1800 K, we can expect that the maximum value the emissivity can get to is $\sim$3 for the assumption of a 1400 K black body. Similarly, the surface covering factor may be $>$1 if the dust is optically thin. However, for the near-IR, this is incompatible with the red SED towards the mid-IR, for which we would expect a Rayleigh-Jeans behaviour if optically thin.

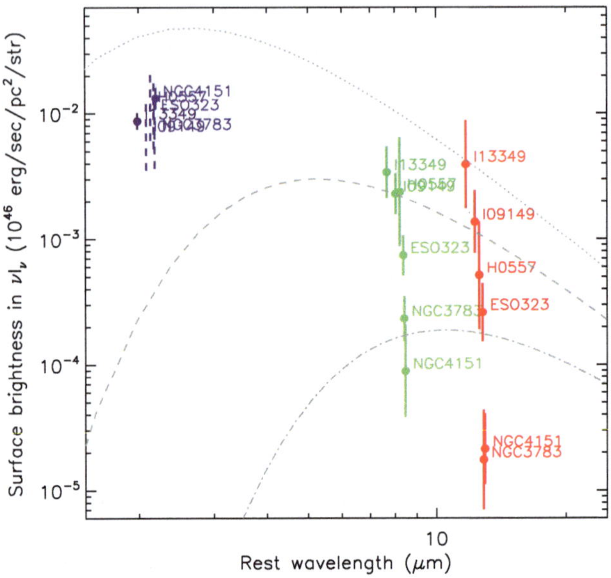

Fig. 6.6 Comparison of the observed surface brightness in the near-IR (*violet*) and mid-IR (*green*: 8 µm; red: 13 µm) for the six AGN in [21]. The *grey lines* are *black bodies* with temperatures of 1400 K (*dotted*), 700 K (*dashed*) and 350 K (*dash-dotted*). See text for further information (Figure from [21])

hand, is surprising, since it firmly limits the volume filling factor of the region to approximately $>(0.1)^{3/2} \sim 0.03$. This value still allows for clumpiness, but it might rule out low-volume filling factor models, as has been put forward in the literature.

The situation is not as clear in the mid-IR. At these wavelengths, the emission is a mix of contributions from the Rayleigh-Jeans tail of hotter dust closer to the AGN and the "black body peaks" of cooler dust. Therefore, it is much more difficult to infer a representative temperature for the size measured with interferometry. [21] compare the surface brightnesses of 700 K and 350 K black body emission to the resolved 8 and 13 µm emission from the half-light radii of their six AGN (see Fig. 6.6). A rather conservative lower limit can be set to the surface emissivity as >0.01, but given the uncertainties in the distribution of the dust, this result should be considered tentative.

In summary, IR interferometry indicates that clumpiness in the inner part of the dust distribution must be considered of the type with volume filling factor > 0.01. With a better mapping of the dust emission using upcoming interferometric instrumentation, it will become possible to put a similar constraint on the cooler dust in the mid-IR that emits most of the energy in the IR.

6.4.5 *The Distribution of the Dust Revealed, But It Is Not Clear What It Means*

One of the key features of IR interferometry is that it spans a rather large range of wavelengths. Aside from broadly covering the near- and mid-IR, structural information can be gathered with spectral resolution between 8 to 13 μm with the VLTI/MIDI instrument. Translated into temperatures, we cover about 270 K to 450 K in νF_ν black-body equivalent. If all temperatures contributed equally, we could expect a difference in size from the shortest to longest wavelength in this window of a factor of 3–4 based on a modified Stefan-Boltzmann law for dust. In reality, however, the dust will not be distributed homogeneously but according to some mass density distribution, which can be considered a distance-depending weighting function for the temperature distribution. Using these simple considerations, and confirming it by full radiative transfer simulations [11], we can come up with the prediction that if the emission is emerging from a compact dust distribution close to the sublimation radius, then the size should not depend too much on wavelength, since the emission is dominated by a small range of hotter temperatures. Conversely, if the dust distribution is shallow, we expect a strong dependence on wavelength as more and more cooler dust at larger distances will contribute to the emission. This will effectively tie the shape of the total IR SED to the wavelength-dependent sizes, with blue SEDs coming from more compact and, thus, smaller distributions, while redder emission originates from more extended distributions (barring obscuration effects) [11].

To test this theoretical relation with observations, we have to remember that the absolute size of the inner boundary of the dust distribution (= sublimation radius) depends on the AGN luminosity. If we want to test the temperature distribution, and in turn the projected mass distribution, we are interested primarily in how the mass is distributed with respect to this inner boundary. Therefore, the absolute sizes we obtain from interferometry should be expressed in units of this inner scaling radius, which can be either inferred from near-IR interferometry or reverberation mapping. Taking this into account, we can indeed identify a dependence of (scaled) mid-IR size on the emission properties of the overall SED. When plotting the mid-IR half-light radii of the type 1 AGN in [21] against the spectral slope in the IR, we find a quite strong dependence of both parameters (see Fig. 6.7). This backs the idea that the wavelength-dependence of the interferometric sizes traces the (projected) distribution of the dust mass [10].

A more puzzling relation is found when comparing the compactness of the dust emission with AGN luminosity. [21] found for their sample of 6 unobscured AGN that the brightness distribution becomes more compact with luminosity (see Fig. 6.8 left). On the luminous end ($\sim 10^{46}$ erg/s), the mid-IR emission seems to be almost entirely emerging from the Wien tail of hot dust in the vicinity of the sublimation radius, while on the lower luminosity side ($\sim 10^{43}$ erg/s) the hot dust is extended beyond 50 times the sublimation radius. Indications of such a trend may also be identified in the type 1s of the more expansive MIDI LP dataset [5], although

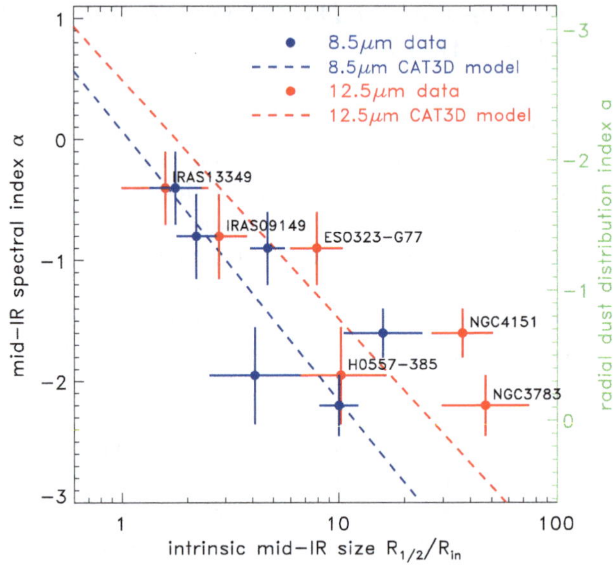

Fig. 6.7 Relation between the observed mid-IR spectral slope (expressed as a power law index) and the 8 μm (*blue*) and 13 μm (*red*) of the type 1 AGN sample in [21]. The *dashed lines* show the relation based on the *CAT3D* radiative transfer model [11] for which the radial dust distribution index was changed (*green scale* on the right y-axis)

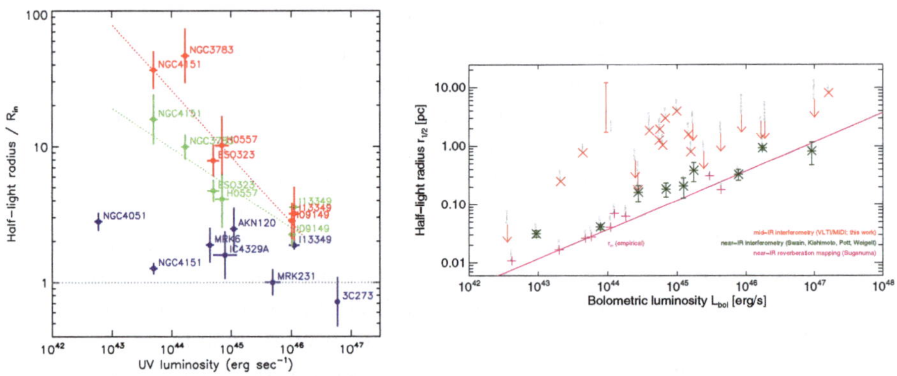

Fig. 6.8 *Left:* Mid-IR half-light radius in units of the inner radius of the dust distribution compared with the luminosity of the six AGN in [21]. The dust distribution becomes more compact at higher luminosity. *Right:* Near-IR and mid-IR sizes of objects in the MIDI Large Programme plotted against AGN luminosity from [5]. A similar trend may be inferred, though at much higher scatter (Figures from [21] and [5])

the picture is messier there (see Fig. 6.8 right). The major problem at this point is the limited number of sources at higher luminosities. If we assume, however, that the trend is real, we have yet to find a physical model to explain it. Note that this "compact when luminous" trend is distinct from the ideas of a receding torus

[25, 32], which has been considered for the effect that higher luminosity AGN might have lower dust covering factors: The receding torus may account for a *vertical* evolution with luminosity, but it does not necessarily require a *radial* evolution of the mass distribution in units of the sublimation radius.

6.4.6 Where Is the Torus After All?

Arguably the most challenging result for our understanding of AGN from IR interferometry comes from detailed observations of the four galaxies NGC 1068 [29], NGC 424 [12], NGC 3783 [13], and Circinus [38]. The mid-IR brightness distribution in each of these sources is mapped with 40 or more individual data points scattered from few 10 m up to 130 m baseline lengths and at all position angles that can be reached with the VLTI. This unprecedented level of detail allowed for precise analysis of the position-angle dependence of the emission source. Surprisingly, all four sources displayed strong elongation in the polar direction of the AGN, i.e. *perpendicular* to the plane of the torus (see Fig. 6.9). While one may argue that radiative transfer and obscuration effects play a role in this appearance for the three type 2 sources, this argument becomes less compelling for NGC 3783, which is a type 1 AGN. Moreover, the elongations in polar direction are substantial, with inferred axis ratios of 3–4. Finally, the emission originating from these polar features dominates the total mid-IR emission in all four sources. Such extreme polar features, in particular in a type 1 source, are very difficult to reconcile with a classical torus model.

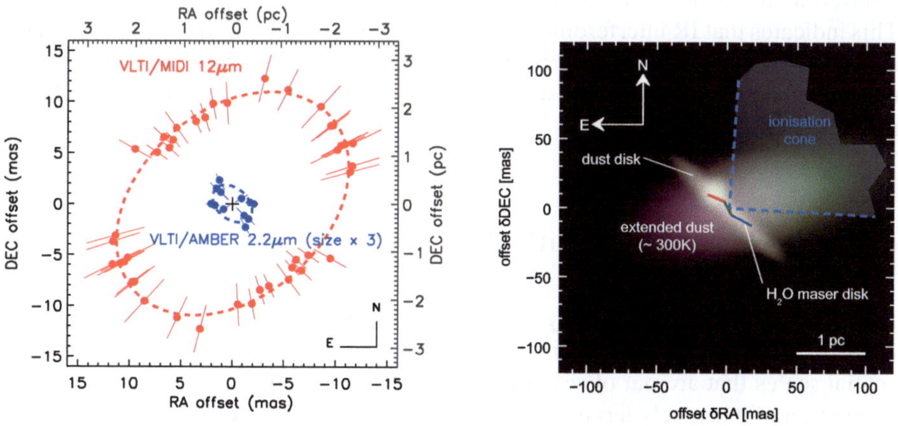

Fig. 6.9 *Left:* Position-angle dependent sizes of the near-iR (*blue*) and mid-IR (*red*) emission in NGC 3783 as observed with VLTI/AMBER and VLTI/MIDI. The mid-IR ellipse is elongated toward the polar region while the near-IR emission is aligned with the plane of the AGN [13]. *Right:* Model image of the dusty environment in the Circinus galaxy based on the mid-IR interferometery data from [38] (See text for further description)

NGC 3783 was also the target of the first successful AGN observations with VLTI/AMBER [40]. A preliminary analysis of the position angle dependence suggests that the hot dust emission as traced by these near-IR interferometry observations is aligned with the equatorial plane where models would locate the torus, i.e. probably perpendicular to the observed mid-IR emission [13]. The covering factor of this near-IR disc has been constrained to $\sim$ 30%. Similarly in Circinus, on top of the 80% of mid-IR flux emerging from the polar region, the bulk of the remaining flux has been found to be in a geometrically thin disc aligned with known maser emission [38]. However, the scale height over most of this disc's extension is substantially smaller than what torus models require.

These two observations suggest that the dusty environment consists of two separate entities: (1) a geometrically thin disc in the equatorial plane of the AGN, potentially dominating the near-IR fluxes, and (2) a strong polar component dominating the overall mid-IR emission. It has been suggested that this polar component represents a dusty wind, which is launched from near the inner region of the disc by the string radiation pressure on dust [12]. Indeed, radiation-hydrodynamical models suggest that such flows can exist [6, 31, 39], but further theoretical and observational work is required to solidify these ideas.

Notwithstanding the final theoretical assessments of the physical origin, these observations strongly suggest that our original idea of a simple torus has to be thrown overboard. The two-component structure offers important new constraints on what models have to reproduce to explain the mass distribution around AGN. Beyond the impact on unification, if the polar emission does indeed originate from a dusty wind, it may be connected to the much larger kpc scale molecular winds that are invoked as a feedback mechanism of AGN onto the galaxy. At least in NGC 1068 and Circinus, we do see that the mid-IR emission extends much beyond interferometric scales and can be traced to at least 100 pc in single-telescope images. This indicates that IR interferometry observes the region of origin of AGN feedback. ALMA may help us making this connection more directly by simultaneously tracing the molecular component of dust clouds at IR interferometry resolution and larger regions.

6.5 Conclusions and Outlook

IR long baseline interferometry has contributed to bring a unique view into the mass distribution around supermassive black holes. It allowed us to access small spatial scales that are out of reach for any single telescope, even for the upcoming generation of extremely large telescopes. The key messages to keep in mind are as follows:

- While in the beginning only the few brightest targets were observable, we are now closing in on 50 different AGN with constraints from interferometry. A whole suite of data sets is available now in the near-IR and mid-IR, ranging from visibilities to closure phases and their differential quantities and monitoring information.

- With the help of spectrally resolved interferometry, it was possible to map the dust brightness distribution around AGN. The dust seems to be clumpy with a volume filling factor of the order 0.01–0.1. The dust undergoes significant evolution in the AGN environment under the influence of the strong, hard radiation field. Finally the dust distribution among the whole AGN sample looks quite diverse, but there are indications of a trend of more compact distributions for higher luminosity AGN. This is yet to be confirmed, however.
- Most surprisingly, the bulk of the pc-scale mid-IR emission in AGN seems to originate from the polar region of the AGN, not from the equatorial plane where the torus is located. A new two-component picture is emerging consisting of a dense dusty disc, in the plane of accretion onto the AGN, and an extended polar wind, where dusty gas is pushed out by the strong radiation pressure from the AGN.

The polar extended emission will become a major field of research for AGN interferometry in the coming years with the new MATISSE (mid-IR) and GRAVITY (near-IR) instruments going online at the VLTI. These instruments will combine all four UT telescopes and enable us to create real images from AGN interferometry data for the first time. MATISSE will also add new capabilities at 3.6 μm and 4.5 μm, which are sweet spots for AGN research. We now have an unprecedented opportunity to pin down the mass distribution around AGN in the dust and gas phase by making use of the high-angular resolution facilities that the European Southern Observatory has to offer in the IR (VLTI) and sub-mm (ALMA).

Another interesting new path for the VLTI was recently suggested [15]. By combining near-IR interferometry with reverberation mapping, is is possible to use simple trigonometric principles to measure precise direct distances to AGN (see Fig. 6.10). As described above, the hot dust around an AGN essentially forms a narrow ring. With near-IR interferometry, we can determine the angular size of this ring. Near-IR reverberation mapping measures the time lag between variability seen in the visual and its dust-reprocessed version in the near-IR. Since the visual light originates from the putative accretion disc, which is at least an order of magnitude smaller than the hot dust ring, the time lag can be converted into a physical size, which corresponds to the physical radius of the hot dust ring. The distance can then be calculated from the ratio of the angular and physical size.

This new "dust parallax" method, being a spin-off of an earlier suggestion for line emission [7], has been showcased for NGC 4151. The distance was determined as $19.0^{+2.4}_{-2.6}$ Mpc, which means that the combined systematical and statistical errors are of the order of 12–13%. If this precision can be held up for other objects, it will be a competitive method for cosmological applications. The systematical uncertainties are dominated by the geometry of the brightness distribution, such as its inclination and orientation, which can be well constrained with the help of the new VLTI instruments. They allow for a better mapping of the near-IR emission source, including phase information to probe asymmetries.

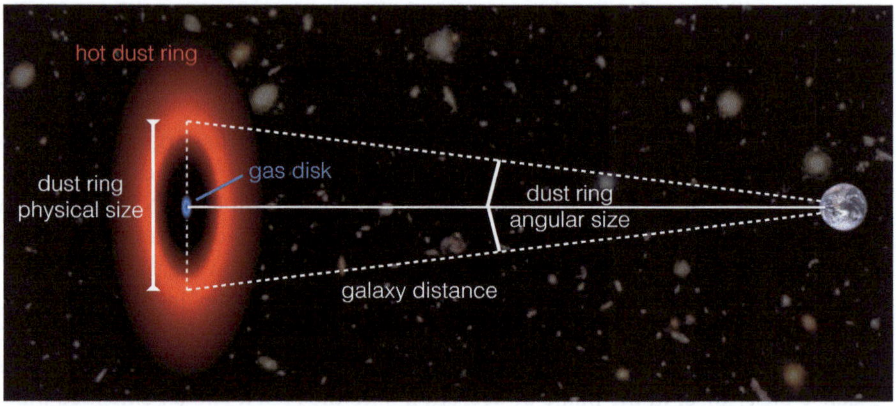

Fig. 6.10 Illustration of the "dust parallax" method to determine geometric distances to AGN [15]. While interferometry measures the angular size of the hot dust ring, its physical size can be inferred from reverberation mapping (See text for details)

One interesting aspect is the possibility to measure geometric distances directly in the Hubble flow at >100 Mpc. Indeed, dozens of AGN with a K-band magnitude between 10.5 and 12 would be suitable for this exercise. If reached with interferometry, we could independently estimate the Hubble constant without invoking the many steps in the cosmological distance ladder and address the current tension between local measurements and inferences from the Cosmic Microwave Backround. Therefore, it may be worth considering to put some additional efforts into increased sensitivity of interferometric instrumentation in the near-IR.

The most impressive recent results [on AGN unification] are due to IR interferometry. Long baseline interferometry is the way of the future.

Hagai Netzer, Annual Review of Astronomy & Astrophysics, 2015

References

1. Antonucci, R.: Unified models for active galactic nuclei and quasars. ARA&A **31**, 473 (1993)
2. Barvainis, R.: Hot dust and the near-infrared bump in the continuum spectra of quasars and active galactic nuclei. ApJ **320**, 537 (1987)
3. Beckert, T., Driebe, T., Hönig, S.F., Weigelt, G.: Probing the dusty environment of the Seyfert 1 nucleus in NGC 3783 with MIDI/VLTI interferometry. A&A **486**, L17 (2008)
4. Burtscher, L., Jaffe, W., Raban, D., Meisenheimer, K., Tristram, K.R.W., Röttgering, H.: Dust emission from a parsec-scale structure in the Seyfert 1 nucleus of NGC 4151. ApJ **705**, L53 (2009)
5. Burtscher, L., Meisenheimer, K., Tristram, K.R.W., et al.: A diversity of dusty AGN tori. Data release for the VLTI/MIDI AGN Large Program and first results for 23 galaxies. A&A **558**, 149 (2013)

6. Dorodnitsyn, A., Kallman, T., & Bisnovatyi-Kogan, G.: AGN obscuration through dusty, infrared-dominated flows. II. Multidimensional, radiation-hydrodynamics modeling. ApJ **747**, 8 (2012)
7. Elvis, M., & Karovska, M.: Quasar parallax: a method for determining direct geometrical distances to quasars. ApJL **581**, L67 (2002)
8. Goulding, A.D., Alexander, D.M., Bauer, F.E., et al.: Deep silicate absorption features in compton-thick active galactic nuclei predominantly arise due to dust in the host galaxy. ApJ **755**, 5 (2012)
9. Hönig, S.F., Beckert, T., Ohnaka, K., Weigelt, G.: Radiative transfer modeling of three-dimensional clumpy AGN tori and its application to NGC 1068. A&A **452**, 459 (2006)
10. Hönig, S.F., Kishimoto, M., Gandhi, P., et al.: The dusty heart of nearby active galaxies. I. High-spatial resolution mid-IR spectro-photometry of Seyfert galaxies. A&A **515**, 23 (2010)
11. Hönig, S.F., & Kishimoto, M.: The dusty heart of nearby active galaxies. II. From clumpy torus models to physical properties of dust around AGN. A&A **523**, 27 (2010)
12. Hönig, S.F., Kishimoto, M., Antonucci, R., et al.: Parsec-scale dust emission from the polar region in the type 2 nucleus of NGC 424. ApJ **755**, 149 (2012)
13. Hönig, S.F., Kishimoto, M., Tristram, K.R.W., et al.: Dust in the polar region as a major contributor to the infrared emission of active galactic nuclei. ApJ **771**, 87 (2013)
14. Hönig, S.F., Gandhi, P., Asmus, D., Mushotzky, R.F., Antonucci, R., et al.: What obscures low-X-ray-scattering active galactic nuclei. MNRAS **438**, 647 (2014)
15. Hönig, S.F., Watson, D., Kishimoto, M., Hjorth, J.: A dust-parallax distance of 19 megaparsecs to the supermassive black hole in NGC 4151. Nature **515**, 528 (2014)
16. Jaffe, W., Meisenheimer, K., Röttgering, H.J.A., Leinert, Ch., Richichi, A., et al.: The central dusty torus in the active nucleus of NGC 1068. Nature **429**, 47 (2004)
17. Kishimoto, M., Hönig, S.F., Beckert, T., Weigelt, G.: The innermost region of AGN tori: implications from the HST/NICMOS type 1 point sources and near-IR reverberation. A&A **476**, 713 (2007)
18. Kishimoto, M., Hönig, S.F., Tristram, K., Weigelt, G.: Possible evidence for a common radial structure in nearby AGN tori. A&A **493**, L57 (2009)
19. Kishimoto, M., Hönig, S.F., Antonucci, R., et al.: Exploring the inner region of type 1 AGNs with the Keck interferometer. A&A **507**, L57 (2009)
20. Kishimoto, M., Hönig, S.F., Antonucci, R., et al.: The innermost dusty structure in active galactic nuclei as probed by the Keck interferometer. A&A **527**, 121 (2011)
21. Kishimoto, M., Hönig, S.F., Antonucci, R., et al.: Mapping the radial structure of AGN tori. A&A **536**, 78 (2011)
22. Kishimoto, M., Hönig, S.F., Antonucci, R., Millan-Gabet, R., Barvainis, R., et al.: Evidence for a receding dust sublimation region around a supermassive black hole. ApJL **775**, L36 (2013)
23. Krolik, J.H., & Begelman, M.: An X-ray heated wind in NGC 1068. ApJL **308**, 55 (1986)
24. Krolik, J.H., & Begelman, M.C.: Molecular tori in Seyfert galaxies – feeding the monster and hiding it. ApJ **329**, 702 (1988)
25. Lawrence, A.: The relative frequency of broad-lined and narrow-lined active galactic nuclei – implications for unified schemes. MNRAS **252**, 586 (1991)
26. López-Gonzaga, N., Jaffe, W., Burtscher, L., Tristram, K.R.W., Meisenheimer, K.: Revealing the large nuclear dust structures in NGC 1068 with MIDI/VLTI. A&A **565**, 71 (2014)
27. Meisenheimer, K., Tristram, K.R.W., Jaffe, W., et al.: Resolving the innermost parsec of Centaurus A at mid-infrared wavelengths. A&A **471**, 453 (2007)
28. Pott, J.-U., Malkan, M.A., Elitzur, M., et al.: Luminosity-variation independent location of the circum-nuclear, hot dust in NGC 4151. ApJ **715**, 736 (2010)
29. Raban, D., Jaffe, W., Röttgering, H., Meisenheimer, K., Tristram, K.R.W.: Resolving the obscuring torus in NGC 1068 with the power of infrared interferometry: revealing the inner funnel of dust. MNRAS **394**, 1325 (2009)
30. Schartmann, M., Meisenheimer, K., Camenzind, M., Wolf, S., Henning, T.: Towards a physical model of dust tori in active galactic nuclei. Radiative transfer calculations for a hydrostatic torus model. A&A **437**, 861 (2005)

31. Schartmann, M., Wada, K., Prieto, M. A., Burkert, A., Tristram, K.R.W.: Time-resolved infrared emission from radiation-driven central obscuring structures in active galactic nuclei. MNRAS **445**, 3878 (2014)
32. Simpson, C.: The luminosity dependence of the type 1 active galactic nucleus fraction. MNRAS **360**, 565 (2005)
33. Suganuma, M., Yoshii, Y., Kobayashi, Y., Minezaki, T., Enya, K., et al.: Reverberation measurements of the inner radius of the dust torus in nearby Seyfert 1 galaxies. ApJ **639**, 46 (2006)
34. Swain, M., Vasisht, G., Akeson, R., Monnier, J., Millan-Gabet, R., et al.: Interferometer observations of subparsec-scale infrared emission in the nucleus of NGC 4151. ApJ **596**, L163 (2003)
35. Tacconi, L.J., Genzel, R., Blietz, M., Cameron, M., Harris, A.I., Madden, S.: The nature of the dense obscuring material in the nucleus of NGC 1068. ApJ **426**, L77 (1994)
36. Tristram, K.R.W., Meisenheimer, K., Jaffe, W., Schartmann, M., Rix, H.-W., et al.: Resolving the complex structure of the dust torus in the active nucleus of the Circinus galaxy. A&A **474**, 837 (2007)
37. Tristram, K.R.W., Raban, D., Meisenheimer, K., Jaffe, W., Röttgering, H., et al.: Parsec-scale dust distributions in Seyfert galaxies. Results of the MIDI AGN snapshot survey. A&A **502**, 67 (2009)
38. Tristram, K.R.W., Burtscher, L., Jaffe, W., Meisenheimer, K., Hönig, S. F., et al.: The dusty torus in the Circinus galaxy: a dense disk and the torus funnel. A&A **563**, 82 (2014)
39. Wada, K.: Radiation-driven fountain and origin of torus around active galactic nuclei. ApJ **758**, 66 (2012)
40. Weigelt, G., Hofmann, K., Kishimoto, M., et al.: VLTI/AMBER observations of the Seyfert nucleus of NGC 3783. A&A **541**, L9 (2012)
41. Wittkowski, M., Kervella, P., Arsenault, R., Paresce, F., Beckert, T., & Weigelt, G.: VLTI/VINCI observations of the nucleus of NGC 1068 using the adaptive optics system MACAO. A&A **418**, L39 (2004)

Chapter 7
Disentangling of Stellar Spectra

Petr Hadrava

7.1 Introduction

Observations and the corresponding theory of binary and multiple stars provide us with a clue to the physics of stars and hence also to the universe beyond the Solar system:

- From observations of a **visual binary**, we can determine its orbital period P, inclination i, eccentricity e and periastron longitude ω, the angular distance α between the components (and hence the ratio of the semi-major axis a to the distance d of the system) and light ratio L_1/L_2 of the components.
- From observations of an **eclipsing binary**, we can also assess P, i, with some limitations also e and ω, and in addition to L_1/L_2 also the ratios of the components' radii $R_{1,2}$ to a.
- From observations of Doppler shifts of lines in spectra of **spectroscopic binaries** we can also obtain P, e and ω. Moreover, we can get the semi-amplitudes of radial-velocity curves $K_{1,2}$, i.e. for a reasonable estimate of i we know the orbital velocities $v_{1,2}$ and we can calculate the absolute size of a and also the masses $M_{1,2}$ of the component stars.

For spectroscopic binaries which are simultaneously eclipsing, we can thus determine the basic physical parameters of the stars – $M_{1,2}$ and $R_{1,2}$, and from colour photometry or spectroscopy also the temperatures $T_{\mathrm{eff}1,2}$ and hence the luminosities $L_{1,2}$ of the component stars and the photometric distance d of the system. From these data, we can calculate the angular separation α, including for stars that we cannot resolve by direct imaging (e.g., $\alpha \leq 0.2\,\mu$ as for the first binary measured in M31 [15]).

P. Hadrava (✉)
Astronomical Institute, Academy of Sciences, Boční II 1401, Praha 4, Czech Republic
e-mail: had@sunstel.asu.cas.cz

H.M.J. Boffin et al. (eds.), *Astronomy at High Angular Resolution*, Astrophysics and Space Science Library 439, DOI 10.1007/978-3-319-39739-9_7

However, the goal of indirect imaging which is usually dependent on some theoretical model of the object (e.g., in this case the model of components radiation and the interstellar absorption) is not to compete with observational techniques of direct imaging, not to say to replace them. To the contrary, the improvement of such techniques is even more desirable in connection with indirect imaging because they may provide a complementary information enabling to verify the theory and to improve the underlying assumptions. For instance, until recently, there were practically no binaries observed as both visual and spectroscopic because the former technique prefers wide orbits while the former close binaries with higher amplitudes of radial velocities. Interferometry, however, enables higher angular resolution, while improved spectroscopic instrumentation and data processing provides better spectral resolution, so that the overlap of these two sets of techniques increases. From a combination of interferometric and spectroscopic observations of (even non-eclipsing) binaries we can determine α, i, $v_{1,2}$, a and hence also the geometric distance d (e.g., [23], which solved the Pleiades distance controversy caused by an error in Hipparcos results).

The observations of binaries can thus provide us with more information than those of single stars, but their interpretation is at the same time more difficult. In particular, the spectra (as well as the colours) of the component stars are blended and can only be distinguished by taking into account their variations with the orbital motion. While for single stars it is usually sufficient to fit their observed spectra with synthetic spectra parameterised with $T_{\rm eff}$, log g (the surface gravity) and abundances of elements X_i, the physics of binaries is more complicated due to their interaction and proximity effects, such as tidal forces or reflection effects. More sophisticated theoretical tools and data-processing methods are thus needed to interpret the observational data. Such methods actually reveal structure within the binary systems and belong thus to the methods of indirect imaging.

7.2 Disentangling of Spectra of Multiple Stars

The classical treatment of observed spectra of spectroscopic binaries consists in measuring radial velocities (RVs), $v_j|_{j=1}^{n}$, of the n component stars (Fig. 7.1). There are various techniques to measure RVs, but they all require some model of the component spectra, at least a simple assumption that the centres of some observed spectral lines correspond to the v_j of a particular component j. The RV curves, i.e. the set of $v_j(t)$ at different times t, are then fitted to find the orbital parameters p of the system. A simultaneous solution of other observational data such as photometry or interferometry is advantageous because these data may better constrain some of the parameters and the set of solved p may be enriched also by some more physical parameters (M, R, $T_{\rm eff}$ etc.). To relate the stellar parameters found with the spectra of individual stars, it is desirable to separate the spectrum F_j of each component j from the observed spectra F of the whole system of n components. This is possible thanks to varying conditions in the blending of components in different exposures

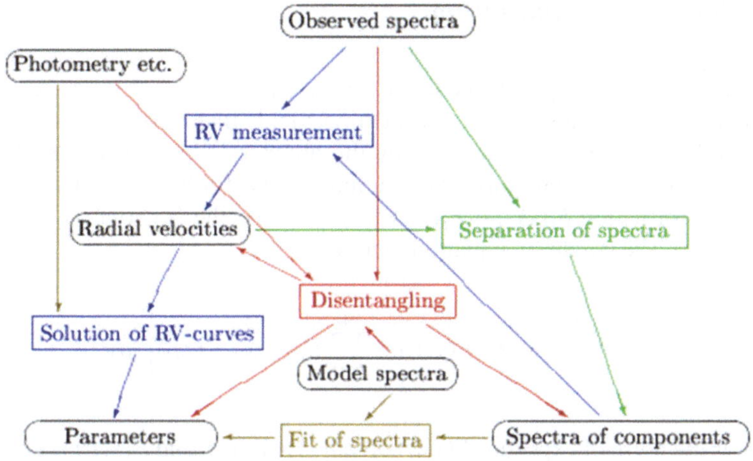

Fig. 7.1 Flow diagram of the standard data processing of spectra of multiple stellar systems and of the disentangling process

with different Doppler shifts (cf. e.g., [2]) or with different light ratios, e.g., during an eclipse (cf. [20]). The information on both the $F_j|_{j=1}^n$ and p is contained in the observed spectra, but it is entangled there because to retrieve one of them we need to know the other.

The basic idea of the method of disentangling spectra (cf. [5, 20]) is to solve simultaneously (or, actually, iteratively) the component spectra $F_j|_{j=1}^n$ and directly the orbital parameters p (instead of the intermediate step of RVs) by fitting all the observed spectra with a model of the form

$$F(x,t;p) = \sum_{j=1}^{n} F_j(x) * \Delta_j(x,t;p) \ , \tag{7.1}$$

where $x \equiv c \ln\lambda/\lambda_0$ is the logarithmic wavelength scale and the broadening function is given by a simple Doppler-shifted Dirac δ-function

$$\Delta_j(x,t;p) = \delta(x - v_j(t;p)) \ . \tag{7.2}$$

This means that the separated component spectra F_j appropriate to the observed object are also used for the treatment of RVs instead of a theoretical model or another template needed in the classical approach.

7.2.1 Fourier Disentangling

The model given by Eq. (7.1) is linear in F_j and non-linear in p. A χ^2-fit of the observed spectra with respect to $F_j(x)$ thus yields linear condition equations of the (generally large) dimension given by the number of bins of the spectra times the number of component stars. Its solution can be further simplified using the Fourier transformation $F_j(x) \rightarrow \tilde{F}_j(y)$ which converts the convolution in Eq. (7.1) into a product. The condition equation then reads

$$0 = \delta\chi^2 = \delta \sum_{l=1}^{N} \frac{1}{\sigma_l^2} \int \left| \tilde{F}(y, t_l) - \sum_{j=1}^{n} \tilde{F}_j(y) \tilde{\Delta}_j(y, t_l; p) \right|^2 dy \, , \tag{7.3}$$

where σ_l is the noise of exposure $F(x, t_l)$ obtained at time t_l and $\tilde{F}(y, t_l)$ is the y-mode of Fourier transform of $F(x, t_l)$. Variation of the right-hand side with respect to $\delta\tilde{F}_k^*(y)$ obviously yields a set of n linear equations for $\tilde{F}_j(y)$, i.e. separated subsets of equations for individual Fourier modes

$$\sum_{j=1}^{n} \sum_{l=1}^{N} \frac{\tilde{\Delta}_k^*(y, t_l; p) \tilde{\Delta}_j(y, t_l; p)}{\sigma_l^2} \tilde{F}_j(y) = \sum_{l=1}^{N} \frac{\tilde{\Delta}_k^*(y, t_l; p) \tilde{F}(y, t_l)}{\sigma_l^2} \, . \tag{7.4}$$

This makes the separation of spectra, i.e. the solution of the linear part of the disentangling, numerically easier and more efficient, compared to the solution in the wavelength domain. The optimisation with respect to δp can be then performed by numerical minimisation of the right-hand side of Eq. (7.3), e.g., using the simplex method.

It is obvious that a necessary (but not sufficient) condition to get a non-singular matrix on the left-hand side of Eq. (7.4) is $N \geq n$ and the vectors $\tilde{\Delta}_k$ should differ for at least n different t_l, i.e. the observations should sufficiently cover orbital phases with different $\tilde{\Delta}_k$. Regarding the intention to minimise the right-hand side of Eq. (7.3) with respect to p, it should be positive, i.e. Eq. (7.4) should be overdetermined and hence $N > n$. A significant over-determinacy is desirable because it can reduce the random noise in the disentangled $F_j(x)$ originating from the observational noise of $F(x, t_l)$.

For a pure orbital Doppler shift, the broadening $\Delta_j(x, t; p)$ is given by Eq. (7.2) and its Fourier transform can be calculated analytically

$$\tilde{\Delta}_j(y, t; p) = s_j(t) \exp(iyv_j(t; p)) \tag{7.5}$$

with $s_j(t) = 1$. It can be seen from here that $\tilde{\Delta}_j(y = 0, t; p) = 1$ and, hence, the matrix on the left-hand side of Eq. (7.4) is singular for the zeroth Fourier mode $y = 0$. This corresponds to the obvious fact (which is also true for the separation of the spectra in the wavelength domain) that the continuum level is insensitive to the orbital motion and consequently it cannot be disentangled from spectroscopy alone. The disentangled component spectra $F_j(x)$ are thus defined up to additive constants, the sum of which (over j) should give the mean value of all observed $F(x, t_l)$ in a particular spectral region. To get a completely separated spectra of each component like a spectrum of a single star, it is necessary to find proportions in which this sum has to be distributed between the components. This information on light ratios of the components can be in principle obtained, e.g., from interferometry (this is an example of complementarity of the direct and indirect imaging). In the case of eclipsing binaries the light ratio is usually obtained from photometry. The light ratio is wavelength dependent and is usually measured in broad wavelength regions. A normalisation of the disentangled spectra from the common continuum of the whole system to the proper continuum of each component should thus take into account the levels of continua influenced by the presence of spectral lines which are different for each component in each spectral region. Another possibility of interpretation of the disentangled spectra is to fit them by model spectra scaled in the intensity with a general unknown linear transformation.

The change of light ratio in the course of an eclipse and possibly also in other orbital phases of tidally distorted or mutually illuminated components of binaries also varies the strengths of lines of all components visible in the spectra. To get a better fit of the observed spectra in this case, line-strength factors $s_j(t)$ are introduced in Eq. (7.5) – cf. [6]. A set of linear equations for the values $s_j(t_l)|_{j=1}^{n}|_{l=1}^{N}$ can be obtained from Eq. (7.3) and solved iteratively with the separation of $F_j(x)$ – the model is bilinear in these two sets of unknown variables. The solution of s_j yields photometric information from the spectra about the luminosity of one component relative to the other one. The decrement of light due to an eclipse depends on the limb darkening which in turn depends on the source function $S_\nu(\tau_\nu)$. In the approximation of plane-parallel stellar atmosphere

$$S_\nu(\tau_\nu) = \sum_k S_{\nu,k}\tau_\nu^k \quad \Rightarrow \quad I_\nu(\tau_\nu = 0, \mu) = \sum_k S_{\nu,k}\mu^k \,, \tag{7.6}$$

which means that the dominant linear term ($k = 1$) in the dependence of the surface intensity $I_\nu \sim (1 - u + u\mu)$ on the directional cosine μ is proportional to the first derivative of S_ν with respect to the monochromatic optical depth τ_ν. In the local thermodynamic equilibrium (LTE) approximation, S_ν is given by the temperature structure of the atmosphere and $dS_\nu/d\tau_\nu$ is proportional to the vertical gradient (in geometric depth) of the temperature and inversely proportional to the monochromatic opacity. In contrary to the frequently adopted approximation that the change of limb darkening is negligible within the narrow line-profile and the value of u found from photometry (dominated by the continuum) can be fixed across the lines, u rapidly decreases from its value at the continuum toward the centre of line (cf. Fig. 7.2). Actually, the absorption lines are present in the spectra just owing to this change. In consequence, the signal contained in spectral lines normalised with respect to the overall continuum of the system may be enhanced not only for the component in foreground but also for the partly eclipsed component when only

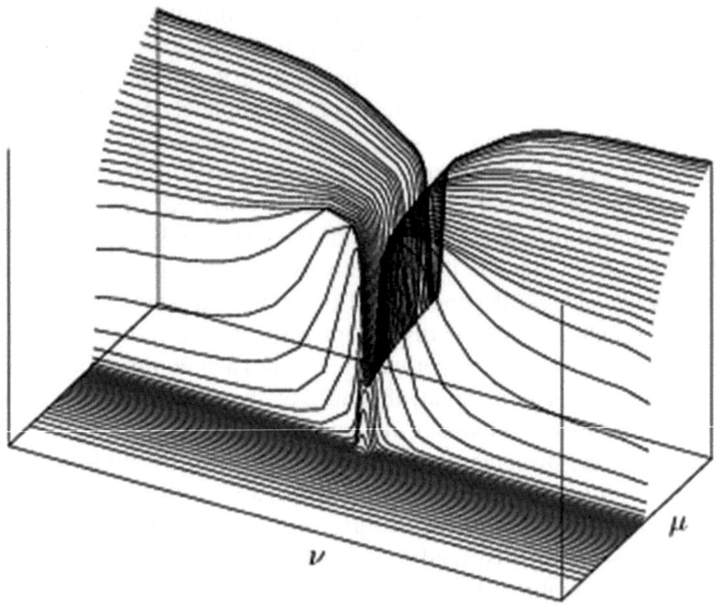

Fig. 7.2 Limb darkening within the profile of Hα line in spherical non-LTE model of stellar atmosphere ($T_{\rm eff}$ = 15,000 K, $\log g = 2.0$; cf. [10])

an edge of its disc is eclipsed which little contributes to the line. The disentangling of the line-strength factors $s_j(t)$ thus enables to map the surface and actually also the depth structure of the eclipsed star. It may also reduce errors in rectification of the observed spectra (i.e. their normalisation with respect to the continuum), which, however, limits the information gain about the structure of the atmosphere. The line-strength factors also enable to disentangle the telluric spectrum [6, 7].

7.2.2 Generalised Disentangling

Taking into account non-LTE or higher order terms in Eq. (7.6), the spectral lines may change during an eclipse, not only in strength, but also in shape. In the approximation of a thin stellar atmosphere, the observed spectral flux $F(x, t)$ is given by an integral of the surface intensity I_j over the visible surface s of all components j,

$$F(x,t) = \sum_{j=1}^{n} \int_s \mu I_j(x, s, \mu, t) * \delta(x - v_j(s,t)) d^2 s \,, \tag{7.7}$$

where μ and v_j are known functions of s. If in analogy with Eq. (7.6) the intensity can be expressed in a form of superposition of several spectral functions F_j^k,

$$I(x, s, \mu, t) = \sum_{k=1} F_j^k(x) f_j^k(s, \mu, t) \ , \tag{7.8}$$

then the disentangling can be generalised from Eq. (7.1) to

$$F(x, t; p) = \sum_{j,k} F_j^k(x) * \Delta_j^k(x, t; p) \ , \tag{7.9}$$

where the broadening functions

$$\Delta_j^k(x, t; p) = \int_s \mu f_j^k(s, \mu, t) \delta(x - v_j(s, t)) d^2 s \tag{7.10}$$

are now dependent not only on the orbital parameters p, but also on other geometrical and physical parameters, as determined by the decomposition in Eq. (7.8).

In particular, if we treat eclipses of a rotating star, its line profile is rotationally broadened and its shape is distorted in the course of a partial eclipse. This so-called "rotational effect" [1, 8, 12, 14, 17, 18] is usually treated as a deviation of measured RVs during the eclipse from an exact Keplerian motion of the component's centres of mass. However, its value depends on the method of measurement and definition of RV of an asymmetric line-profile. The line-profiles are usually modelled in the form of Eq. (7.1), i.e. as a convolution of the profiles of a non-rotating star with a rotational broadening function. The rotational broadening, however, depends on the limb darkening – it has a semi-circular shape for a rigidly rotating sphere with uniform intensity and a parabolic shape for intensity linearly darkened to zero at the disc edge (see Fig. 7.3). As can be seen in Fig. 7.2, the line-profile generally varies across the disc (instead of a simple scaling) and, following Eq. (7.6), it should be expressed as a sum of contributions with different limb darkenings. A generalisation from Eqs. (7.1) to (7.9) for the two or more modes k of the limb darkening can thus fit the observed spectra better. The eclipse then enables to map the variations of line-profiles and hence the atmospheric structure if the quality of

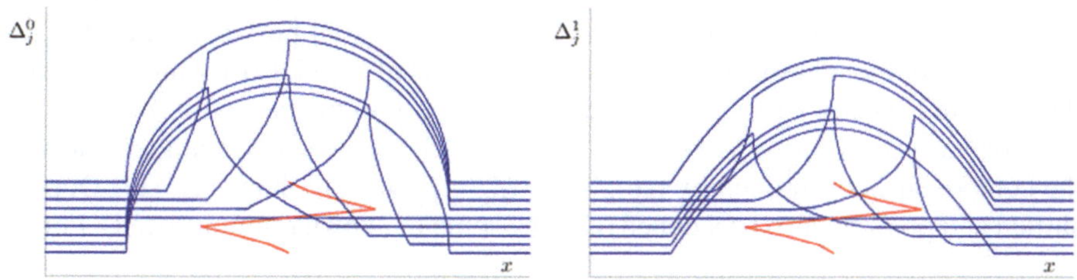

Fig. 7.3 Rotational broadening during an eclipse for $u = 0$ (*left*) and $u = 1$ (*right panel*; cf. [8])

the spectroscopic data (S/N, resolution, phase coverage) is sufficient. In addition to the synthetic spectra, more detailed results from model atmospheres are needed for interpretation of such results (as well as the other moments of $I_\nu(\mu)$). In any case, even if the data are not sufficient to distinguish between different modes k, it is preferable to fit the data with a proper model of shape of Δ_j than to modify v_j only in Eq. (7.2) because we can get a better fit of the data and more reliable parameters from the eclipse (e.g. the radii, rotational velocities or even inclinations of rotational axes, cf. [1]).

In addition to the Doppler shift according to Eq. (7.2), the broadening functions in the generalised disentangling can thus also include intrinsic variations of line profiles caused by various reasons – either geometric (e.g., eclipses, tidal distortion, spots) or physical. A frequent example of the latter case is pulsation. In the case of radial pulsations, the broadening functions for individual limb-darkening modes can be calculated explicitly (as well as their Fourier transforms, cf. [11]). This enables to disentangle directly the pulsation velocities of Cepheids and other radial pulsators and to avoid the need of the so-called projection factor which relates the effective RV measured in a particular way to the pulsation velocity. The line-strength factors $s_j(t_l)$ introduced in Eq. (7.5) as the simplest generalisation of Eq. (7.2), which are different for different lines in Cepheids depending on the influence of the temperature variations during the pulsation period on the rate of a particular transition, can reveal the temperature. In combination with photometric observations of luminosity variations or interferometric observations of changes of angular diameter, the disentangling of spectra of Cepheids may thus provide primary method of photometric or geometric determination of distance known as the Baade–Wesselink calibration of the luminosity–period relation.

7.2.3 *Constrained Disentangling*

As mentioned in the introduction, the aim of separating the spectra of the components is to compare them with theoretical models and to find the parameters of the component's atmospheres (cf. the "Fit of spectra" in Fig. 7.1). This comparison is usually performed by means of a least-squares fit similarly to the separation itself. Provided we have a grid of model spectra, which we believe correspond to some of the component stars, or a code for their synthesis, we can use them as template(s) by which the disentangling is constrained to a subspace of all possible solutions. The sum of the component spectra on the right-hand side of Eq. (7.1) by which we model the observed spectra $F(x, t; p)$ then generally splits into the part with the unknown component spectra $F_j(x)|_{j=1}^{m}$ (where $m < n$), which are to be disentangled and with the components constrained by templates $G_j|_{j=m+1}^{n}$ that are, however, still Doppler shifted and possibly also broadened (e.g., by some unknown rotation speed). For the separation of F_j we thus solve a set of linear equations

$$\sum_{j=1}^{m} F_j(x) * \Delta_j(x,t;p) = F(x,t;p) - \sum_{j=m+1}^{n} G_j(x) * \Delta_j(x,t;p) \tag{7.11}$$

of the order of m only, while the template-constrained components G_j are subtracted from the observations on the right-hand side (they are the additional input to the disentangling in Fig. 7.1). The lower number of unknown spectra makes the solution less sensitive to errors on the observed data – on the other hand, the number of unknown parameters p in broadenings at both F_j and G_j usually increases a bit. For instance, the systemic velocity γ is uncertain for the standard disentangling and can only be determined when a template G_j is accepted. The line strengths of G_j are usually given relatively to the continuum level of the component j, while in the observed spectra F they are normalised by the sum of continua. Free line-strength factors s_j are thus to be solved. They reduce at the same time the uncertainty in the levels of continua discussed in Sect. 7.2.1.

As it was already mentioned and shown in Fig. 7.1, it is advantageous also in the classical approach to treat the spectroscopy together with other types of data because these may better constrain some of the parameters. The simplest way to utilise such an additional information in the disentangling is to fix the values of parameters which we know well from the other data and to converge only those, which can better be determined from the spectroscopy. However, the constraint has often the form of a hypersurface $f(p) = 0$ in the space of parameters p. For instance, the times of primary and secondary eclipses in binaries depend on the time of periastron passage, eccentricity and periastron longitude (slightly also on the inclination), but cannot determine any of these parameters unless information about some of them is available from elsewhere, e.g., from the RV-curve. The minimising of χ^2 in the p-space should thus be constrained to the hypersurface or to the cross-section of the hypersurfaces if more different constraints $f_k(p) = 0$ are given by various observations. Such a minimisation can be performed using the method of Lagrange multipliers λ_k, i.e. instead of Eq. (7.3), we minimise its more complex variant

$$0 = \delta \left\{ \sum_{l=1}^{N} \frac{1}{\sigma_l^2} \int |\tilde{F}(y,t_l) - \sum_{j=1}^{m} \tilde{F}_j(y)\tilde{\Delta}_j(y,t_l;p) - \sum_{j=m+1}^{n} \tilde{G}_j(y)\tilde{\Delta}_j(y,t_l;p)|^2 dy \right.$$
$$\left. + \sum_k \lambda_k f_k^2(p) \right\} . \tag{7.12}$$

If the constraints $f_k = 0$ are observational, they are not sharp, but due to the observational error of a probabilistic nature only. The functions $f_k^2(p)$ can thus be chosen as $(O - C)^2$, where $C = C(p)$ is a model of the observations, O, and the multipliers $\lambda_k = \sigma_k^{-2}$ are given by errors on these observations. The disentangling thus turns into a simultaneous solution, i.e. χ^2-fitting of all available data. From the point of view of accuracy and conservation of information, it is preferable to fit the data in their original form than to reduce them into some intermediate values, such as RVs instead of the source spectra or separations and positions angles instead of visibilities for spectroscopic or interferometric binaries. On the other hand, such extracted data preserved in the literature are often the only information available, hence it is still worth to enrich, e.g., the disentangling of some spectra by simultaneous solution of RV-curves from other exposures.

7.2.4 Numerical Representation

The observed and solved spectra as well as their Fourier transforms were treated in the previous sections as continuous functions of x or y, respectively. In practice, however, the spectra are usually observed and registered using a chip with discrete pixels and represented in the data-processing from the rough data to the final separated component spectra as a finite set of values in discrete bins. It means that the space of RVs in which the spectra and the broadening functions are represented is quantised. Nevertheless, the position of a smooth line stretched over several bins can be determined with a sub-pixel resolution from asymmetry of the signal in surrounding bins. A sub-pixel resolution in RVs enables to determine the orbital parameters with a higher precision, but this, however, requires representing with a similar resolution the operator $\delta(x - v_j)$ in Eq. (7.2) or in its Fourier transform Eq. (7.5). For an integer multiple of the bin size Δ_x, the convolution with the shifted δ-function is given by multiplication with a unit off-diagonal matrix in the x-representation and by $\exp(iyv_j)$ in the y-representation, where the v_j must be rounded to the integer multiple of Δ_x to keep the periodicity in the finite set of y-values. The change of signal in a sub-pixel shift for $v < \Delta_x$ can be found from its Taylor expansion, in which its derivatives are approximated from the differences in the neighbouring bins (cf. [9]), i.e.

$$\begin{aligned}\delta(x - v) \simeq \delta(x) - \frac{v}{2\Delta_x}(\delta(x + \Delta_x) - \delta(x - \Delta_x)) + \\ + \frac{v^2}{2\Delta_x^2}(\delta(x + \Delta_x) - 2\delta(x) + \delta(x - \Delta_x)) + o(v^3)\end{aligned} \tag{7.13}$$

and its Fourier transform

$$\exp(iyv_j) \simeq 1 + \frac{iv}{\Delta_x}\sin(y\Delta_x) + \frac{v^2}{\Delta_x^2}(\cos(y\Delta_x) - 1) + o(v^3)\ . \tag{7.14}$$

The Fourier disentangling with this sub-pixel resolution can be performed using the author's code KOREL, which is now available as an on-line service of the Virtual

Observatory (cf. [21]). A question frequently asked by its users is what signal-to-noise ratio (S/N) of the input spectra is needed for applicability of the method of disentangling. A general simple answer is "any, but the better is the input S/N, the better is the S/N of the disentangled component spectra and the better is the accuracy of the disentangled parameters". However, the problem is more complex. The precision of the results also depends on the degree of the overdetermination, because the noise in disentangled spectra decreases as $\sqrt{N}$ with the number N of the input spectra. But other important parameters are the spectral resolution, phase coverage and information content of the input spectra. The useful signal for disentangling are the spectral lines, i.e. deviations of the spectrum from the continuum, while the S/N compares the noise with the continuum level. To clarify this problem, a more detailed analysis is presented in the Appendix.

7.3 Disentangling of Spectra of Interacting Binaries

The method of spectra disentangling assumes that the observed system consists of several discrete sources with some general unknown spectra, the changes of which can be parameterised by means of the broadening functions Δ_j. In the case of interacting binaries with mass exchange, there is a significant contribution of radiation from the circumstellar matter continuously moving around the component stars with a complicated velocity field. The methods of Doppler tomography are used in these cases to map the distribution of the emission sources in the velocity space (cf. [13]). The intrinsic spectrum of an infinitesimal volume of the source is mostly supposed to be a Dirac δ-function in frequency at least in the vicinity of each line used for the Doppler imaging. This simplifying assumption is obviously false. (E.g., the presence of other lines or absorption features in P-Cygni profiles violate the mirror symmetry in the opposite phase supposed for the standard Doppler tomography.) It would thus be desirable to disentangle intrinsic spectra in each point of the velocity space (or even better in each point of the geometric space). This problem, however, does not have a unique solution, because a narrow line of source spread over a wide region of the velocity space results in the same spectra as a wide line from a narrow source.

Nevertheless, it is possible to approximate the smooth distribution of the circumstellar matter in the interacting binaries by several blobs of gas and to disentangle their spectra and (as a free parameter) also their positions in the velocity space. As an example, disentangling of Cyg X-1 is shown in Fig. 7.4 (cf. [22] for details). Although this source is variable on a wide range of time-scale, which violates the underlying assumptions of both the Doppler tomography as well as the disentangling, both techniques are able to reveal main features of the circumstellar matter (cf. [4, 19]).

Even in the cases when the assumptions of both these Doppler techniques are satisfied, the resolved distribution of the radiation sources in the velocity space does not yield their distribution in space and the consequent angular resolution.

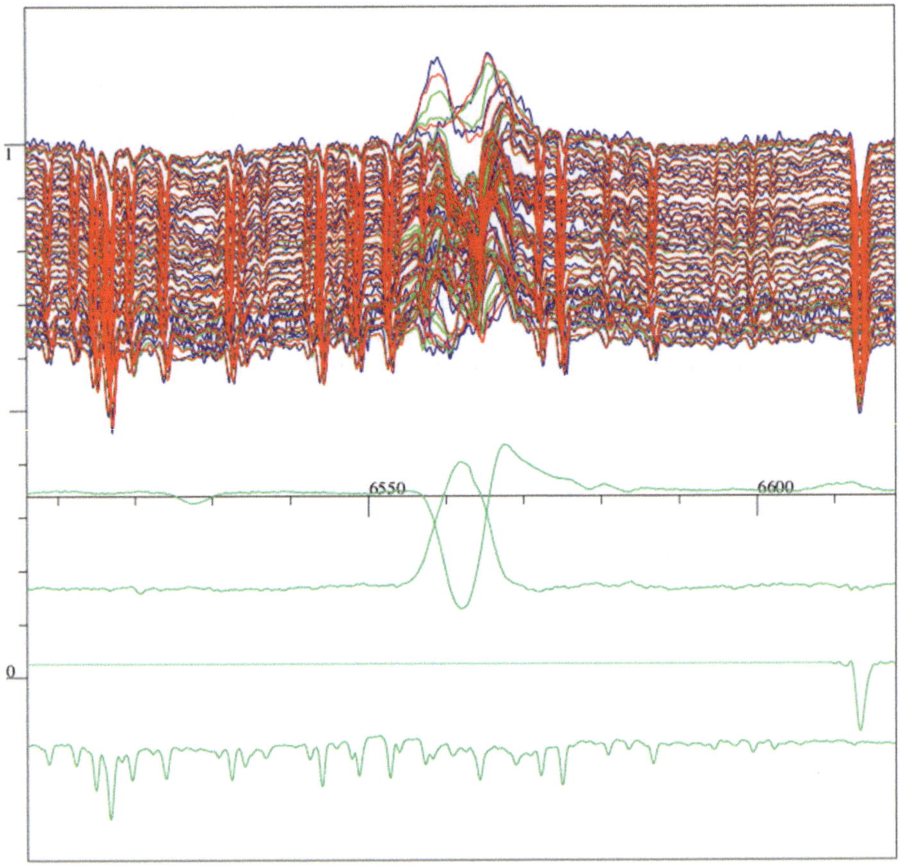

Fig. 7.4 Disentangled Hα profile of Cyg X-1. The input observed spectra (*blue lines* in the upper part of the panel) are overplotted by their reconstruction (*red lines*) from the disentangled component spectra (bottom four *green lines*): P-Cyg profile of the supergiant, emission profile of the circumstellar matter, diffuse interstellar band (the disentangled DIB is constrained by a template constant across the Hα) and telluric lines. The emission is shifted in phase with respect to the unseen black-hole secondary and its strength is anti-correlated with the X-ray emission

Generally there is no one-to-one correspondence between these spaces due to the non-uniformity of the velocity field. To get an information about the spatial structure of the source, the interpretation must be constrained by an appropriate theoretical model (based e.g. on radiation hydrodynamics, cf. [3]) of the source and its radiation. The method of synthetic Dopplerograms is promising for this purpose (cf. [4, 16]).

A direct comparison of the observations with synthetic spectra based on sophisticated models of the sources may be more precise because it may avoid discrepancies caused by the simplifying assumptions of the methods of indirect imaging. However, these techniques are useful for indicating the dominant features in the observed phenomena and their achievements as well as failures are intuitive in suggesting the way in which the theory should proceed.

Acknowledgements This work has been supported by the Czech Science Foundation (GAČR 14-37086G) – Albert Einstein Center for Gravitation and Astrophysics.

Appendix: Bayesian Estimation of Parameters Errors

The errors of parameters of a stellar system determined by disentangling of spectra can be estimated using Bayesian statistics. The errors are caused mainly by the noise in the observed spectra but they are also influenced by the phase distribution of the observations and by the sensitivity of the available data to a particular parameter. Usually we "know" in advance only that the values of parameters p can be expected in some "reasonable" regions (e.g., hundreds of km/s for orbital RVs, days or from hours to months for the orbital period, the line strengths and widths corresponding to the spectral types and rotational broadening, etc.). If previous studies of the same system exist their results can be taken as a more specific limitation of the possible range of p (and a first estimate of the new solution), which we want to verify or improve using a set of new data. If the old data are available they can be included together with the new ones into the new solution, otherwise the reliability of the old results must be estimated and they can be included into the new solution as constraints to the solution in the space of parameters p. Having the set of new observations, i.e. the spectra $F(x)$, we can find the "posterior" probability $P(p|F)$ of p agreeing with the new data according to Bayes' theorem

$$P(p|F) = \frac{P(F|p)P(p)}{P(F)} , \tag{7.15}$$

where $P(p)$ is the "prior" probability of p (i.e. either a smooth characteristic of the wide range of "reasonable" values or the constraint resulting from the old data), $P(F|p)$ is the "liability" that for a chosen value of p the particular shape of $F(x)$ will be detected and $P(F)$ is a general probability do detect the signal $F(x)$. The best new solution $p_{\max}$ can be then defined as the maximum of $P(p|F)$ in the p-space or we can find a mean value

$$\langle p \rangle = \int pP(p|F)dp \tag{7.16}$$

(if $P(p|F)$ is normalised to unity, i.e., $\int P(p|F)dp = 1$). In both cases the result depends on our choice of measures in the p space. The errors of resulting parameters p can be estimated from the behaviour of $P(p|F)$ around the found solution. Because p is generally multidimensional (and different dimensions can be of different nature), it is not sufficient to attribute some error bar to each component of p separately as it is a common habit. If $P(p|F)$ is sufficiently smooth around its maximum, it can be expanded into a Taylor series up to quadratic terms in variation of δp and the non-diagonal components of the corresponding quadratic form determine the correlation of the parameters. Generally, however, there may be several local maxima in the p-space, which may be treated as different solutions.[1] The errors of these different solutions must be then determined from a local behaviour of $P(p|F)$. Generally, the information about the studied system is more completely characterised by mapping of $P(p|F)$ than by a simple list of found values of p and their errors.

Errors of Line Strengths and Radial Velocities

We shall illustrate the calculation of $P(p|F)$ first on a toy model of measurements of one spectrum of a single star. Let us assume that the observed spectrum $F(x)$ is given by

$$F(x) = f(x,p) + \delta F(x) \,, \tag{7.17}$$

where $\delta F(x)$ is a random observational noise. We fit $F(x)$ by a model $f(x,p)$ dependent on a parameter (or a set of parameters) p by minimising the residual error

$$S(p) = \int_D (F(x) - f(x,p))^2 dx \tag{7.18}$$

integrated over the whole observed region D of x. The equation(s) for p thus reads

$$0 = \frac{\partial S(p)}{\partial p} = 2 \int \frac{\partial f(x,p)}{\partial p} (f(x,p) - F(x)) dx \,. \tag{7.19}$$

The random fluctuations δF in the observed signal $F(x)$ blur this condition and result in a deviation δp of the solution from its correct value. These variations are related by the condition

$$\delta p \frac{\partial}{\partial p} \int \frac{\partial f(x,p)}{\partial p} (f(x,p) - F(x)) dx = \int \frac{\partial f(x,p)}{\partial p} \delta F(x) dx \,. \tag{7.20}$$

[1] A typical example is the epoch of a periodic RV curve, where for each solution there is an infinite set of solutions differing by an integer multiple of the period. These solutions are usually taken as equivalent. The errors and correlation of the epoch with the period depend on which of these solutions we choose. The period itself may have several solutions due to aliasing in quasi-periodically sampled data or due to an interference of data obtained in two time-remote observational seasons.

For example, let the spectrum be rectified, i.e. normalised to the continuum, which means $f = 1 - \varphi$, where $\varphi(x)$ corresponds to the spectral lines. We shall assume first that the only unknown free parameter p is the strength of lines, i.e., the model has the form

$$f(x,p) = 1 - p\varphi_1(x) , \tag{7.21}$$

where φ_1 is a pattern of the line-profile(s) imprinted into the observed spectrum $F(x)$ with an unknown amplitude (e.g., due to uncertainty in element abundances, due to a contamination of the signal by light of another star or due to instrumental error in subtracting dark signal). Equation (7.19) is then a simple linear equation with solution

$$p = \frac{\int (1 - F(x))\varphi_1(x)dx}{\int \varphi_1^2(x)dx} \tag{7.22}$$

and in agreement with Eq. (7.20) its variations are given by

$$\delta p = -\frac{\int \varphi_1(x)\delta F(x)dx}{\int \varphi_1^2(x)dx} . \tag{7.23}$$

The integrals in these equations are actually summations over K individual pixels (each one of size $\Delta_x = D/K$) in real observations,

$$\int \varphi_1(x)\delta F(x)dx = \Delta_x \sum_{i=1}^{K} \varphi_1(x_i)\delta F(x_i) , \tag{7.24}$$

$$\int \varphi_1^2(x)dx = \Delta_x \sum_{i=1}^{K} \varphi_1^2(x_i) . \tag{7.25}$$

Let us assume that the probability distribution of the noise δF of the signal $F(x)$ in each pixel can be approximated as a Gaussian with standard deviation σ, i.e., the probability of its value δF in one pixel x_i is

$$P(\delta F(x_i)) \simeq \exp(-\frac{\delta F(x_i)^2}{2\sigma^2}) \tag{7.26}$$

and, hence, the statistical mean value is $\langle \delta F^2 \rangle = \sigma^2$. In addition, the noise in different pixels is supposed to be statistically independent, i.e.,

$$\langle \delta F(x_i)\delta F(x_j) \rangle = \sigma^2 \delta_{ij} . \tag{7.27}$$

The probability distribution of error δp is then also Gaussian[2] with

$$\langle \delta p^2 \rangle = \frac{\Delta_x^2}{\left[\int \varphi_1^2(x)dx\right]^2} \langle [\sum_{i=1}^{K} \varphi_1(x_i)\delta F(x_i)]^2 \rangle = \frac{\Delta_x \sigma^2}{\int \varphi_1^2(x)dx} . \tag{7.28}$$

The squared error of p thus decreases inversely proportionally with the number of pixels covering the profile $\varphi(x)$. It can be seen from Eqs. (7.22) and (7.23) that the strength p and its uncertainty δp are predominantly determined by the parts of the spectrum where the lines φ_1 of the model are deep.

If the unknown free parameter p is the Doppler shift of the spectrum, the model has the form

$$f(x) = 1 - \varphi_0(x - p) , \tag{7.29}$$

where $\varphi_0(x)$ is a model line-profile in laboratory wavelength scale. The residual noise

$$S(p) = \int (\varphi_0(x - p) - \varphi(x) + \delta F(x))^2 dx \tag{7.30}$$

is then no more a simple quadratic function of p and Eq. (7.19), which reads now

$$0 = \frac{\partial S(p)}{\partial p} = -2 \int \frac{d\varphi_0(x - p)}{dx} (\varphi_0(x - p) - \varphi(x) + \delta F(x))dx, \tag{7.31}$$

may have several solutions for p corresponding to coincidences of some improperly identified observed lines with wrong lines in the model. The uncertainty of p in each of these solutions can be estimated from the depth and width of the local minimum of the residual noise. Equation (7.20) reads now

$$\delta p \int \left(\frac{d\varphi_0}{dx}\right)^2 dx = \int \frac{d\varphi_0}{dx} \delta F(x)dx , \tag{7.32}$$

where we have skipped the term $\frac{d^2\varphi_0}{dx^2}(\varphi_0 - \varphi)$ which should vanish at the correct value $p = p_0$ where $\varphi = \varphi_0$ (this need not be true at a false minimum). This equation shows that the error δp of RV is dominated by the noise δF in the steepest parts of the line profile. Analogously to Eq. (7.28), the mean squared value of this

---

[2] If a quantity x has a probability distribution $\exp(-(x - x_0)^2/(2\alpha^2))$ and quantity y a distribution $\exp(-(y - y_0)^2/(2\beta^2))$ then the linear combination $ax + by$ has a mean value $\langle ax + by \rangle = ax_0 + by_0$ and the quadratic error $\langle (a(x - x_0) + b(y - y_0))^2 \rangle = a^2\alpha^2 + b^2\beta^2$. Similarly for a sum of more statistically independent quantities we get recursively $\langle \sum_{i=1}^n a_i x_i \rangle = \sum_{i=1}^n a_i x_{i0}$ and $\langle (\sum_{i=1}^n a_i(x_i - x_{i0}))^2 \rangle = \sum_{i=1}^n a_i^2 \alpha_i^2$.

error is now

$$\langle \delta p^2 \rangle = \frac{\Delta_x \sigma^2}{\int (\frac{d\varphi_0}{dx})^2 dx} \,. \tag{7.33}$$

The difference between Eqs. (7.28) and (7.33) is due to the fact that the fit of line strengths is most sensitive to the part of the spectrum where the line is deepest, while the Doppler shift is most sensitive to the wings where the line profile is steepest.

Errors in Multidimensional Space of Parameters

We can optimise the fit of the observed spectrum simultaneously with respect to the line-strength, Doppler shift and some additional parameters. Generally, if the model spectrum $\varphi_0(x, p)$ is a function of several parameters $p_i|_{i=1}^m$, we get for them a set of m generally non-linear equations for p

$$0 = \int \frac{\partial \varphi_0}{\partial p_i}(1 - F(x) - \varphi_0(x, p))dx \,. \tag{7.34}$$

Linearising, we arrive at the set of equations

$$M_{ij}\delta p_j = \delta R_i \tag{7.35}$$

for variations δp_j caused by variations $\delta F(x)$ of the observed spectrum. The matrix M_{ij} is given here by

$$M_{ij} \equiv \int \left(\frac{\partial \varphi_0}{\partial p_i}\frac{\partial \varphi_0}{\partial p_j} - \frac{\partial^2 \varphi_0}{\partial p_i \partial p_j}(1 - F(x) - \varphi_0(x, p)) \right) dx \tag{7.36}$$

and the right-hand side by

$$\delta R_i \equiv - \int \frac{\partial \varphi_0}{\partial p_i}\delta F(x)dx = \Delta_x \sum_{k=1}^{K} \frac{\partial \varphi_0(x_k)}{\partial p_i}\delta F(x_k) \,. \tag{7.37}$$

Depending on the form of the model φ_0, some variation δF can be compensated by a change of different p_i, so that their variations may be correlated

$$\langle \delta p_j \delta p_k \rangle = M_{ji}^{-1} M_{kl}^{-1} \langle \delta R_i \delta R_l \rangle \,. \tag{7.38}$$

Assuming that the noise $\delta F(x)$ is statistically independent at different pixels x_k, x_l, we have

$$\langle \delta R_i \delta R_l \rangle = \Delta_x^2 \sum_{k,l=1}^{K} \frac{\partial \varphi_0(x_k)}{\partial p_i} \frac{\partial \varphi_0(x_l)}{\partial p_j} \langle \delta F(x_k) \delta F(x_l) \rangle =$$
$$= \Delta_x \int \frac{\partial \varphi_0}{\partial p_i} \frac{\partial \varphi_0}{\partial p_j} \langle \delta F(x)^2 \rangle dx \,. \tag{7.39}$$

If the noise $\langle \delta F(x)^2 \rangle = \sigma^2$ is statistically the same in all pixels and if the second term in the integral in Eq. (7.36) is negligible compared to the first one (which is the case if the fit is good and $F(x) \simeq 1 - \varphi_0(x, p)$), then

$$\langle \delta R_i \delta R_l \rangle = \Delta_x \sigma^2 M_{il} \tag{7.40}$$

and

$$\langle \delta p_j \delta p_k \rangle = \Delta_x \sigma^2 M_{jk}^{-1} \,. \tag{7.41}$$

As an example, let us investigate a fit by a simple Gaussian profile

$$f(x, p) = 1 - \varphi_0(x, p) = 1 - p_1 \exp\left(-\frac{(x - p_3)^2}{p_2^2}\right) \,. \tag{7.42}$$

Equation (7.34) then reads

$$0 = \int (x - p_3)^k \varphi_0 (1 - F - \varphi_0) dx \,, \tag{7.43}$$

where $k = 0, 2, 1$ in conditions for p_1, p_2, p_3, resp. The variation of φ_0 reads

$$\delta \varphi_0 = \left(\frac{1}{p_1} \delta p_1 + 2 \frac{(x - p_3)^2}{p_2^3} \delta p_2 + 2 \frac{x - p_3}{p_2^2} \delta p_3 \right) \varphi_0 \,, \tag{7.44}$$

so that neglecting the second term in Eq. (7.36), the matrix M has the form

$$M_{ij} \simeq \int \frac{\partial \varphi_0}{\partial p_i} \frac{\partial \varphi_0}{\partial p_j} dx = \sqrt{\frac{\pi}{2}} \begin{pmatrix} p_2 & \frac{p_1}{2} & 0 \\ \frac{p_1}{2} & \frac{3p_1^2}{4p_2} & 0 \\ 0 & 0 & \frac{p_1^2}{p_2} \end{pmatrix} \tag{7.45}$$

and hence according to Eq. (7.41) the correlation matrix of parameter errors reads

$$\langle \delta p_j \delta p_k \rangle \simeq \sqrt{\frac{2}{\pi}} \Delta_x \sigma^2 \begin{pmatrix} \frac{3}{2p_2} & -\frac{1}{p_1} & 0 \\ -\frac{1}{p_1} & \frac{2p_2}{p_1^2} & 0 \\ 0 & 0 & \frac{p_2}{p_1^2} \end{pmatrix} . \quad (7.46)$$

It means that the errors of the depth p_1 and the width p_2 of the line-profile given by Eq. (7.42), which are due to the part of the perturbation δF symmetric with respect to the centre of the line, are anti-correlated, while the error in the position p_3 of the line centre is not correlated with them and its squared value is half of that of the width.

Results of a Monte Carlo simulation of this example can be seen in Fig. 7.5. The red line shows the profile (cf. Eq. (7.42)) for $p_1 = 0.2$, $p_2 = 0.2$ and $p_3 = 0.0$. The black line has added a randomly generated noise with $\sigma = 0.04$ in 1024 pixels in the displayed interval of $x \in (-1, +1)$, i.e., $\Delta_x = 2^{-9} \simeq 0.00195$. The blue line gives the best fit to this particular simulated measurement, which was found for values of parameters $p_1 \simeq 0.2002$, $p_2 \simeq 0.2085$ and $p_3 = 0.0029$. The figure included in the bottom left corner shows the scatter of p_1 and p_2 around their true values for 1000 different choices of δF (the drawn parts of coordinates correspond to ± 0.01). Similarly the histogram in the right corner shows the distribution of p_3 (the width of each histogram column is 0.001). If the noise σ is decreased then

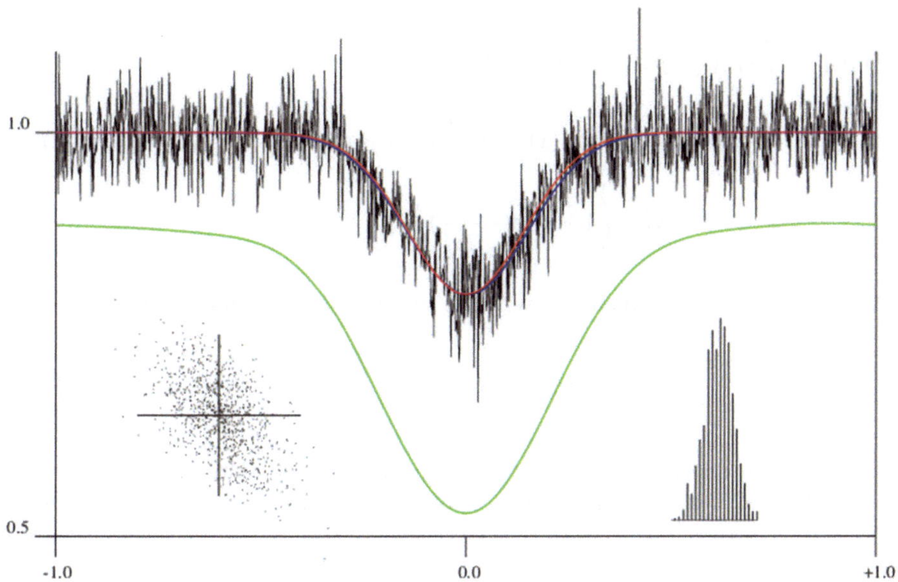

Fig. 7.5 Gaussian line-profile (*red line*) with a simulated noise (*black line*) and its best fit (*blue line*; see text for a detailed description)

in agreement with Eq. (7.38) or its approximation (Eq. (7.41)) also the uncertainty $\langle \delta p_j \delta p_k \rangle$ of the parameters p is reduced. This agrees with the Bayes theorem (Eq. (7.15)) according to which the probability $P(p|F)$ of a larger difference of the true value of p from its best fit to particular data $F(x)$ decreases with decreasing probability $P(F|p)$ that the random noise will mimic a wrong spectrum.

For real observed spectra we do not know the exact value of σ (which can only be estimated from the level of signal integrated during the exposure), while we may have a rough estimate only of Δ_x and, generally, we do not know the explicit form of the matrix M_{ij} and its inversion. However, we can estimate these values from the value S_0 of the residuum $S(p)$ in its minimum and from variations of $S(p)$ with respect to variations of p. If the spectrum is correctly fitted by a proper model, the residuum S_0 should be given by the noise[3] only, i.e.,

$$S_0 \simeq \int \delta F^2(x)dx = D\sigma^2 = K\Delta_x\sigma^2 \,. \tag{7.47}$$

The value of Δ_x and hence for known D also the number K of statistically independent pixels of noise can be estimated from auto-correlation of the residual noise. In our example, the numerical simulation results in $S_0/\Delta_x \simeq 1.73$, giving $\sigma \simeq 0.0411$ in good agreement with the value 0.04 for which the noise was generated. The residuum $S(p_3)$ is drawn by the green line in Fig. 7.5 as a function of p_3 for $p_{1,2}$ fixed to their best values (the zero level is shifted to the bottom of the panel for $S(p_3)$). Its behaviour can be estimated substituting φ_0 from Eqs. (7.42) into (7.30) also for φ,

$$\begin{aligned} S(p_3) &\simeq \int (\varphi_0(x-p_3) - \varphi_0(x) + \delta F(x))^2 dx \\ &\simeq \int \delta F^2(x)dx + \int \varphi_0^2(x)dx + \int \varphi_0^2(x-p_3)dx - 2\int \varphi_0(x-p_3)\varphi_0(x)dx \\ &= S_0 + \sqrt{2\pi}p_1^2 p_2 \left(1 - \exp(-\frac{p_3^2}{2p_2^2})\right) \,. \end{aligned} \tag{7.48}$$

The residual $S(p_3)$ thus increases with the square of δp_3 around its minimum

$$S(\delta p_3) \simeq S_0 + \sqrt{\frac{\pi}{2}} \frac{p_1^2}{p_2} \delta p_3^2 \,, \tag{7.49}$$

but it approaches saturation at level $S_\infty = S_0 + \sqrt{2\pi}p_1^2 p_2$ for large δp_3 (in agreement with numerical results $S_\infty/\Delta_x \simeq 12.5$ in our example). This saturation corresponds

[3]In the discrete representation of N input spectra, the residual noise after disentangling of m components is given by the sum $S_0 \simeq (N-m)K\sigma^2$. However, it should be noted that both the signal and its noise may be rescaled by the Fourier transform and hence (due to the use of FFT) the residuum on output from KOREL is KS_0.

to the possibility that the line visible in the spectrum is actually a random fluctuation of the noise, while the real line is in the distant wavelength δp_3, but it is hidden in another noise fluctuation. The probability of this "hidden" solution is proportional to the width D of the wavelength interval. This may be in principle infinite, so that the probability of this solution may be high, even for a small value of σ. The limitation of D to the interval $(-1, +1)$ in our example has a character of the prior $P(p_3)$ which is chosen equal to zero out of this interval.

Regarding the assumption of the Gaussian noise (7.26), the liability $P(\delta F|p)$ of the noise $\delta F(x_i)|_{i=1}^{K}$ in all pixels is

$$P(\delta F|p) \simeq \exp\left(-\frac{\sum_i \delta F(x_i)^2}{2\sigma^2}\right) = \exp\left(-\frac{S_0}{2\Delta_x\sigma^2}\right) . \tag{7.50}$$

Following Bayes' theorem (Eq. (7.15)), the probability of p for known F (and a 'flat' prior $P(p)$) reads

$$P(p|F) \simeq \exp\left(-\frac{S(p)}{2\Delta_x\sigma^2}\right) \simeq \exp\left(-\frac{KS(p)}{2S_0}\right) . \tag{7.51}$$

We thus obtain from Eq. (7.49) the probability distribution of radial velocity δp_3 of the "visible" line

$$P(\delta p_3|F) \simeq \exp\left(-\frac{KS(\delta p_3)}{2S_0}\right) \sim \exp\left(-\sqrt{\frac{\pi}{2}}\frac{p_1^2}{2\Delta_x\sigma^2 p_2}\right), \tag{7.52}$$

which yields, according to the definition (Eq. (7.16)), a mean value of squared velocity shift

$$\langle\delta p_3^2\rangle = \frac{\int \delta p_3^2 P(\delta p_3|F) d\delta p_3}{\int P(\delta p_3|F) d\delta p_3} \simeq \sqrt{\frac{2}{\pi}}\frac{\Delta_x\sigma^2 p_2}{p_1^2}, \tag{7.53}$$

in agreement with component $\{j,k\} = \{3,3\}$ of Eq. (7.46).

It is worth noting that the error $\sqrt{\langle\delta p_3^2\rangle}$ on the radial velocity p_3 is proportional to the ratio σ/p_1 of the noise to the line depth. Regarding the problem of desirable S/N-ratio mentioned in Sect. 7.2.4, we can see from this relation that there is no limitation. However, the higher the S/N-ratio is, the more precise is the determination of the disentangled parameters. From the observational and instrumental point of views, the S/N of spectra refers to the ratio of the noise to the overall spectral flux, i.e. the signal means the continuum. However, it is obvious that the signal which yields an information about RV and other spectral features is the modulation of the flux by spectral lines. It is thus desirable to optimise in the observations the ratio of line depths to the (photon) noise. The RV error is also proportional to the square root of the line width p_2 which indicates that a higher precision can be achieved from sharper lines, although they are covered by smaller number of pixels and are thus more influenced by fluctuations in photon counts.

In the disentangling technique, we do not have explicitly given in Eq. (7.17) the template spectrum $f(x,p)$, like in Eqs. (7.21) or (7.42), but a more complicated expression like the right-hand side of Eqs. (7.1) or (7.9). The component spectra $F_j(x)$ whose imprints we want to find in the observations are now also unknown parameters that can be retrieved with some errors only. In analogy to Eq. (7.28), this error is proportional to the noise σ of the input spectra. However, usually the separation of the component spectra is overdetermined, as explained in Sect. 7.2.1 and hence, following the relations in the Footnote 2, the amplitude of the noise of the disentangled spectra decreases inversely proportionally to the square root of the number of observed spectra. The dependence of the model $f(x,p)$ on nonlinear parameters p in the broadening functions $\Delta_j(x,t;p)$ is more complicated and generally cannot be expressed explicitly like in the above given toy models. However, it is possible to map the distribution of the residual noise numerically as a function of these parameters.

References

1. Albrecht, S.: The long history of the Rossiter – McLaughlin effect and its recent applications. IAUS **282**, 379–384 (2012)
2. Bagnuolo, W.G., Jr., Gies, D.R.: Tomographic separation of composite spectra – the components of the O-star spectroscopic binary AO cassiopeiae. ApJ **376**, 266–271 (1991)
3. Čechura, J., Hadrava, P.: Stellar wind in state transitions of high-mass X-ray binaries. A&A **575**, A5 (2015)
4. Čechura, J., Vrtilek, S.D., Hadrava, P.: Interpreting the X-ray state transitions of Cygnus X-1. MNRAS **450**, 2410–2422 (2015)
5. Hadrava, P.: Orbital elements of multiple spectroscopic stars. A&AS **114**, 393–396 (1995)
6. Hadrava, P.: Relative line photometry of eclipsing binaries. A&AS **122**, 581–584 (1997)
7. Hadrava, P.: Disentangling telluric lines in stellar spectra. A&A **448**, 1149–1152 (2006)
8. Hadrava, P.: Line profiles during eclipses of binary stars. ASPC **370**, 164–168 (2007)
9. Hadrava, P.: Notes on the disentangling of spectra. I. Enhancement in precision. A&A **494**, 399–402 (2009)
10. Hadrava, P., Kubát, J.: Limb darkening and line-profile variations in eclipsing binaries. ASPC **288**, 149–152 (2003)
11. Hadrava, P., Šlechta, M., Škoda, P.: Notes on disentangling of spectra. II. Intrinsic line-profile variability due to cepheid pulsations. A&A **507**, 397–404 (2009)
12. Holt, R.J.: Spectroscopic determination of stellar rotation. Astron. Astro-Phys. **12**, 646 (1893)
13. Marsh, T.R., Horne, K.: Images of accretion discs. II – Doppler tomography. MNRAS **235**, 269–286 (1988)
14. McLaughlin, D.B.: Some results of a spectrographic study of the Algol system. ApJ **60**, 22–31 (1924)
15. Ribas, I., Jordi, C., Vilardell, F., et al.: First determination of the distance and fundamental properties of an eclipsing binary in the andromeda galaxy. ApJ **635**, L37–L40 (2005)
16. Richards, M.T., Cocking, A.S., Fisher, J.G., et al.: Images of gravitational and magnetic phenomena derived from two-dimensional back-projection Doppler tomography of interacting binary stars. ApJ **795**, 160 (2014)

17. Rossiter, R.A.: On the detection of an effect of rotation during eclipse in the velocity of the brigher component of beta Lyrae, and on the constancy of velocity of this system. ApJ **60**, 15–21 (1924)
18. Schlesinger, F.: The Algol-variable δ Librae. Publ. Allegh. Obs. **1**, 123–134 (1910)
19. Sharova, O.I., Agafonov, M.I., Karitskaya, E.A., et al.: Doppler tomography in 2D and 3D of the X-ray binary Cyg X-1 for June 2007. IAUS **282**, 201–202 (2012)
20. Simon, K.P., Sturm, E.: Disentangling of composite spectra. A&A **281**, 286 (1994)
21. Škoda, P., Hadrava, P.: Fourier disentangling using the technology of virtual observatory. ASPC **435**, 71–75 (2010)
22. Yan, J.Z., Liu, Q.Z., Hadrava, P.: Optical spectroscopic observations of Cyg X-1 = HDE 226868. AJ **136**, 631–640 (2008)
23. Zwahlen, N., North, P., Debernardi, Y., et al.: A purely geometric distance to the binary star Atlas, a member of the Pleiades. A&A **425**, L45–L48 (2004)

Chapter 8
Velocity Fields in Stellar Atmospheres Probed by Tomography

Alain Jorissen, Sophie Van Eck, and Kateryna Kravchenko

8.1 Introduction

In this Chapter, we describe recent progress made in the implementation of a tomographic technique used to probe velocity fields as a function of depth in stellar atmospheres. A first basic version of the technique was described in Jorissen et al. [19]. It must be stated right away that the word *tomography* is used here with a meaning close to its etymological roots ("write/display cuts"), which differs from the broader meaning in use within the astronomical community (reconstruction of a structure using projections taken under different angles, using inverse methods; see Chaps. 10 and 12). There is no inverse method involved here.

The concept originates from the recognition by Schwarzschild [30] that the doubling of the Hα line observed in Cepheid variables of the W Vir type follows a specific time sequence that may be accounted for by a shock front passing through the line-formation zone. Line doubling has since been observed in many other lines, both in the infrared and in the optical, as will be detailed below. For Mira variables, Merrill [27] was the first to suggest that the observed spectral changes may be explained by the presence of a shock wave moving outward. This idea has since been investigated by a number of authors ([4] and references therein).

By looking at lines forming at different depths in the atmosphere, one may locate the shock front: its depth will correspond to the depth of the layer where double-peaked lines form. By considering a temporal sequence of spectra, it is then possible to follow the propagation of the shock front across the stellar atmosphere, and even derive its velocity (provided, however, that the depth may be known in *geometrical* units rather than in *optical-depth* units, as we shall discuss in Sects. 8.2 and 8.4).

A. Jorissen (✉) • S. Van Eck • K. Kravchenko
Institut d'Astronomie et d'Astrophysique, Université libre de Bruxelles, Brussel, Belgium
e-mail: alain.jorissen@ulb.ac.be; svaneck@ulb.ac.be; kateryna.kravchenko@ulb.ac.be

H.M.J. Boffin et al. (eds.), *Astronomy at High Angular Resolution*, Astrophysics and Space Science Library 439, DOI 10.1007/978-3-319-39739-9_8

The derivation of the velocity of the shock wave passing through the atmosphere is crucial in the study of the pulsation and mass-loss process for long-period variables (LPVs) of Mira type. There has been a long-standing controversy about the shock speed in LPVs [12], which hampers the construction of reliable theoretical models of shock waves in LPVs [10]. The shock speed should be correctly reproduced by state-of-the-art, dynamical models [18] of Mira stars, which still involve a few free parameters (for instance the mixing length, the turbulent viscosity parameter and the amplitude and velocity of the piston used as inner boundary condition to trigger the pulsations).

In pulsating atmospheres like those of LPVs or Cepheids, what has since been known as the Schwarzschild scenario [2, 30] states that, when the shock front is still located underneath the line-forming region, lines forming at all depths are redshifted, since matter in the line-forming region is falling down. Then the shock reaches the deepest optically-thin layer, just above continuum-formation, where the faintest lines form. Next to the existing red component, those lines then exhibit a blue component, corresponding to rising matter. As time passes and the shock front propagates to upper layers, stronger and stronger lines start exhibiting a blue component. At the same time, the red component has totally vanished from the lines forming deepest in the atmosphere, since the full extent of the deepest line-forming region is now occupied by rising matter. Finally, when the shock wave has gone through the entire photosphere, all lines (and not only the faintest any longer) exhibit a blue component only. This outward motion of the shock front has indeed been revealed by the tomographic method applied to Mira variables [19, 20].

The Schwarzschild scenario thus predicts that in a pulsating stellar atmosphere, any given line will appear double when the shock front crosses the layer where it forms. Before that time, the line was redshifted, and after that time, the line is blueshifted. As the shock wake decelerates, the blue peak progressively turns into a red peak. An S-shaped velocity curve ensues (Fig. 8.1), as seen on either infrared lines [16, 17] or optical lines [2, 4]. There is a time shift between the S-shaped velocity curve and the light curve, and this time shift depends on the formation depth of the considered line (compare infrared and optical lines in Fig. 8.1). The S-shaped velocity curve is thus another manifestation of the Schwarzschild scenario.

Line doubling in Mira variables was recorded as well using the infrared second-overtone ($\Delta\upsilon = 3$) rotation-vibration CO line system around 1.6–1.8 μm [17]. This spectral window coincides with the minimum of the continuum opacity. Therefore, lines in that infrared wavelength range originate from rather deep layers, and most of these lines are strong, whereas the same layer gives rise in the optical to weak lines only. Nevertheless, these weak optical lines are used by the tomographic method originally designed by [3], and combined with stronger optical lines forming higher up in the atmosphere, they offer a way to probe a larger range of optical depths than is possible in the infrared. The tomographic method, and its recent improvements, will be discussed in Sect. 8.2. Results obtained by the tomographic method applied to pulsating long-period variables and supergiants are then reviewed in Sect. 8.3.

So far, the method has only been used with optical depths as proxy for geometrical depths. The knowledge of the relation between geometrical and optical

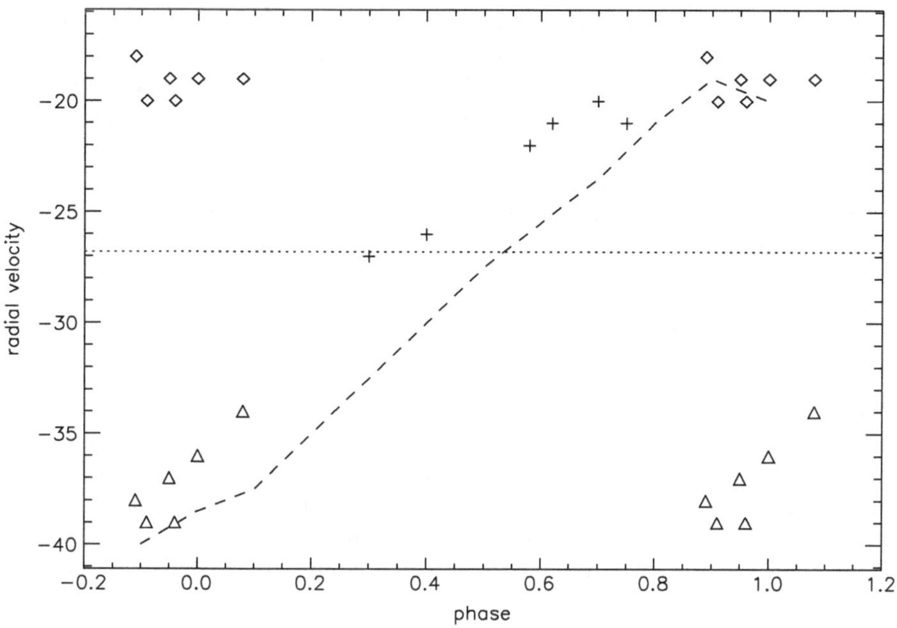

Fig. 8.1 Velocity variations exhibited by optical lines in the spectrum of the carbon Mira variable S Cep (*crosses*: single component; *triangles*: blue component; *diamonds*: red component), as a function of the light phase. *The dotted line* indicates the centre-of-mass velocity from [28]. The velocities between phases 0.8 and 1.1 are duplicated to illustrate the velocity behaviour through an entire light cycle. The *dashed line* is an eye fit to the radial-velocity curve of S Cep obtained by [16] from infrared CN lines ($\Delta\upsilon = 2$) around 2 μm (their Fig. 1). Note the phase shift between the S-shaped curves for the optical and infrared lines (Reproduced with permission from [4])

depths would open the way to derive directly the shock-front velocity. We discuss that issue in Sect. 8.4.

8.2 Method

The tomographic method involves two steps. The first step rests on our ability to construct reliable synthetic spectra of late-type giant stars [15], and from those, to identify the depth of formation of any given spectral line. Then, to cope with the fact that the spectra of late-type stars are very crowded, particularly in the optical domain, the information on the average shape of lines forming at a given depth in the atmosphere is extracted from a cross-correlation technique. This co-addition of the profiles of many different lines through the cross-correlation process results as well in an increase of the signal-to-noise ratio for a global property like the cross-correlation function (CCF), as compared to individual line profiles. This way, the signal-to-noise ratio in the CCF is increased by a factor $\sqrt{n}$, where n is the number of co-added lines. We now discuss the two steps of the tomographic method in turn.

The information relating to the line doubling (velocity shift and line shape) is in fact distributed among a large number of spectral lines, and can be summed up into an average profile, or more precisely into a CCF. If the correlation of the stellar spectrum with a mask involves many lines, it is possible to extract the relevant information from very crowded spectra, making it possible to use spectra with relatively low signal-to-noise ratios (∼10). The CCF writes

$$\mathrm{CCF}(V_r) = \int_{\lambda_1}^{\lambda_2} s\,[\lambda - \Delta\lambda(\lambda, V_r)]\ m(\lambda)\ \mathrm{d}\lambda, \tag{8.1}$$

where $\Delta\lambda(\lambda, V_r) = \lambda\ V_r/c$, c is the speed of light, $s(\lambda)$ is the observed spectrum, $m(\lambda)$ is the template (a binary template has been adopted, being 1 at the location of the spectral lines of interest, and 0 elsewhere), and λ_1 and λ_2 are the boundaries of the spectral range covered by the observed spectrum. The radial velocity V_r for the considered mask is then obtained from the CCF minimum.

We refer the reader to reference [5] for a detailed description of the CCF mathematical properties.

The mask $m(\lambda)$ used in Eq. 8.1 contains all lines forming in a given range of optical depths. We use an optical-depth scale at a reference wavelength of 5000 Å. Typically, the width of the bins is $\Delta \log \tau_{5000} = 0.5$ in the range $-3.5 \leq \log \tau_{5000} \leq 0.0$, the number of lines ranging from about 3000 in the innermost layer to a few hundreds in the outermost layer (for the spectral range 400–800 nm, and excluding spectral windows strongly contaminated by telluric lines).

The original implementation of the method [3] relied on the Eddington-Barbier approximation, stating that in a grey atmosphere, the flux emerges from the layer located at optical depth $\tau_\lambda = 2/3$. Thus a line at wavelength λ was supposed to form at optical depth $\tau_\lambda = 2/3$. An appropriate model atmosphere is then used to convert this monochromatic optical depth $\tau_\lambda = 2/3$ into a reference optical depth (at 5000 Å for instance), which is then a proxy for geometrical depth. As shown by [25], the Eddington-Barbier approximation gives however a valid depth of line formation only for sufficiently strong lines. In the recent implementation of the method, we use instead the contribution function[1] to the line depression in the flux (C_U; see [1]), in order to correctly assess the depth of line formation:

$$C_U(\tau_\lambda, \lambda) = \int_0^1 \frac{\kappa_l}{\kappa_l + \kappa_c}\ (I_c - S_l)\ \mathrm{e}^{-\tau_\lambda/\mu}\mathrm{d}\mu, \tag{8.2}$$

[1]The contribution function gives the relative contribution of the different atmospheric layers to an observed quantity, which can be the emergent intensity or flux, or the line depression in the surface intensity or flux.

where κ_c and κ_l are the continuum and line opacities, respectively; τ_λ is the total (line + continuum) optical depth; I_c is the continuum intensity, S_l is the source function for the lines (all these variables being λ-dependent) and $\mu = \cos\theta$, where θ is the polar angle between the direction of the light pencil and the normal to the atmospheric layers.

The contribution function is thus a two-dimensional surface (top panel of Fig. 8.2). A crest line may be traced on that surface, and when that crest line is projected on the (λ, τ_{5000}) plane, it defines the so-called depth function $\log \tau_{5000} = x(\lambda)$ (middle panel of Fig. 8.2): for a given wavelength λ, the depth function specifies the value of $\log \tau_{5000}$ corresponding to the maximum of the contribution function. The depth function thus provides the optical depth (at the reference wavelength of 5000 Å) at which a line (if any) present at wavelength λ forms. For the sake of conciseness, we will from now on use the short-hand notation $x \equiv \log \tau_{5000}$.

Different masks C_i are then constructed from the collection of N_i lines located at wavelengths $\lambda_{i,j}$ $(1 \le j \le N_i)$ such that $x_i \le x(\lambda_{i,j}) < x_{i+1} = x_i + \Delta x$, where $x_0 = -3.5$, and $\Delta x = 0.5$ is an optimal value to keep enough lines in any given mask without losing too much resolution in terms of depth (middle panel of Fig. 8.2). A natural limit to the resolution that can be achieved on the optical depth is provided by the width of the contribution function (CF) [1], and according to the Nyquist-Shannon theorem of elementary signal theory, there is no advantage in taking Δx smaller than half the CF width of typical lines probed by the mask. Each mask C_i should then probe lines forming at depths in the range $x_i, x_i + \Delta x$ in the atmosphere. These masks are used as templates to correlate with the observed spectra of the target stars.

As an illustration, the masks have been applied to a spectrum of Z Oph (a Mira with a photometric period of 349 d) obtained on 1998, August 5–6 (see Table 3 of [4] for details) with ELODIE[2] (Fig. 8.3). The CCF shapes change from blue-peaked only in the innermost layer to red-peaked only in the outermost layer, with a double-peaked CCF in the intermediate layers, in accordance with the presence of a shock front in the latter. This procedure thus provides the velocity field as a function of (optical) depth in the atmosphere, whereas it has been checked that for static atmospheres (i.e., non-pulsating red giants), CCFs are single-peaked for all masks and the CCF minimum yields the same radial velocity in all masks. The correctness of the information delivered by the CCF is demonstrated by Fig. 8.4: the CCF shape indeed corresponds to the shape of the lines probed by the corresponding mask.

[2]The high-resolution spectrograph at the *Observatoire de Haute Provence*. See Sect. 8.3.1 for details.

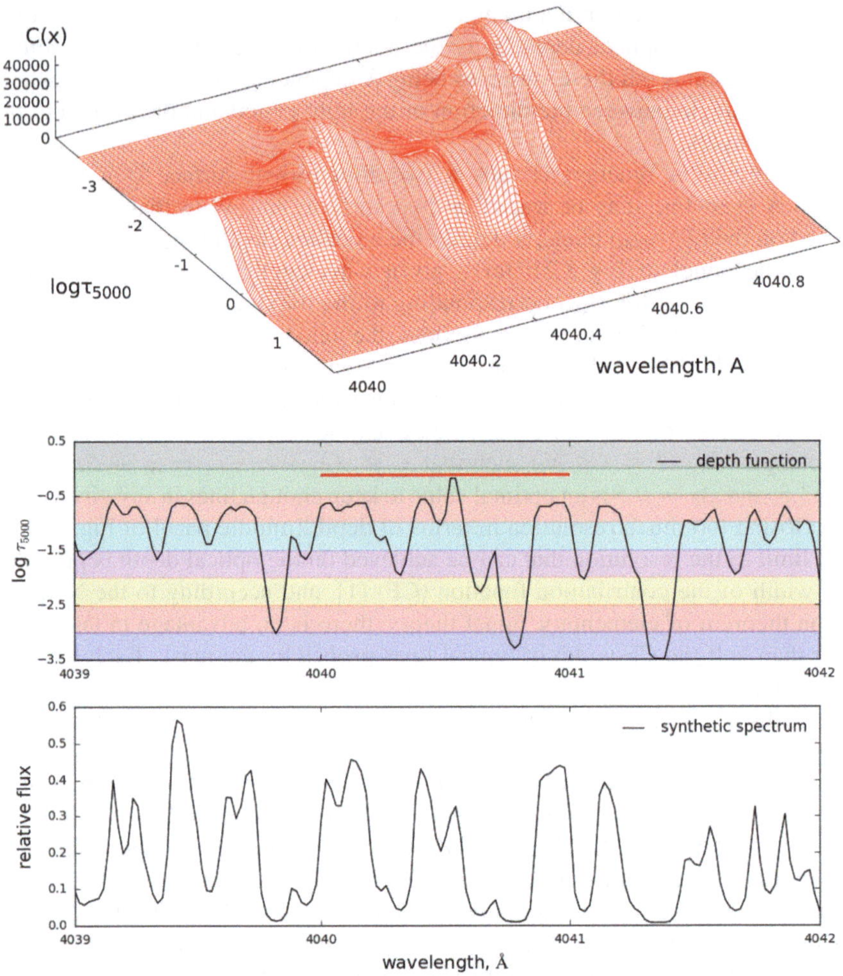

Fig. 8.2 *Top panel*: The contribution function to the line depression in flux, in the spectral window $404.0 \leq \lambda(\mathrm{nm}) \leq 404.1$, for a supergiant model with $T_{\mathrm{eff}} = 3700\,\mathrm{K}$ and $\log g = -0.35$. *Middle panel*: The depth function corresponding to the crest line of the contribution function in the $(\log \tau_{5000}, \lambda)$ plane. *The horizontal line* identifies the spectral window of the displayed contribution function. The different shaded regions delineate the $\log \tau_{5000}$ ranges corresponding to the different masks. For the sake of comparison, note that in the infrared around 1.7 μm, all lines form in the limited range $-1 \leq \log \tau_{5000} \leq 0$. *Bottom panel*: The corresponding synthetic spectrum

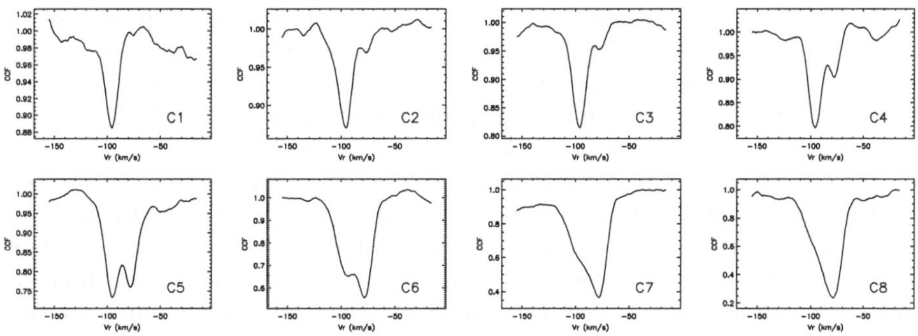

Fig. 8.3 The CCFs of the Mira Z Oph at phase 0.08 (1998, August 05–06) obtained with the tomographic masks. The set of tomographic masks used for Z Oph was constructed from a synthetic spectrum at $T_{\rm eff} = 3500\,\rm K$ and $\log g = 0.9$ (see [3] for details). Note how the shape of the CCFs evolve from the innermost layer (involving ascending matter only, hence *C1* exhibits a single blue peak) to the outermost layer (involving mostly matter falling in, hence *C8* exhibits predominantly a red peak). This spatial sequence of line doubling reflects the presence of a shock wave in the line-forming region, with the shock front being centered on the layer probed by the mask *C5*

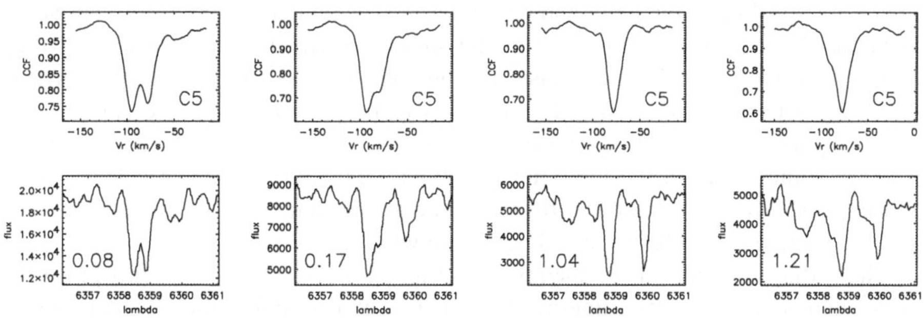

Fig. 8.4 Comparison of the λ 6358.69 Fe I line as seen in the spectrum of Z Oph at phases 0.08, 0.17, 1.04 and 1.21 (*bottom row*, from *left* to *right*), with the CCF obtained with the tomographic mask *C5* (*upper row*), to which belongs the λ 6358.69 Fe I line. Z Oph is warm enough at maximum light (K3). Therefore, its spectrum is not too crowded; clean, almost unblended, lines as the one displayed here may be isolated

8.3 Results

8.3.1 Application to the Mira Variables RT Cyg and RY Cep

A long-term monitoring of the Mira stars RT Cyg ($P = 190\,\rm d$; $6.0 \leq V \leq 13.1$) and RY Cep ($P = 145\,\rm d$; $8.6 \leq V \leq 13.6$) has been performed with the fibre-fed echelle spectrograph ELODIE [6]. The spectrograph ELODIE is mounted on the 1.93-m telescope of the Observatoire de Haute Provence (France), and covers the full range from 390.6 to 681.1 nm in one exposure at a resolving power of 42,000.

For RT Cyg, a monitoring in August-September 1999 covered phases −0.20 to 0.16 around maximum light, with 32 spectra obtained during this phase range (corresponding to an average resolution of $\Delta\phi = 0.01$; Fig. 8.5). For RY Cep, the monitoring covered a *full* light cycle extending from August 2001 (phase −0.17) to February 2002 (phase 0.92), and 40 spectra were obtained (Fig. 8.6).

The sequences of CCFs presented in Figs. 8.5 and 8.6 clearly follow the temporal evolution predicted by the Schwarzschild scenario, with a single red component progressively transforming into a single blue component around maximum light. There are moreover clear phase lags between the different layers – this transformation occurring at later phases in outer layers. Thus, the Schwarzschild scenario holds for Mira variables, and this conclusion definitely points towards the velocity stratification associated with the shock wave as the cause of the double absorption lines observed in Mira variables, as opposed to complex radiative processes (e.g., radiative release of thermal energy into the post-shock layer or temperature inversion) [13, 22].

Radial velocities have been extracted from the CCFs of Figs. 8.5 and 8.6 to yield the curves displayed in Figs. 8.7 and 8.8. We just stress here the noteworthy features of the radial-velocity curve of RY Cep, displayed in Fig. 8.8 for the three masks C3 (innermost), C4 (intermediate) and C6 (outermost): (i) the red peaks (corresponding to matter falling in) are observed at the same velocity in all three masks; they disappear at increasingly later phases as one considers layers closer to the surface, since the outermost layers are the last ones to be penetrated by the shock wave which suppresses the infalling component (i.e., red peak); (ii) the maximum outward velocity is the same for the three masks (about −247 km/s), and is reached around phase −0.1. After that, matter will decelerate in the innermost layers first, as they lose the impetus provided by the shock wave which is moving away.

The radial-velocity curve displayed in Fig. 8.8 corresponds to a full pulsation cycle, and at least in the innermost mask C3, it resembles the S-shape curves obtained by [17] (see also Fig. 8.1) for the CO $\Delta\upsilon = 3$ infrared lines. These 1.6–1.8 μm lines, emerging from a spectral window close to the continuum-opacity minimum, form deep in the atmosphere ($-1 \leq \log \tau_{5000} \leq 0$), and it is therefore not surprising that the S-pattern is best seen in the innermost C3 mask which probes in fact the same optical-depth range. The tomographic method reveals how that S-shaped velocity curve gets distorted further out in the atmosphere.

8.3.2 Other Pulsating Variables

Doubling of absorption lines, sometimes with a documented occurrence of the Schwarzschild phenomenon, has been reported in several kinds of pulsating stars (RR Lyrae: [7, 9]; β Cephei: [26]; BL Herculis: [8, 11]; RV Tauri: [12]), but the tomographic technique has, so far, never been applied to any of them.

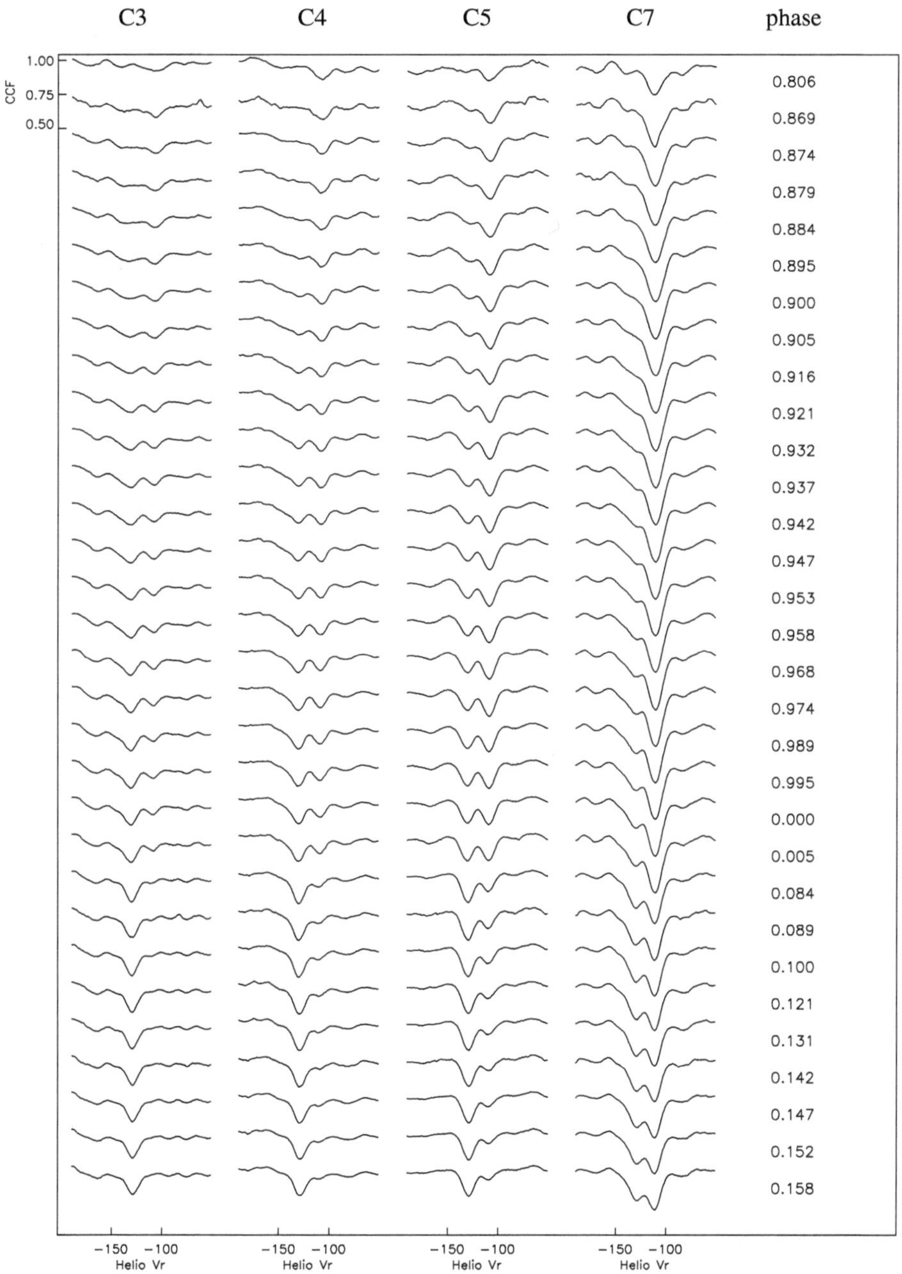

Fig. 8.5 Sequence of cross-correlation profiles of RT Cyg obtained with the tomographic masks *C3, C4, C5* and *C7* (from left to right) in August-September 1999, around maximum light. The labels beside each CCF refer to the corresponding visual phase based on the AAVSO ephemeris (Mattei, 1999, priv. comm)

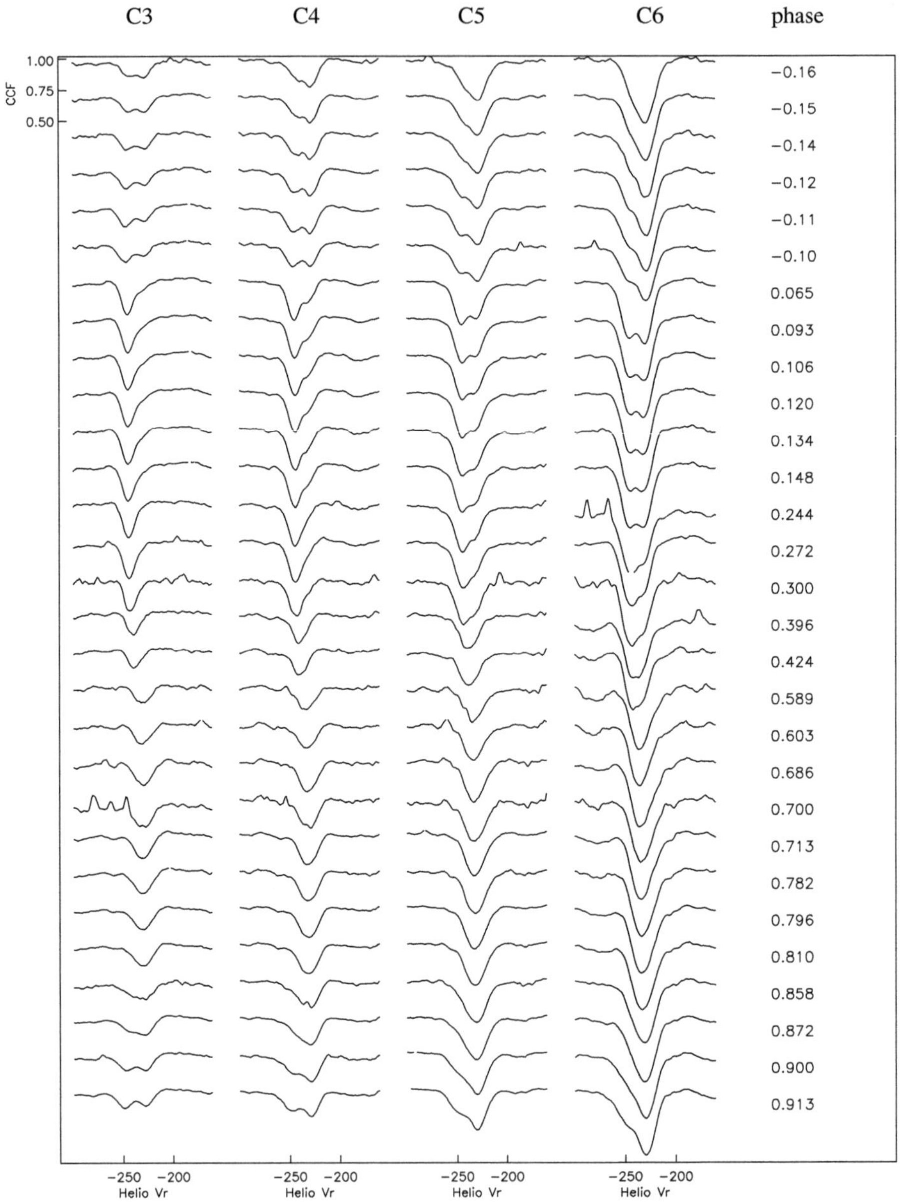

Fig. 8.6 Same as Fig. 8.5 for RY Cep with the tomographic masks *C3* (inner), *C4, C5* and *C6* (outer), from left to right. The labels beside each CCF refer to the corresponding visual phase based on the AFOEV monitoring

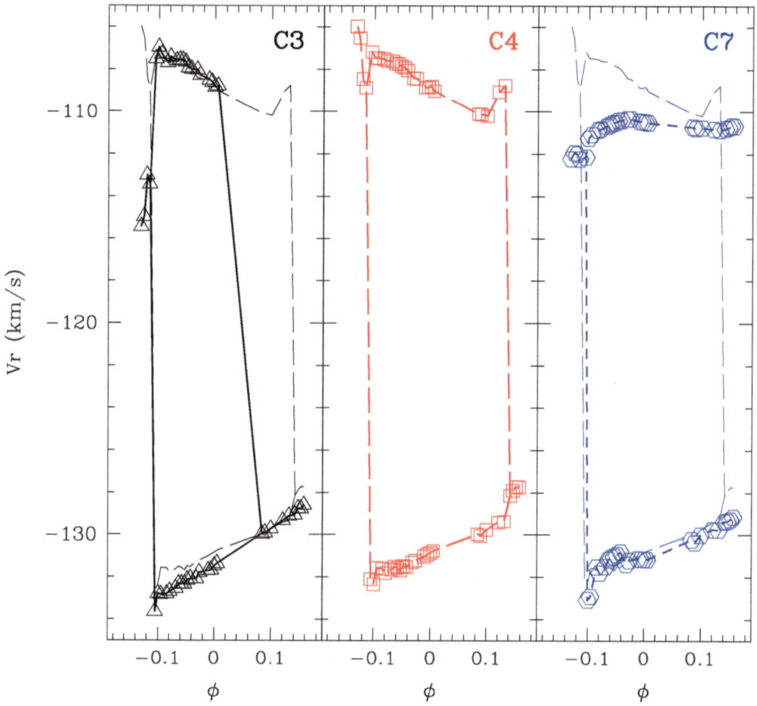

Fig. 8.7 Radial velocity curves obtained for RT Cyg with the tomographic masks *C3* (*leftmost panel – solid line – triangles*), *C4* (*middle panel – long-dashed line – squares*), and *C7* (*rightmost panel – short-dashed line – circles*). The *C4* radial-velocity curve (*long-dashed line*) has been duplicated in the *left-* and *rightmost panels* to allow an easy comparison

8.3.3 Supergiants

The tomographic method has been applied to supergiants by [21], using the first generation masks from [3]. No line doubling was recorded. However, applying the new generation masks (those obtained with the method of Sect. 8.2) on the supergiant μ Cep, one of the brightest red supergiant ($M_{\rm bol} = -9.1$, initial mass larger than 25 $M_\odot$ [24]), reveals another story.

A preliminary analysis of μ Cep's visual light curve led to a primary period around 860 d [23] and a long secondary period around 4400 d. More recent data (Fig. 8.9) do not seem to confirm, however, these earlier conclusions, even if there are clear variations. These photometric periodicities could be ascribed to pulsation, shock waves, large convective cells, circumstellar material or to the interplay between several of these factors. We argue below that the development of convective cells seems to play an important role in the observed photometric pattern. The effective temperature of μ Cep has been estimated at 3700 K [24] or 3750 K [21], and the gravity at $\log g = -0.36$ [21]. A series of 66 high-resolution spectra of

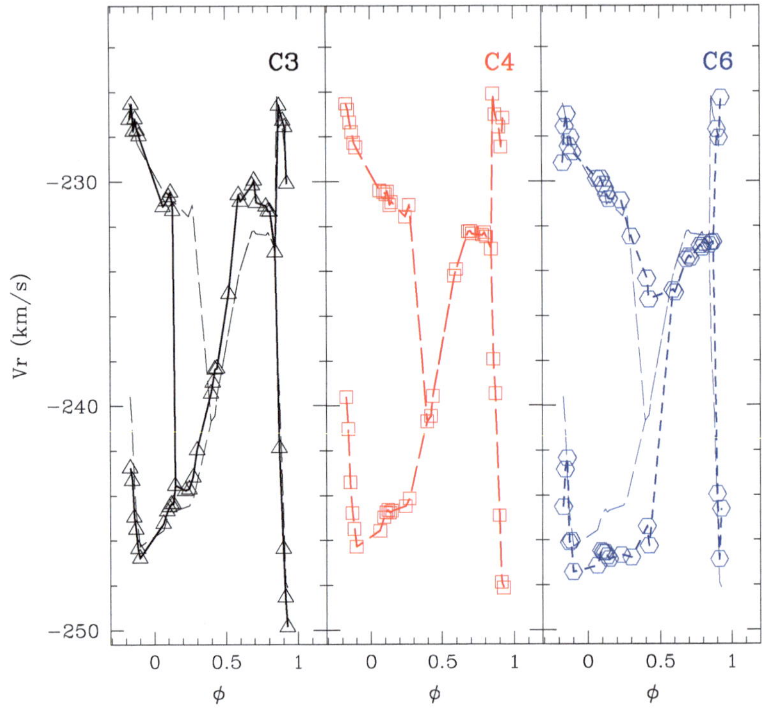

Fig. 8.8 Same as Fig. 8.7 for RY Cep, using the tomographic masks *C3*, *C4*, and *C6*

μ Cep has been obtained with the HERMES spectrograph [29] operated on the MERCATOR telescope (La Palma). These 66 spectra cover a time span of 1500 d (2011–2015), so nearly two photometric cycles.

Episodes of line doubling, like that displayed in Fig. 8.10, are observed, but are restricted to the innermost layers, and do not follow a Schwarzschild scenario. As shown on Figs. 8.9 and 8.10, these double peaks correspond to the apparition of a component blue-shifted by 10–30 km/s (rising matter thus) with respect to the pre-existing, red peak. The blue and red peaks, in the layers where they co-exist, are equally strong. Comparison with the AAVSO light curve reveals that such episodes of line doubling *always* appear during the rising part of the light curve (Fig. 8.9). A likely explanation is that these phases correspond to the emergence at the base of the photosphere of a new convective cell with rising, hot matter, responsible for the brightening and the blue-shifted velocity component. Interestingly, the velocity difference between the red and blue peaks is correlated with the amplitude of the excursion of the rising light curve, larger excursions corresponding to larger values of $Vr(\text{red}) - Vr(\text{blue})$.

Another striking result emerging from the application of the tomographic method to the supergiant μ Cep is the fact that the CCF depth varies with time, with the same periodicity as the light change, but with some phase lag: a hysteresis loop ensues, the

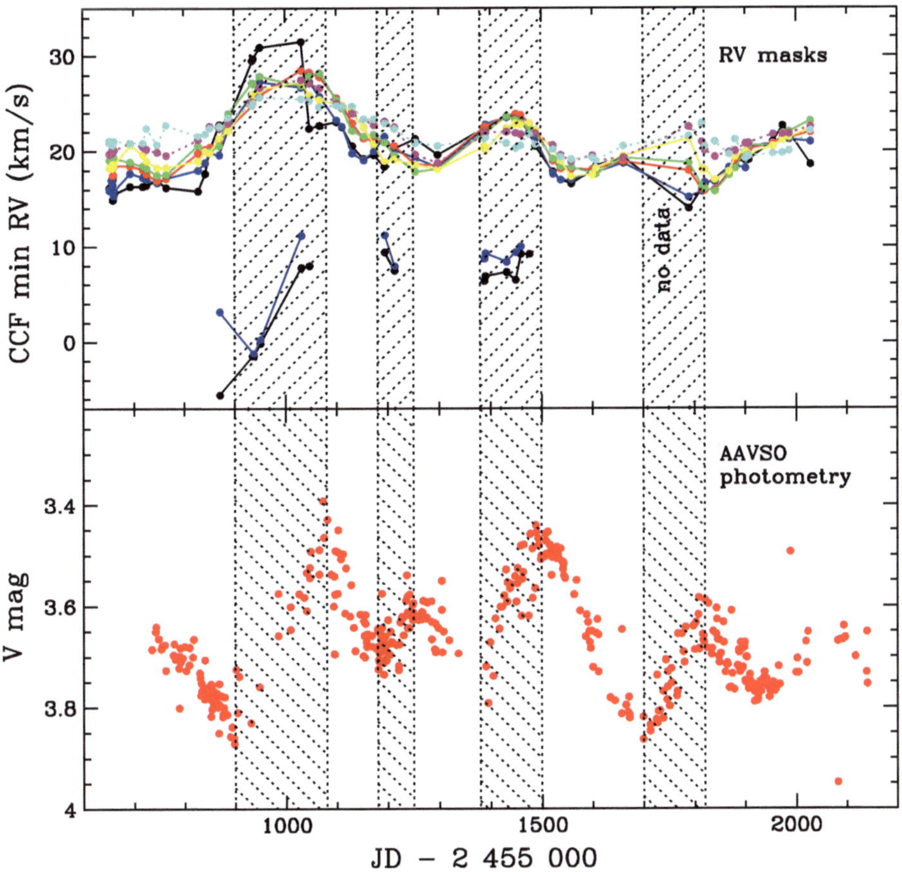

Fig. 8.9 *Top panel*: The velocity variations of the supergiant μ Cep, in the different masks. The inner masks exhibit the largest velocity amplitudes. The incomplete curves in the lower part of the upper panel correspond to the *blue peaks* appearing in the innermost masks during the rising part of the light curve (*shaded areas*). *Bottom panel*: The AAVSO light curve. Line doubling appears on the rising portion of the light curve

lines being deeper when the star is brighter. A similar hysteresis loop exists between the CCF depth and the radial velocity (Fig. 8.11). This property was already noted for Betelgeuse [14], in a temperature-velocity plane, and interpreted as a signature of convective cells. Indeed, hot rising (blue-shifted) matter progressively cools down before going down (red-shifted). Various line-depth ratios (like V I λ 6251.83/Fe I λ 6252.57) were used as temperature indicators by [14]. Since we observe a similar loop in the CCF depth-velocity plane (Fig. 8.11), it is likely that the CCF depth is to some extent temperature-sensitive. However, this is much more difficult to prove for a global quantity like the CCF depth than for line-depth ratios.

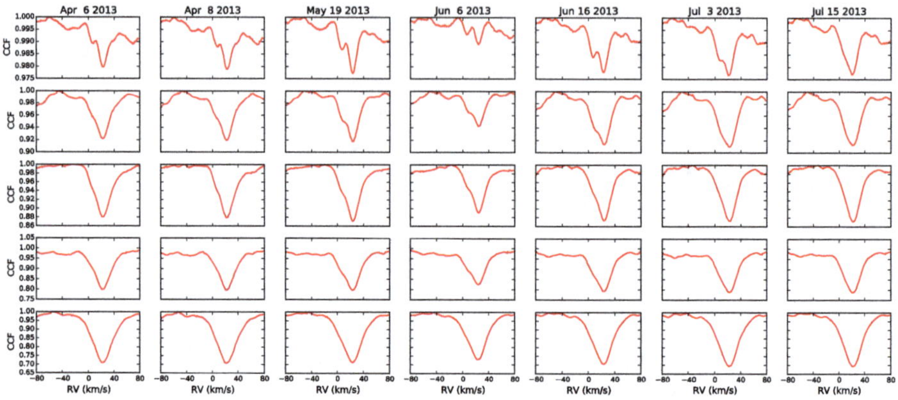

Fig. 8.10 An excerpt of the tomographic sequence of the red supergiant μ Cep over the period April–July 2013 (JD 2456389–2456489). The mask probing the innermost layer is at the top. Note that the line doubling appearing in deep layers does not follow a Schwarzschild scenario

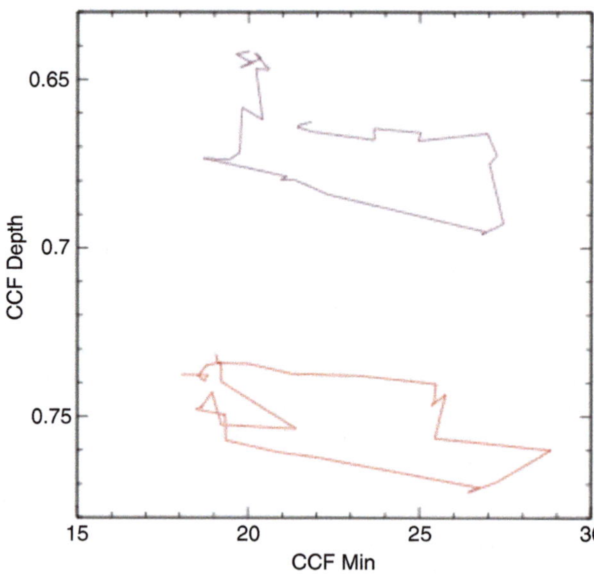

Fig. 8.11 Hysteresis in the CCF depth – CCF min plane (in km/s) for the red supergiant μ Cep for a series of spectra covering roughly a full photometric cycle. A counter-clockwise loop is followed in this diagram, corresponding to the two outermost numerical masks (the outermost is on top), during the course of a photometric cycle

To summarise, in the supergiant star μ Cep, line doubling systematically occurs

- on the rising part of the light curve;
- in the innermost masks;
- is never seen in the outermost masks;
- along a hysteresis loop in the velocity-temperature plane.

All these arguments suggest that convective cells rather than shock waves are responsible for the line doubling observed in red supergiants. They behave thus in a very different manner than the pulsating variables described in the previous sections.

8.4 Future Prospects: Transforming Optical Depths to Geometrical Depths to Access the Shock Velocity

So far, the tomographic method has only been used with optical depths as proxy for geometrical depths. The knowledge of the relation between geometrical and optical depths would open the way to derive directly the shock-front velocity. Although such a relation is provided by model atmospheres, it may be sensitive upon ill-defined parameters like the stellar mass. It seems therefore preferable to derive the relation between geometrical and optical depths empirically, by combining tomographic and interferometric data. In order to calibrate the reference-optical-scale τ_{5000} in terms of geometrical depth, it is necessary to measure the stellar radius at the wavelengths corresponding to each tomographic mask. To do so, high-resolution AMBER data have been acquired, and first results should be available soon.

Acknowledgements The authors thank B. Plez, A. Chiavassa, and H. Van Winckel for useful discussions. SvE is FNRS research associate. Based on observations obtained with the Mercator Telescope and the HERMES spectrograph, supported by the Fund for Scientific Research of Flanders (FWO), the Research Council of K.U.Leuven, the Fonds National de la Recherche Scientifique (F.R.S.-FNRS), Belgium, the Royal Observatory of Belgium, the Observatoire de Genève, Switzerland and the Thüringer Landessternwarte Tautenburg, Germany.

References

1. Albrow, M.D., Cottrell, P.L.: Contribution functions and the depths of formation of spectral lines in Cepheids. MNRAS **278**, 337 (1996)
2. Alvarez, R., Jorissen, A., Plez, B., Gillet, D., Fokin, A.: Envelope tomography of long-period variable stars. I. The Schwarzschild mechanism and the Balmer emission lines. A&A **362**, 655–665 (2000)
3. Alvarez, R., Jorissen, A., Plez, B., Gillet, D., Fokin, A., Dedecker, M.: Envelope tomography of long-period variable stars II. Method. A&A **379**, 288–304 (2001)

4. Alvarez, R., Jorissen, A., Plez, B., Gillet, D., Fokin, A., Dedecker, M.: Envelope tomography of long-period variable stars III. Line-doubling frequency among Mira stars. A&A **379**, 305–322 (2001)
5. Baranne, A., Mayor, M., Poncet, J.L.: CORAVEL – a new tool for radial velocity measurements. Vistas Astron. **23**, 279–316 (1979)
6. Baranne, A., Queloz, D., Mayor, M., Adrianzyk, G., Knispel, G., Kohler, D., Lacroix, D., Meunier, J.P., Rimbaud, G., Vin, A.: ELODIE: a spectrograph for accurate radial velocity measurements. A&AS **119**, 373–390 (1996)
7. Chadid, M., Gillet, D.: Observation of the metallic line doubling phenomenon in the variable star RR Lyrae. A&A **308**, 481–488 (1996)
8. Fokin, A.B., Gillet, D.: The shock wave propagation effects in BL Herculis. II. Nonlinear model and theoretical profile of Hα. A&A **290**, 875–884 (1994)
9. Fokin, A.B., Gillet, D.: Theoretical interpretation of the metallic line doubling phenomenon in the variable star RR Lyrae. A&A **325**, 1013–1024 (1997)
10. Fox, M.W., Wood, P.R.: Shock waves in Mira variables. II – theoretical models. ApJ **297**, 455–475 (1985)
11. Gillet, D., Burki, G., Chatel, A., Duquennoy, A., Lebre, A.: The shock wave propagation effects in BL Herculis. 1: the hydrogen alpha profile. A&A **286**, 508–514 (1994)
12. Gillet, D., Lafon, J.P.J., David, P.: Radiative shocks in atomic and molecular stellar atmospheres. III – the shock wave velocity problem in Mira stars. A&A **220**, 185–196 (1989)
13. Gillet, D., Maurice, E., Bouchet, P., Ferlet, R.: The shock-induced variability of emission profiles in S CAR. A&A **148**, 155–164 (1985)
14. Gray, D.F.: Mass motions in the photosphere of Betelgeuse. AJ **135**, 1450–1458 (2008)
15. Gustafsson, B., Edvardsson, B., Eriksson, K., Jørgensen, U.G., Nordlund, Å., Plez, B.: A grid of MARCS model atmospheres for late-type stars. I. Methods and general properties. A&A **486**, 951–970 (2008)
16. Hinkle, K.H., Barnbaum, C.: Infrared velocities of long period variables: the carbon Mira S Cep. AJ **111**, 913 (1996)
17. Hinkle, K.H., Lebzelter, T., Scharlach, W.W.G.: Infrared velocities of long period variables: CO $\Delta(\upsilon) = 3$ in four Miras and five SR variables. AJ **114**, 2686 (1997)
18. Ireland, M.J., Scholz, M., Wood, P.R.: Dynamical opacity-sampling models of Mira variables – I. Modelling description and analysis of approximations. MNRAS **391**, 1994–2002 (2008)
19. Jorissen, A., Alvarez, R., Plez, B., Gillet, D., Fokin, A.: Tomography of the atmosphere of long-period variable stars. In: Astro Tomography. Lecture Notes in Physics, vol. 573, pp. 301–306. Springer, Berlin (2001)
20. Jorissen, A., Alvarez, R., Plez, B., Gillet, D., Fokin, A.: Tomography of the atmosphere of long-period variable stars. In: Journées Mira 2002, Université de Montpellier II (published only as arXiv:astro-ph/0301361) (2003)
21. Josselin, E., Plez, B.: Atmospheric dynamics and the mass loss process in red supergiant stars. A&A **469**, 671–680 (2007)
22. Karp, A.H.: Hydrodynamic models of a Cepheid atmosphere. III – line spectrum and radius determinations. ApJ **201**, 641–652 (1975)
23. Kiss, L.L., Szabó, G.M., Bedding, T.R.: Variability in red supergiant stars: pulsations, long secondary periods and convection noise. MNRAS **372**, 1721–1734 (2006)
24. Levesque, E.M., Massey, P., Olsen, K.A.G., Plez, B., Josselin, E., Maeder, A., Meynet, G.: The Effective temperature scale of galactic red supergiants: cool, but not as cool as we thought. ApJ **628**, 973–985 (2005)
25. Magain, P.: Contribution functions and the depths of formation of spectral lines. A&A **163**, 135–139 (1986)
26. Mathias, P., Gillet, D., Fokin, A.B., Cambon, T.: On the origin of shock waves in the beta Cephei star BW Vulpeculae. A&A **339**, 525–530 (1998)
27. Merrill, P.W.: Red stars. PASP **67**, 199 (1955)
28. Neri, R., Kahane, C., Lucas, R., Bujarrabal, V., Loup, C.: A (12) CO (J=1–0) and (J=2–1) atlas of circumstellar envelopes of AGB and post-AGB stars. A&AS **130**, 1–64 (1998)

29. Raskin, G., van Winckel, H., Hensberge, H., Jorissen, A., Lehmann, H., Waelkens, C., Avila, G., de Cuyper, J.P., Degroote, P., Dubosson, R., Dumortier, L., Frémat, Y., Laux, U., Michaud, B., Morren, J., Perez Padilla, J., Pessemier, W., Prins, S., Smolders, K., van Eck, S., Winkler, J.: HERMES: a high-resolution fibre-fed spectrograph for the Mercator telescope. A&A **526**, A69 (2011)
30. Schwarzschild, M.: Transactions of the IAU, vol. VIII, pp. 811–812. Cambridge University Press, Cambridge (1952)

Chapter 9
Eclipse Mapping: Astrotomography of Accretion Discs

Raymundo Baptista

9.1 Context and Motivations

In weakly-magnetic cataclysmic variables (CVs) a late-type star overfills its Roche lobe and feeds a companion white dwarf (WD) via an accretion disc, which usually dominates the ultraviolet and optical light of the system [60]. The temperatures in CV discs may vary from 5000 K in the outer regions to over 50,000 K close to the disc centre, and the surface density may vary by equally significant amounts over the disc surface. Therefore, the spectrum emitted by different regions of the accretion disc may be very distinct. Furthermore, aside of the component stars and the accretion disc, additional sources contribute to the integrated light from the binary: a bright spot (BS) or stream forms where the gas from the mass-donor star hits the outer edge of the disc, tidally-induced spiral structures might develop in the outer regions of extended discs [8, 59], and a fraction of the transferred mass may also be ejected from the binary in a wind from the disc surface [9, 38]. Because what one directly observes is the combination of the spectra emitted from these diverse regions and sources, the interpretation of CVs observations is usually plagued by the ambiguity associated with composite spectra. The most effective way to overcome these difficulties is with spatially resolved data, in which the light from the different sources might be disentangled. However, at typical sizes of less than a solar radius and distances of hundreds of parsecs, accretion discs in CVs are seen at angular diameters of tens of micro-arcseconds, and spatially resolving them is well beyond the reach of current direct imaging interferometric techniques, both at optical and radio wavelength ranges. Thus, resolving an accretion disc in a CV is presently possible only via indirect imaging.

R. Baptista (✉)
Departamento de Física, UFSC, Campus Trindade, 88040-900, Florianópolis, Brazil
e-mail: raybap@gmail.com

H.M.J. Boffin et al. (eds.), *Astronomy at High Angular Resolution*, Astrophysics and Space Science Library 439, DOI 10.1007/978-3-319-39739-9_9

Developed in the 1980s by Horne [33], the Eclipse Mapping Method is an indirect imaging technique which provides spatially resolved observational constraints of accretion discs in CVs on angular scales of micro-arcseconds. It assembles the information contained in the shape of the eclipse into a map of the accretion disc surface brightness distribution. The following sections describe the technique and provide a set of examples aiming at illustrating the wealth of possible applications.

9.2 Principles and Inner Workings

The three basic assumptions of the standard eclipse mapping method are: (i) the surface of the secondary star is given by its Roche equipotential, (ii) the brightness distribution is constrained to the orbital plane, and (iii) the emitted radiation is independent of the orbital phase. While assumption (i) is reasonably robust and always employed, the others are simplifications that do not hold in all situations and may be relaxed with three-dimensional (3D) eclipse mapping implementations (see Sect. 9.5.4).

A grid of intensities centred on the WD, the eclipse map, is defined in the orbital plane. One usually adopts the distance from the disc centre to the internal Lagrangian point, R_{L1}, as the length scale. With this definition the primary lobe has about the same size and form for any reasonable value of the binary mass ratio q ($=M_2/M_1$, where M_2 and M_1 are the masses of the mass-donor star and the WD, respectively) [33]. If the eclipse map is an N points flat, square grid of side λR_{L1}, each of its surface element (pixel) has an area $(\lambda R_{L1})^2/N$ and an associated intensity I_j. The solid angle comprised by each pixel as seen from the earth is then

$$\theta^2 = \left[\frac{(\lambda R_{L1})^2}{N}\frac{1}{d^2}\right] \cos i \; , \tag{9.1}$$

where d is the distance to the system. The value of λ defines the area of the eclipse map while the choice of N sets its spatial resolution.

The model eclipse light curve $m(\phi)$ is derived from the intensities in the eclipse map,

$$m(\phi) = \theta^2 \sum_{j=1}^{N} I_j V_j(\phi) \; . \tag{9.2}$$

The eclipse geometry $V_j(\phi)$ specifies the fractional visibility of each pixel as a function of orbital phase and may include fore-shortening and limb darkening factors [35, 50, 65]. The fractional visibility of a given pixel is obtained by dividing the pixel into smaller tiles and evaluating the Roche potential along the line of sight for each tile to see if the potential falls below the value of the equipotential that defines the Roche surface of the mass-donor star. If so, the tile is occulted. The fractional visibility of the pixel is then the sum of the visible tiles divided by the number of tiles.

The eclipse geometry is determined by the inclination i, the binary mass ratio q and the phase of inferior conjunction ϕ_0 [33, 35]. These parameters set the shape and extension of the projected shadow of the mass-donor star in the orbital plane. As the binary phase ϕ changes during eclipse, the resulting shadow rotates around the L1 point, progressively covering/uncovering different parts of the accretion disc. This creates a grid of criss-crossed ingress/egress arcs in the orbital plane. A pixel with coordinates (x, y) within the region covered by this grid disappears and reappears from eclipse at a particular pair of binary phases (ϕ_i, ϕ_e).[1] Thus, the eclipse geometry sets the connection between the image space (the eclipse map) and the data space (the light curve).

Figure 9.1 shows an example of eclipse geometry and depicts the connection between the image and data spaces. Information about the location of the occulted brightness sources is embedded in the shape of the eclipse light curve. The width of the eclipse increases as the light source moves closer to the L1 point, whereas it is displaced towards negative (positive) phases if the light source is in the leading (trailing) disc side. For example, the phase width and range of the eclipse in the upper right panel of Fig. 9.1 tells us that the corresponding light source (the blue pixel in the left-hand panel of Fig. 9.1) is located in the leading, near side of the disc. Figure 9.2 illustrates the simulation of the eclipse of a fitted brightness distribution while showing the comparison between the resulting model light curve and the data light curve. The geometry in this case is $i = 81°$ and $q = 0.5$. The left-hand panels show the data light curve (small dots) and the model light curve (solid line) as it is being drawn at five different orbital phases along the eclipse (indicated in the upper right corner). The right-hand panels show the best-fit disc brightness distribution and how it is progressively covered by the dark shadow of the mass-donor star during the eclipse. The slope of the light curve steepens whenever a bright source is occulted or reappears from eclipse. The two asymmetric arcs of the brightness distribution of this example are occulted at distinct phases, leading to a V-shaped eclipse with two bulges at ingress and at egress. The flux at phase $\phi = 0$ does not go to zero because a sizeable fraction of the disc remains visible at mid-eclipse.

Given the eclipse geometry, a model light curve can be calculated for any assumed brightness distribution in the eclipse map. A computer code then iteratively adjusts the intensities in the map (treated as independent parameters) to find the brightness distribution the model light curve of which fits the data eclipse light curve within the uncertainties (Fig. 9.3). Because the one-dimensional data light curve cannot fully constrain a two-dimensional map, additional freedom remains to optimise some map property. A maximum entropy (MEM) procedure [56, 57] is

[1]Pixels outside the region covered by the grid of criss-crossed arcs are never eclipsed, and have no corresponding ingress/egress phases (ϕ_i, ϕ_e). Accordingly, there is no information about the surface brightness distribution of these unocculted regions in the shape of the eclipse.

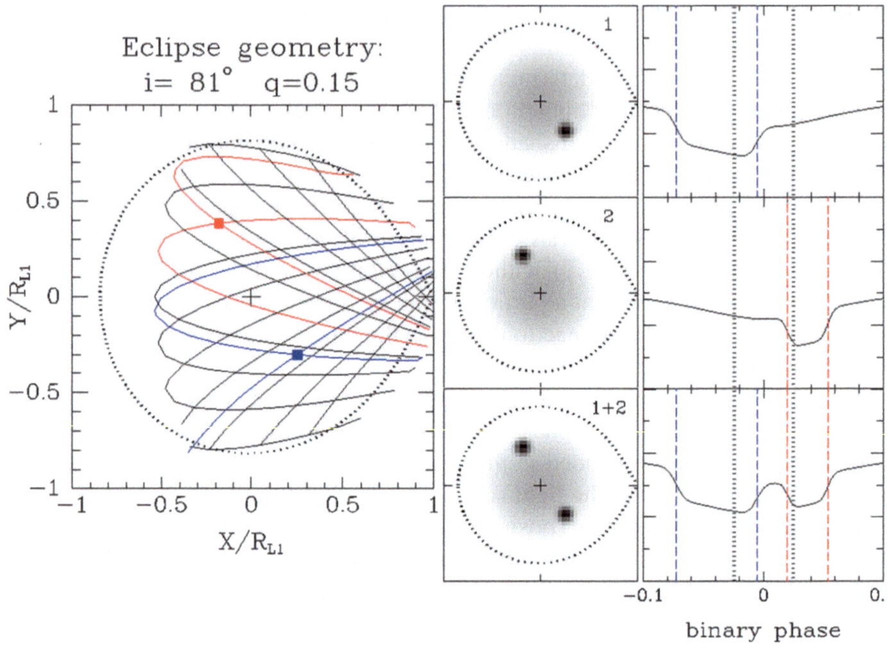

Fig. 9.1 *Left-hand panel*: A grid of criss-crossed ingress/egress arcs for the eclipse geometry $i = 81°$, $q = 0.15$, and an eclipse map with $\lambda = 2$. The ingress/egress arcs for the *blue* and *red pixels* are shown in the respective colour. A *dotted line* depicts the primary Roche lobe; a cross marks the disc centre. The mass-donor star is to the right of the panel and the stars rotate counter-clockwise (equivalently, the observer and the mass-donor star shadow rotate clockwise). *Middle-panels*: grayscale brightness distributions with a faint, extended disc plus bright, narrow Gaussian spots at the positions of the *blue* and *red pixels* in the *left panel*. *Right-hand panels*: the model eclipse light curves obtained by convolving the eclipse geometry in the *left panel* with the brightness distributions in the *middle panels*. *Vertical dotted lines* mark the ingress/egress of the disc centre; *blue/red vertical dashed lines* depict the ingress/egress phases of the *blue/red pixels*

used to select, among all possible solutions, the one that maximises the entropy of the eclipse map with respect to a smooth default map.

The entropy of the eclipse map p with respect to the default map q is defined as

$$S = -\sum_{j=1}^{N} p_j \ln\left(\frac{p_j}{q_j}\right) , \tag{9.3}$$

where p and q are written as

$$p_j = \frac{I_j}{\sum_k I_k} \quad , \quad q_j = \frac{D_j}{\sum_k D_k} . \tag{9.4}$$

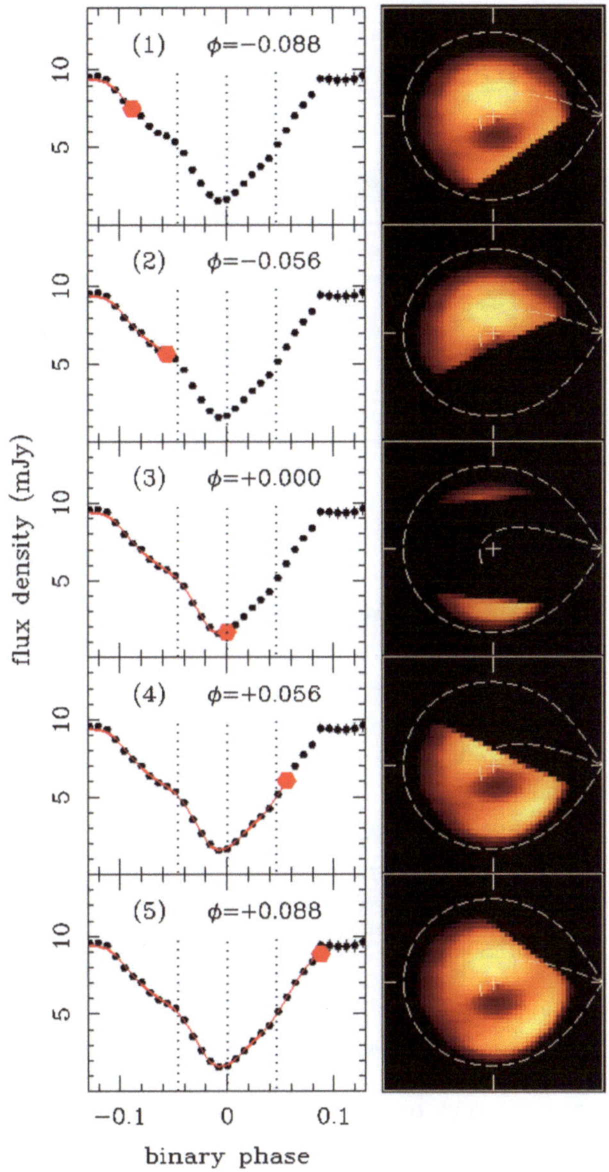

Fig. 9.2 Simulation of a disc eclipse ($i = 81°, q = 0.5$). *Left-hand panels*: data light curve (*dots*) and model light curve (*solid line*) at five different binary phases (indicated in the *upper right corner*). *Vertical dotted lines* mark mid-eclipse and the ingress/egress phases of the disc centre. *Right-hand panels*: visible parts of the best-fit eclipse map (in a false colour blackbody logarithmic scale) at the corresponding binary phases. *Dashed lines* show the primary Roche lobe and the ballistic trajectory of the gas from the mass-donor star; crosses mark the centre of the disc. The mass-donor star is to the right of each panel; the stars and the accretion disc gas rotate counter-clockwise

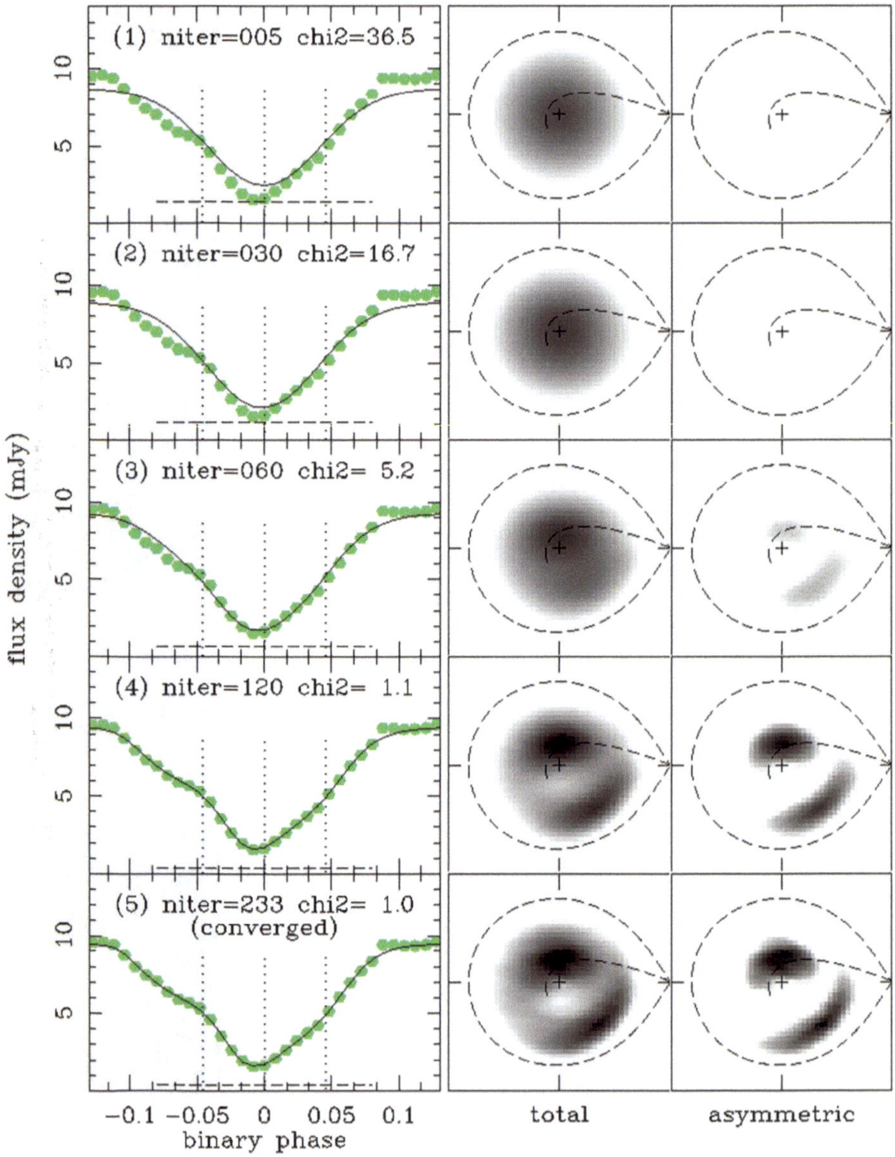

Fig. 9.3 Example of the iterative fit of a data light curve with a MEM eclipse mapping code [15]. *Left-hand panels*: data (*points*) and model (*solid line*) light curves at five different stages of the iteration process towards convergence. *Vertical dotted lines* mark mid-eclipse and the ingress/egress of the disc centre. An *horizontal dashed line* depicts the uneclipsed component of the total flux. *niter* and *chi2* give the current iteration number and χ^2 value, respectively. *Middle panels*: corresponding eclipse maps in a logarithmic grayscale; darker regions are brighter. The notation is the similar to that of Fig. 9.2. *Right-hand panels*: the asymmetric component of the eclipse maps in the *middle panels*

Other functional forms for the entropy appear in the literature [35, 60]. These are equivalent to (9.3) when p and q are written in terms of proportions.

The default map D_j is generally defined as a weighted average of the intensities in the eclipse map,

$$D_j = \frac{\sum_k \omega_{jk} I_k}{\sum_k \omega_{jk}} , \tag{9.5}$$

where the weight function ω_{jk} is specified by the user. A priori information about the disc (e.g., axi-symmetry) is included in the default map via ω_{jk}. Prescriptions for the weight function ω_{jk} and their effect in the reconstructions are discussed in Sect. 9.3. In the absence of any constraints on I_j, the entropy has a maximum $S_{max} = 0$ when $p_j = q_j$, or when the eclipse map and the default map are identical.

The consistency of an eclipse map may be checked using the χ^2 as a constraint function,

$$\chi^2 = \frac{1}{M} \sum_{\phi=1}^{M} \left(\frac{m(\phi) - d(\phi)}{\sigma(\phi)} \right)^2 = \frac{1}{M} \sum_{\phi=1}^{M} r(\phi)^2 , \tag{9.6}$$

where $d(\phi)$ is the data light curve, $\sigma(\phi)$ are the corresponding uncertainties, $r(\phi)$ is the residual at the orbital phase ϕ, and M is the number of data points in the light curve. Alternatively, the constraint function may be a combination of the χ^2 and the R-statistics [15],

$$R = \frac{1}{\sqrt{M-1}} \sum_{\phi=1}^{M-1} r(\phi)\, r(\phi+1) , \tag{9.7}$$

to minimise the presence of correlated residuals in the model light curve [14]. For the case of uncorrelated normally distributed residuals, the R-statistics has a Gaussian probability distribution function with average zero and unity standard deviation. Requiring the code to achieve an $R = 0$, is equivalent to asking for a solution with uncorrelated residuals in the model light curve.

The final MEM solution is the eclipse map that is as close as possible to its default map as allowed by the constraint imposed by the light curve and its associated uncertainties [35, 50]. In mathematical terms, the problem is one of constrained maximisation, where the function to maximise is the entropy and the constraint is a consistency statistics that measures the quality of the fitted model to the data light curve. Different codes exist to solve this problem. Many of the eclipse mapping codes are based on the commercial optimisation package MEMSYS [56]. Alternative implementations using conjugate-gradients algorithms [14, 15], CLEAN-like algorithms [58] and genetic algorithms [20] are also used.

Rutten et al. [52] found that the entropy function can be a useful tool to signal and to isolate the fraction of the total light which is not coming from the accretion disc plane. They noted that when the light curve is contaminated by the presence of additional light (e.g., from the mass-donor star) the reconstructed map shows a spurious structure in its outermost region. This is because the eclipse mapping method assumes that all the light is coming from the accretion disc, in which case the eclipse depth and width are correlated in the sense that a steeper shape corresponds to a deeper eclipse. The addition of an uneclipsed component in the light curve (i.e., light from a source other than the accretion disc) ruins this correlation. To account for the extra amount of light at mid-eclipse and to preserve the brightness distribution derived from the eclipse shape the algorithm inserts the additional light in the region of the map which is least affected by the eclipse. Since the entropy measures the amount of structure in the map, the presence of these spurious structures is flagged with lower entropy values. The correct offset level may be found by comparing a set of maps obtained with different offsets and selecting the one with highest entropy. Alternatively, the value of the zero-intensity level can be included in the mapping algorithm as an additional free parameter to be fitted along with the intensity map in the search for the MEM solution [9, 49]. A detailed discussion on the reliability and consistency of the estimation of the uneclipsed component can be found in [16].

Figure 9.3 gives an example of the convergence process of a MEM reconstruction. The code starts from a flat map and quickly evolves towards an axi-symmetric Gaussian map which reproduces the gross features of the data light curve. However, at this point the model light curve is not a good match to the data ($\chi^2 = 16.7$), failing to reproduce the double stepped ingress/egress bulges of the eclipse shape. Several more iterations are required in order to match these features in the data light curve, which demands building two asymmetric bright arcs in the leading and trailing sides of the disc. The reconstruction converges for a final unity χ^2 value after 233 iterations.

9.3 Performance and Limitations

A crucial aspects of eclipse mapping is the selection of the weight function for the default map, ω_{jk}, which allows the investigator to steer the MEM solution towards a determined type of disc map. The choice $\omega_{jk} = 1$ results in a uniform default map which leads to the *most uniform eclipse map* consistent with the data. This happens not to be a good choice for eclipse mapping, because it results in a map severely distorted by criss-crossed artefacts [19, 33, 58] as the flux of point sources is spread along their ingress and egress arcs (Fig. 9.4a). This led to the adoption of a weight function which sets the default map as an axi-symmetric average of the eclipse map, thereby yielding the *most nearly axi-symmetric map* that fits the data [33]. It suppresses the azimuthal information in the default map while keeping the radial structure of I_j on scales greater than a radial blur width Δ_R. This seems a reasonable choice for accretion disc mapping because one expects the disc material to be roughly in Keplerian orbits, so that local departures from axi-symmetry will

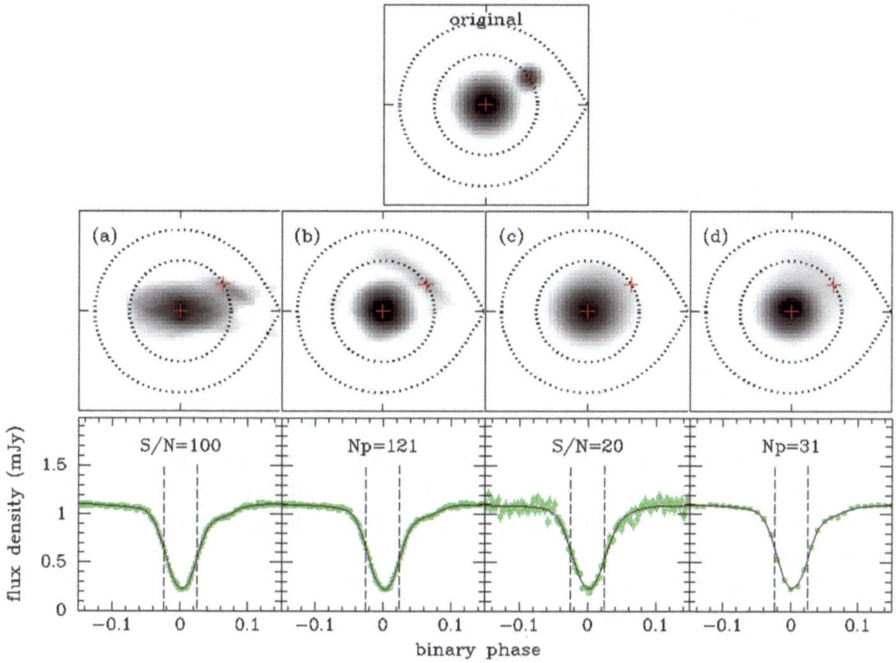

Fig. 9.4 Effects of the default map, phase resolution (= number of data points in the light curve) and signal-to-noise ratio (S/N) of the input data on the eclipse mapping reconstruction. The *top panel* shows an artificial, test brightness distribution in a logarithmic grayscale. The notation is similar to that of Fig. 9.1. The *middle panels* show reconstructions of the test brightness distribution for different combination of parameters: (**a**) uniform default map on a light curve with $N = 121$ data points and $S/N = 100$; (**b**) default of limited azimuthal smearing (Eq. 9.8), $Np = 121$ and $S/N = 100$; (**c**) default of Eq. 9.8, $Np = 121$ and $S/N = 20$; and (**d**) default of Eq. 9.8, $Np = 31$ and $S/N = 100$. The *lower panels* show the corresponding data (*points*) and model (*solid line*) light curves. *Vertical dashed lines* mark the ingress/egress phases of the disc centre

tend to be smeared away by the strong shear. This is a commonly used option and is also known as the default map of full azimuthal smearing.

The full azimuthal smearing default results in rather distorted reproduction of asymmetric structures such as a bright spot at the disc rim. In this case, the reconstructed map exhibits a lower integrated flux in the asymmetric source region, the excess being redistributed as a concentric annulus about the same radial distance. By limiting the amount of azimuthal smearing it is possible to alleviate this effect and to start recovering azimuthal information in the accretion disc. Two prescriptions in this regard were proposed. Rutten et al. [47] limited the amount of azimuthal smearing by averaging over a polar Gaussian weight function of *constant angle* Δ_θ along the map,[2]

$$\omega_{jk} = \exp\left[-\frac{1}{2}\left(\frac{(R_j - R_k)^2}{\Delta_R^2} + \frac{\theta_{jk}^2}{\Delta_\theta^2}\right)\right] \tag{9.8}$$

while Baptista et al. [16] chose to use a polar Gaussian function of *constant arc length* Δ_s through the map,

$$\omega_{jk} = \exp\left[-\frac{1}{2}\left(\frac{(R_j - R_k)^2}{\Delta_R^2} + \frac{s_{jk}^2}{\Delta_s^2}\right)\right] . \tag{9.9}$$

The reconstructions in Fig. 9.4b–d used the default function of Eq. 9.8. In particular, the reconstructions in Fig. 9.4a, b were obtained from the same input light curve. This comparison illustrates that the default of limited azimuthal smearing provides a much better reconstruction of the central brightness distribution while still allowing to recover the location of a bright spot at disc rim, although with some azimuthal smearing.

The concept of default map was extended with the introduction of a double default function [11, 19, 58],

$$D_j = (D1_j)^n\,(D2_j)^m \tag{9.10}$$

(where $D1_j$ and $D2_j$ are separate default functions and $n+m = 1$), together with the idea of a negative default function (e.g., $m < 0$). While a positive default function steers the MEM solution *towards* its intrinsic property (e.g., axi-symmetry), a negative default function may be used to drive the MEM solution *away* from its intrinsic property (e.g., the presence of the undesired criss-crossed arcs) [58]. The combination of a positive axi-symmetric default function with a negative criss-crossed arcs default function was key to allow the recovery of accretion disc spiral structures with eclipse mapping [11, 30]. The eclipse maps shown in Figs. 9.2, 9.3 and 9.5 were obtained with this double default setup.

The quality of an eclipse mapping reconstruction is tied to the quality of the input data light curve. Specifically, the ability to recover brightness sources and the spatial resolution of an eclipse map depend on the phase (or, time) resolution and the signal-to-noise ratio (S/N) of the data light curve. Baptista and Steiner [14] provide an expression to compute the spatial resolution of the eclipse map which matches the phase resolution of the data light curve. Figure 9.4 shows the effects of (i) degrading the phase resolution and (ii) lowering the S/N of the data light curve on the quality of the resulting eclipse map. As a reference for comparison, the eclipse map in Fig. 9.4b was obtained from a light curve of good phase resolution ($Np = 121$ data

[2] R_j and R_k are the distances from pixels j and k to the centre of the disc; θ_{jk} is the azimuthal angle between pixels j and k; s_{jk} is the arc-length between pixels j and k.

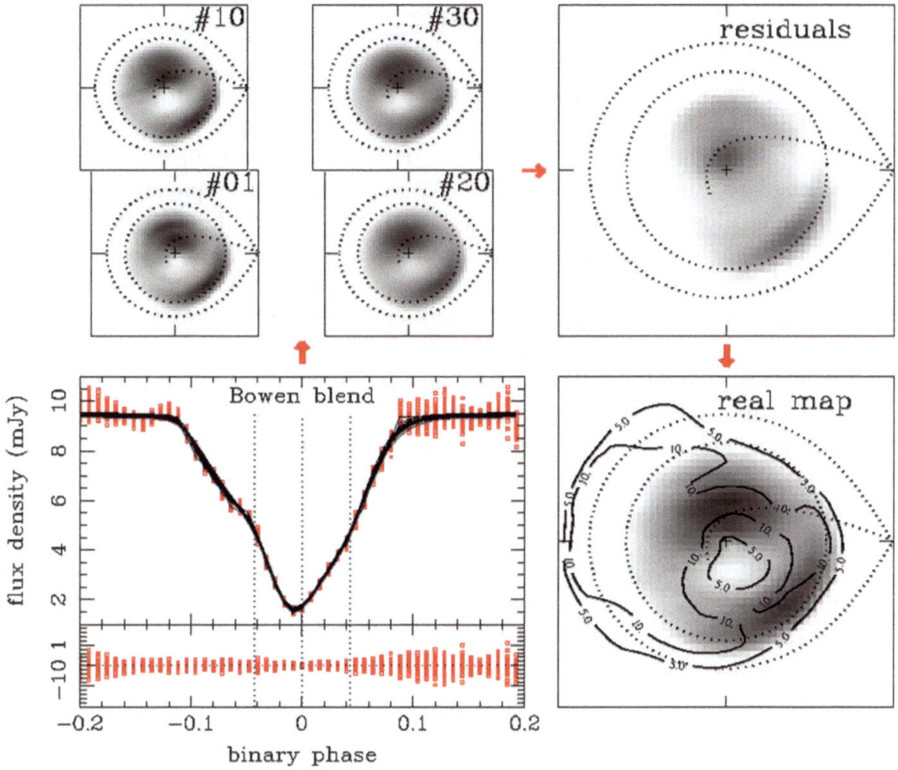

Fig. 9.5 Computing the statistical significance of an eclipse map from the uncertainties in the data light curve. *Lower left panels*: A set of 50 randomised light curves (*dots*) and corresponding eclipse mapping models (*solid lines*) superimposed in phase. The *bottom panel* shows these light curves after subtraction of the real data light curve. The scatter depicts the uncertainty in the flux at each phase. *Top left panels*: a sample of 4 of the eclipse maps from the randomised light curves, in a logarithmic grayscale. The notation is similar to that of Fig. 9.4. *Top right panel*: the map of the standard deviations with respect to the mean intensity (the "residuals" map). *Lower right panel*: the eclipse map from the real data light curve. Contour lines for $S/N = 5$ and 10 are overplotted; features in the eclipse maps are statistically significant at or above the $5\,\sigma$ confidence level

points) and high signal-to-noise ($S/N = 100$). Figure 9.4c shows that degrading the S/N reduces the ability to identify fainter brightness sources. Because of the reduced S/N, the model light curve is not forced to follow the egress shoulder that signals the presence of the bright spot at disc rim. As a consequence, the eclipse map has only a weak asymmetry at the position of the bright spot. Figure 9.4d shows that the effect of reducing the phase resolution in the light curve is to reduce the spatial resolution of the eclipse map, with an increase in the radial/azimuthal blur effects (i.e., point sources would look increasingly out of focus with decreasing phase resolution). The effects of low S/N seem more dramatic than those of low phase resolution.

9.4 Error Propagation Procedures

Because the maximum-entropy eclipse mapping is a non-linear inversion method, it is not possible to compute the uncertainties in the eclipse map directly from the uncertainties in the data light curve. Hence, statistical uncertainties of an eclipse map are usually computed with a Monte Carlo error propagation procedure [52]. This is illustrated in Fig. 9.5. A set of N ($\geq$20) artificial light curves is created from the original data in which the flux at each phase is varied around the true value according to a Gaussian distribution with standard deviation equals to the uncertainty at that point. The artificial light curves are fitted with the eclipse mapping code to generate a set of randomised eclipse maps. These are then combined to produce a "residuals" map by taking the pixel-to-pixel standard deviation with respect to the mean intensity. This yields the statistical uncertainty at each pixel. For $N = 20$, the uncertainty of the standard deviation is <20 %, sufficient to illustrate the confidence limits of the eclipse map, whereas $N \geq 200$ is needed in order to bring the uncertainty in the standard deviation down to 7 % [30]. Uncertainties obtained with this procedure may be used to estimate the errors in the derived radial brightness temperature and intensity distributions. A map of the statistical significance (or the inverse of the relative error) is obtained by dividing the true eclipse map by the "residuals" map [5, 11, 30].

An alternative error propagation procedure involves the use of simulations with the Bootstrap technique (e.g., [6]).

9.5 Applications

This section focuses on selected applications of the eclipse mapping method which illustrates some of the possible scientific problems that have been and may be addresses with it. The reader is referred to [2, 34, 35, 65] for more comprehensive reviews of the results obtained with this technique.

9.5.1 Spectral Mapping: Spatially-Resolved Disc Spectra

The eclipse mapping method is capable of delivering spatially-resolved spectra of accretion discs when the technique is applied to time-resolved eclipse spectrophotometric data, providing the best example of its ability to disentangle the light from different emission sources in the binary. The time-series of spectra is divided up into numerous ($\sim$100) spectral bins and light curves are extracted for each bin. The light curves are then analysed to produce a series of monochromatic eclipse maps covering the whole spectrum. Finally, the maps are combined to obtain the spectrum for any region of interest on the disc surface [47].

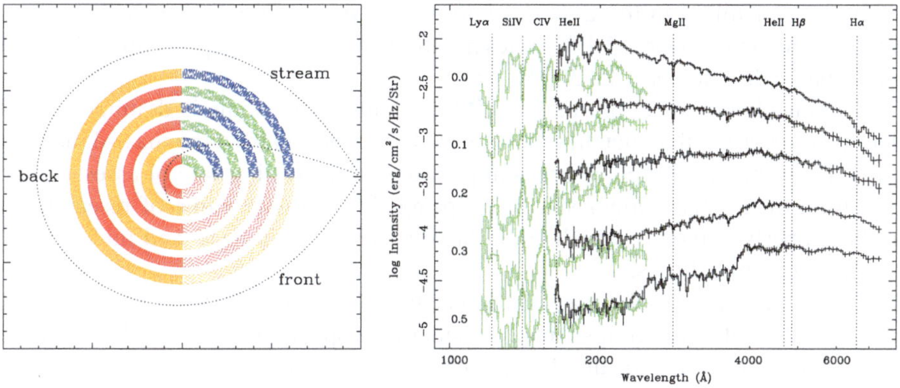

Fig. 9.6 An example of spectral mapping, from HST/FOS UV-optical eclipse spectroscopy of the nova-like variable UX UMa. *Left panel*: a schematic diagram of the eclipse map, with the annular regions used to extract spatially resolved spectra. The disc is divided into three major azimuthal regions (the back side, the front side, and the quarter section containing the gas stream trajectory), and into a set of 6 concentric annuli with radius increasing in steps of 0.1 R_{L1} and of width 0.05 R_{L1}. *Right panel*: Spatially resolved spectra of the back region of the UX UMa accretion disc in two occasions (August 1994 in *green*; November 1994 in *black*). The spectra were computed for the set of concentric annular sections shown in the *left panel*. The most prominent line transitions are indicated by *vertical dotted lines* (Reprinted with permission from [10])

The spectral mapping analysis of nova-like variables [10, 13, 29, 47, 49] shows that their inner accretion disc is characterised by a blue continuum filled with absorption bands and lines which cross over to emission with increasing disc radius (Fig. 9.6). The continuum emission becomes progressively fainter and redder as one moves outwards, reflecting the radial temperature gradient. These high mass-accretion ($\sim 10^{-8} - 10^{-9}\,M_\odot\,\text{year}^{-1}$) discs seem hot and optically thick in their inner regions and cool and optically thin in their outer parts. The spectrum of the infalling gas stream in UX UMa [10] and UU Aqr [13] is noticeably different from the disc spectrum at the same radius, suggesting the existence of gas stream "disc-skimming" overflow that can be seen down to $R \simeq (0.1 - 0.2)\,R_{L1}$.

In contrast, spatially-resolved spectra of low-mass accretion ($\sim 10^{-9} - 10^{-11}\,M_\odot\,\text{year}^{-1}$), quiescent dwarf nova show that the lines are in emission at all disc radii [53]. The observed differences between the spectra of the disc hemisphere farther from and closer to the L1 point might be interpreted in terms of chromospheric emission from a disc with non-negligible opening angle (i.e., a limb-brightening effect).

The spectrum of the uneclipsed component both for the nova-like systems and quiescent dwarf novae shows strong emission lines and the Balmer jump in emission indicating that the uneclipsed light has an important contribution from optically thin gas (e.g., [13]). The lines and optically thin continuum emission are most probably emitted in a vertically extended disc chromosphere + wind [9, 38]. This additional source of radiation may be responsible both for flattening the ultraviolet spectral slope and for filling in the Balmer jump of the optically thick disc spectrum, and might explain the historical difficulties in fitting integrated disc spectrum with stellar atmosphere disc models [25, 39, 41, 61].

The uneclipsed spectrum of UX UMa at long wavelengths is dominated by a late-type spectrum that matches the expected contribution from the mass-donor star [49]. Thus, the uneclipsed component provides an interesting way of assessing the spectrum of the mass-donor star in eclipsing CVs. This is a line of research yet to be properly explored, which could largely benefit from near- and mid-infrared spectroscopy of eclipses.

Spectral mapping is also useful to isolate the spectrum of the WD, represented by the central pixel of the eclipse map. A stellar atmosphere model fit to the extracted white dwarf spectrum for the dwarf nova V2051 Oph yields the WD temperature and an independent estimate of the distance to the binary [53].

9.5.2 Time-Lapse Mapping: Dwarf Nova Outbursts

Eclipse maps give snapshots of the accretion disc at a given time. Time-resolved eclipse mapping may be used to track changes in the disc structure, e.g., to assess variations in mass accretion rate or to follow the evolution of the surface brightness distribution through a dwarf nova (DN) outburst cycle.

DN outbursts are explained in terms of either (i) the time dependent response of a viscous accretion disc to a burst of mass transfer from the donor star (the mass transfer instability model, MTIM, e.g., [17]), or (ii) a limit-cycle behaviour driven by a thermal-viscous disc-instability (the disc-instability model, DIM, e.g., [24, 40]), in which matter progressively accumulates in a low viscosity disc (quiescence) until a critical surface density is reached at a given radius, causing a heating wave to switch the disc to a high viscosity regime (outburst) that allows the gas to diffuse rapidly inwards and onto the white dwarf. There is a set of distinct predictions from these models that might be critically tested with time-lapse eclipse mapping [3].

An example of time-lapse eclipse mapping is given in Fig. 9.7. Eclipse maps covering the full outburst cycle of the dwarf nova EX Dra [7] show the formation of a one-armed spiral structure at the early stages of the outburst and reveal how the disc expands during the rise until it fills most of the primary Roche lobe at maximum light. During the decline stage, the disc becomes progressively fainter until only a small bright region around the WD is left at minimum light. The evolution of the radial temperature distribution shows the presence of an outward-moving heating wave during rise and of an inward-moving cooling wave in the decline.

The temperatures of the EX Dra outbursting disc are above those expected for the hot, ionised outbursting disc gas in the DIM framework [40, 60], and the cooling wave decelerates as it travels inwards [7]. Since these results are consistent with both DIM and MTIM, they offer no power to discriminate between these models. However, the same eclipse mapping study reveals that the radial temperature distribution in quiescence follows the $T \propto R^{-3/4}$ law of opaque steady-state discs

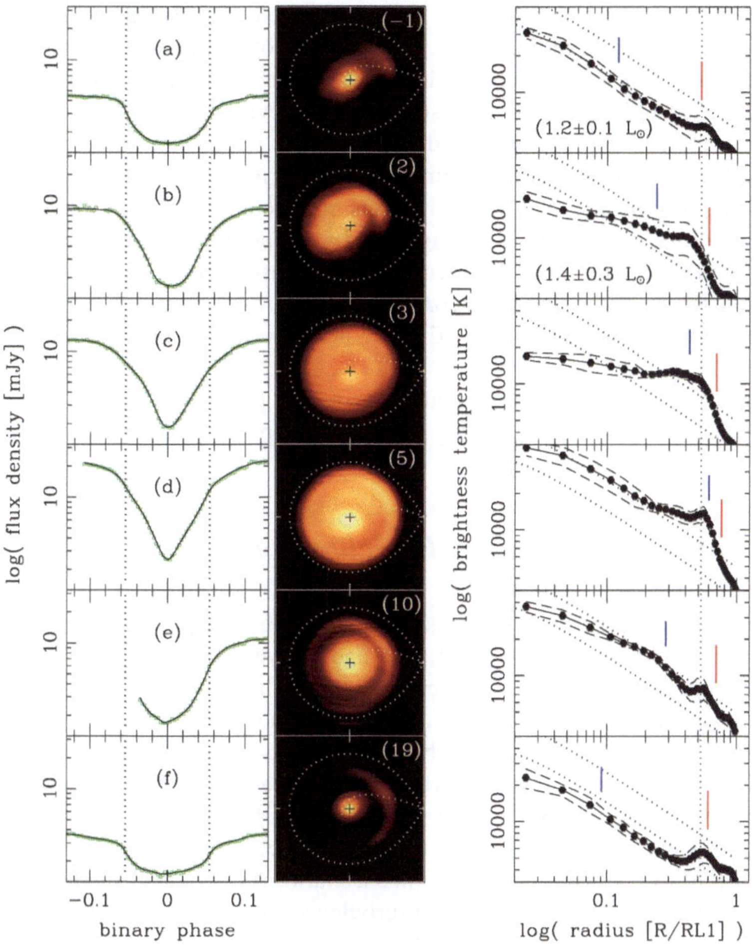

Fig. 9.7 Time-lapse eclipse mapping of the dwarf nova EX Dra along its outburst cycle. *Left-hand panels*: Data (*dots*) and model (*solid line*) light curves in (**a**) quiescent, (**b**) early rise, (**c**) late rise, (**d**) outburst maximum, (**e**) early decline, and (**f**) late decline stages. *Vertical dotted lines* mark ingress/egress phases of disc centre. *Middle panels*: eclipse maps in a false colour blackbody logarithmic grayscale. The notation is similar to that of Fig. 9.2. The numbers in parenthesis indicate the time (in days) elapsed since outburst onset. *Right-hand panels*: azimuthal-averaged radial brightness temperature distributions for the eclipse maps in the *middle panels*. *Dashed lines* show the 1-σ limit on the average temperature for a given radius. A *dotted vertical line* depicts the radial position of the bright spot in quiescence; vertical ticks mark the position of the outer edge of the disc (*red*) and the radial position at which the disc temperature falls below 11,000 K (*blue*). Steady-state disc models for mass accretion rates of log $\dot{M}= -8.0$ and $-9.0\,M_\odot$ year^{-1} are plotted as *dotted lines* for comparison. Numbers in parenthesis list the integrated disc luminosity (Reprinted with permission from [7])

(Fig. 9.7a), and that this quiescent state is reached only 2 d after the end of the outburst. Both results indicate that the viscosity of its quiescent disc is as large as in outburst – in contradiction with DIM and in agreement with predictions of the MTIM. In addition, the early rise map shows evidence of enhanced gas stream emission, indicative of an enhanced mass transfer rate at this early outburst stage (Fig. 9.7b). The integrated disc luminosity at early rise ($1.4 \pm 0.3\, L_{\odot}$) is comparable to that in quiescence ($1.2 \pm 0.1\, L_{\odot}$), and is not enough to support the idea that the enhanced mass transfer could be triggered by an increased irradiation of the mass donor by the accretion disc. This led to the conclusion that the observed enhanced mass transfer at early rise is not a consequence of the ongoing outburst, but its cause – suggesting that the outbursts of EX Dra are powered by bursts of mass transfer [3].

Time-lapse eclipse mapping studies were also performed for the dwarf novae Z Cha [63], OY Car [23, 51], IP Peg [19], V2051 Oph [12], and HT Cas [26, 36].

9.5.3 Flickering Mapping: Revealing the Disc Viscosity

Flickering is the intrinsic brightness fluctuation seen in light curves of T Tau stars [31], mass-exchanging binaries [1, 22] and active galactic nuclei [27]. Optical studies suggest there are two different sources of flickering in CVs: (i) the stream-disc impact region (possibly because of unsteady mass inflow or post-shock turbulence [55, 62]) and/or (ii) turbulent inner disc regions (possibly because of magneto-hydrodynamic (MHD) turbulence, unsteady WD accretion or events of magnetic reconnection at the disc atmosphere [21, 28, 37]). With its ability to spatially-resolve and to disentangle different sources, flickering mapping has been a useful tool to confirm and extend this scenario. As an added bonus, if the disc-related flickering is caused by MHD turbulence, one may infer the radial run of the disc viscosity parameter α_{ss} [54] from the relative flickering amplitude, (σ_D/D) [28],

$$\alpha_{ss}(R) \simeq 0.23 \left[\frac{R}{50\,H}\right] \left[\frac{\sigma_D(R)}{0.05\,D(R)}\right]^2 , \tag{9.11}$$

where H is the disc scale height.

From a large, uniform ensemble of light curves of a given CV it is possible to separate the steady-light component (the average flux in a given phase bin), low- and high-frequency flickering amplitudes (the scatter with respect to the average flux, computed with the ensemble [18] and single [21] methods) as a function of binary phase, to derive corresponding maps of surface brightness distributions from their eclipse shapes and, thereafter, to compute the radial run of the relative amplitude of the disc flickering component [5, 6].

Flickering mapping of the DN V2051 Oph reveals that the low-frequency flickering arises mainly in an overflowing gas stream and is associated with the mass

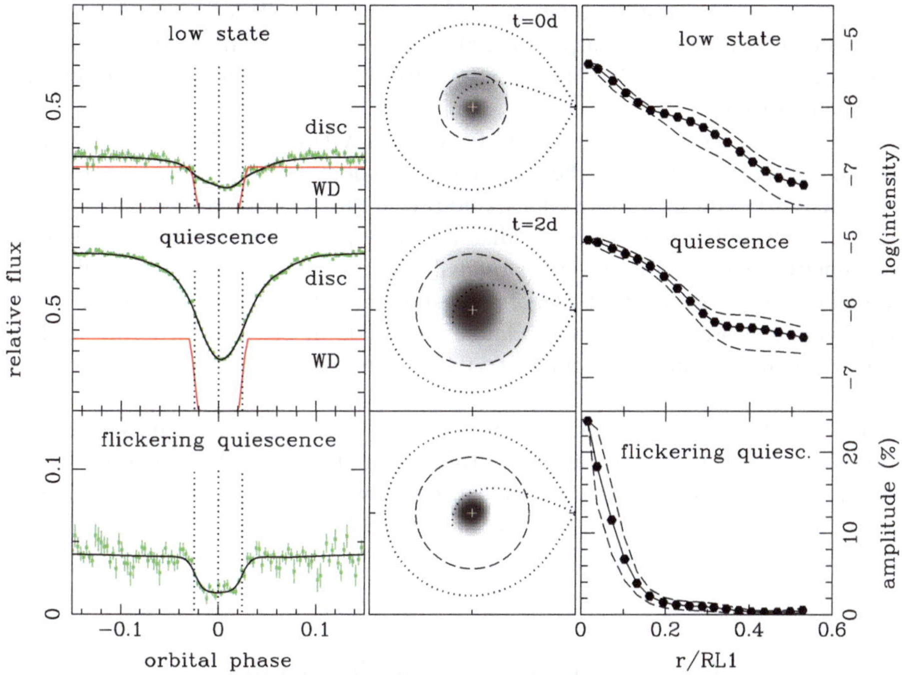

Fig. 9.8 *Left-hand panels*: data (*green dots*) and eclipse mapping model (*black lines*) disc light curves, and the extracted light curve of the WD (*red lines*). *Top* and *middle panels* show, respectively, the low state and quiescence steady-light component, while the *lower panel* shows the high-frequency ($f > 3.3\,mHz$) flickering curve. *Middle panels*: the corresponding eclipse maps in a logarithmic grayscale. The notations is similar to that of Fig. 9.1. *Dashed circles* mark the outer disc radius in each case. *Right-hand panels*: azimuthal-averaged radial intensity (*top*, *middle*) and flickering amplitude in percent (*bottom*) distributions for the eclipse maps in the *middle panels*. *Dashed lines* show the 1-σ limit on the average value for a given radius

transfer process. The high-frequency flickering originates in the accretion disc and has a relative amplitude of a few percent, independent of disc radius and brightness level, leading to large values $\alpha_{ss} \simeq (0.1 - 0.2)$ at all disc radii [5].

Figure 9.8 shows the results of the eclipse mapping analysis of an extensive data set of optical light curves of the dwarf nova HT Cas [3]. These observations frame a 2 d transition from a low state (largely reduced mass transfer rate) back to quiescence, allowing the application of both time-lapse and flickering mapping techniques to derive independent estimates of α_{ss}. In the low state, the gas stream hits the disc at the circularisation radius R_{circ} [60], and the accretion disc has its smallest possible size. The disc fast viscous response to the onset of mass transfer, increasing its brightness and expanding its outer radius at a speed $v \simeq +0.4\,\mathrm{km\,s^{-1}}$, implies $\alpha_{ss} \simeq 0.3 - 0.5$. The newly added disc gas reaches the WD at disc centre soon after mass transfer recovery ($\sim$1 d), also implying a large disc viscosity parameter, $\alpha_{ss} \simeq 0.5$. Flickering mapping reveal a minor, low-frequency BS-stream flickering component in the outer disc, plus a main disc flickering component the amplitude of which rises sharply towards disc centre (Fig. 9.8), leading to a radial dependency $\alpha_{ss}(R) \propto R^{-2}$ with $\alpha_{ss} > 0.1$ for $R < R_{circ}$ – in agreement with the time-lapse results.

A similar analysis was performed for the DN V4140 Sgr [4]. Eclipse mapping in quiescence indicate that the steady-light is dominated by emission from an extended disc with negligible contribution from the WD, suggesting that efficient accretion through a high-viscosity disc is taking place. Flickering maps show an asymmetric source at disc rim (BS-stream flickering) and an extended central source (disc flickering) several times larger in radius than the WD. Unless the thin disc approximation breaks down, the relative amplitude of the disc flickering leads to large α_{ss}'s in the inner disc regions ($\simeq$0.2 – 1.0), which decrease with increasing radius.

In contrast, in the nova-like UU Aqr optical flickering arises mainly in tidally-induced spiral shocks in its outer disc [6]. Assuming that the turbulent disc model applies, its disc viscosity parameter increases outwards and reaches $\alpha_{ss} \sim 0.5$ at the position of the shocks, suggesting that they might be an effective source of angular momentum removal of disc gas.

9.5.4 3D Eclipse Mapping: Disc Opening Angle and Superhumps

Standard eclipse mapping assumes a simple flat, geometrically thin disc model. Real discs may however violate this assumption in the limit of high $\dot{M}$. Disc half-opening angles of $\beta \geq 4°$ are predicted for $\dot{M} \geq 5 \times 10^{-9}\,M_{\odot}\,year^{-1}$ [43]. At large inclinations ($i \geq 80°$) this may lead to artificial front-back asymmetries in the eclipse map because of the different effective areas of surface elements in the near and far sides of a flared disc as seen by an observer on Earth. Furthermore, the assumption that the emitted radiation is independent of orbital phase implies that any out-of-eclipse brightness change (e.g., orbital modulation due to BS anisotropic emission) has to be removed before the light curve can be analysed.[3]

A step to overcome these limitations is to go beyond the standard assumptions allowing the eclipse mapping surface to become three-dimensional. This leads to 3D eclipse mapping [50]. For example, with the inclusion of a disc rim in the eclipse mapping method, the out-of-eclipse modulation can be modelled as the fore-shortening of an azimuthal-dependent brightness distribution in the disc rim [19]. This procedure allows one to recover the azimuthal (phase) dependency of the BS emission. Moreover, motivated by the front-back asymmetry that appeared in the flat-disc map and by the difficulties in removing the asymmetry with the assumption

[3]This is usually done by fitting a spline function to the phases outside eclipse, dividing the light curve by the fitted spline, and scaling the result to the spline function value at phase zero (e.g., [9]).

of an uneclipsed component, [45] introduced a flared disc in their eclipse mapping of ultraviolet light curves of Z Cha at outburst. They found that the asymmetry vanishes and the disc is mostly axi-symmetric for a disc half-opening angle of $\beta = 6°$.

In 3D eclipse mapping, the mapping surface usually consists of a grid of $N1$ pixels on a conical surface centred at the WD position and inclined at a half-opening angle β with respect to the orbital plane, plus a circular rim of $N2$ pixels orthogonal to the orbital plane at a distance R_d ($<R_{L1}$) from disc centre. An *entropy landscape* technique [48] may be used to cope with the extra degree of freedom that comes along with the additional geometry parameters β and R_d. Simulations [4] show that if an eclipse mapping reconstruction is performed with the wrong choice of β and R_d, the code develops artefacts in the brightness distribution in order to compensate for the incorrect parameters. Eclipse maps with these spurious, additional structures have lower entropy than the map with the correct choice of β and R_d. Therefore, the best-fit geometry can be found by searching the space of parameters for the pair of (β, R_d) values of highest entropy (Fig. 9.9, left-hand panels).

An example of 3D eclipse mapping is shown in Fig. 9.9, where the technique was applied to study the evolution of the accretion disc surface brightness of the DN V4140 Sgr in a superoutburst [4]. The entropy landscape analysis indicates that the accretion disc is geometrically thin both in outburst ($\beta = 1.0° \pm 0.5°$) and in quiescence ($\beta = 0.5° \pm 0.5°$); it fills the primary Roche lobe in outburst and progressively shrinks to about half this size in quiescence. They also find that the disc is elliptical in outburst and decline, with an eccentricity $e = 0.13$. At both outburst stages, the disc orientation is such that superhump maximum occurs when the mass-donor star is aligned with the bulge of the elliptical disc. This lends observational support for the tidal resonance instability model of superhumps [32, 42, 64].

Additional observational support for the presence of elliptical precessing discs in CVs come from the eclipse mapping analyses of light curves of the DN Z Cha in superoutburst [44] and of the permanent superhumper V348 Pup [46].

9.6 Summary

Eclipse mapping is a unique, powerful technique to investigate:

- Accretion disc spectra (allowing one to separate the disc atmosphere emission at different distances from disc centre, and to isolate the spectrum of the WD, BS, a possibly outflowing disc wind, and even the faint, red mass-donor star);
- Accretion disc structures (such as gas stream overflow, tidally-induced spiral shocks, elliptical precessing discs, and magnetic accretion curtains);
- Time evolution of accretion discs (tracing the mass and angular momentum redistribution throughout dwarf nova outbursts, or to follow brightness changes caused by lighthouse effects in discs with fast spinning, magnetic WDs);

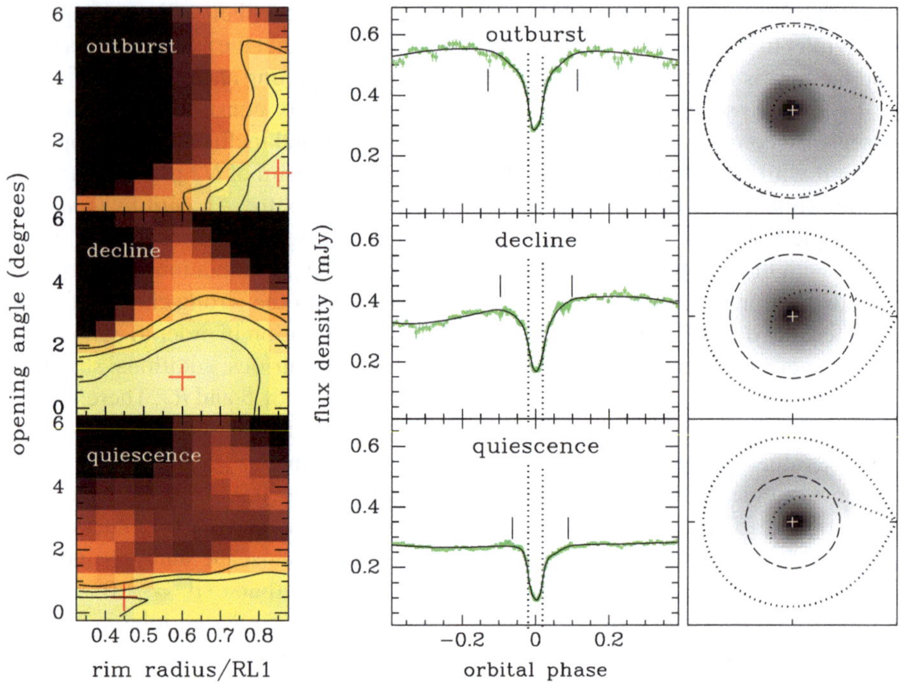

Fig. 9.9 3D eclipse mapping of V4140 Sgr. *Left panels*: entropy landscape for the outburst (*top*), decline (*middle*) and quiescence (*bottom*) light curves in a blackbody false-colour scale. Regions in *yellow* have higher entropy. A *red cross* marks the combination of disc half-opening angle β and rim radius R_d of highest entropy. *Middle panels*: data (*dots*) and model (*solid line*) light curves for the pair of (β, R_d) values of highest entropy in each case. *Vertical dotted lines* mark the ingress/egress phases of the WD, while vertical tick marks depict the phases of eclipse ingress/egress. *Right panels*: corresponding eclipse maps in a logarithmic grayscale. The notation is similar to that of Fig. 9.2. *Dashed circles* show the best-fit disc rim radius (Reprinted with permission from [4])

- Accretion disc viscosity (either via flickering mapping or by measuring the velocity of transition waves during dwarf nova outbursts with time-lapse mapping).

Acknowledgements RB acknowledges financial support from CNPq/Brazil through grant no. 308 946/2011-1.

References

1. Augusteijn, T., Karatasos, K., Papadakis, M., et al.: Coordinated X-ray and optical observations of Scorpius X-1. A&A **265**, 177 (1992)
2. Baptista, R.: Eclipse mapping of accretion discs. In: Boffin, H.M.J., Steeghs, D., Cuypers, J. (eds.) Astrotomography: Indirect Imaging Methods in Observational Astronomy. Lecture Notes in Physics, vol. 573, p. 307. Springer, Berlin (2001)

3. Baptista, R.: Accretion disk evolution in dwarf novae through outbursts: disk instability and mass-transfer instability cases. Mem. S. A. It. **83**, 530 (2012)
4. Baptista, R., Borges, B., Oliveira, A.S.: in prep.
5. Baptista, R., Bortoletto, A.: Eclipse mapping of the flickering sources in the dwarf nova V2051 Ophiuchi. AJ **128**, 411 (2004)
6. Baptista, R., Bortoletto, A.: A two-armed pattern in flickering maps of the nova-like variable UU Aquarii. ApJ **676**, 1240 (2008)
7. Baptista, R., Catalán, M.S.: Changes in the structure of the accretion disc of EX Draconis through the outburst cycle. MNRAS **324**, 599 (2001)
8. Baptista, R., Harlaftis, E.T., Steeghs, D.: Eclipse maps of spiral shocks in the accretion disc of IP Pegasi in outburst. MNRAS **314**, 727 (2000)
9. Baptista, R., Horne, K., Hilditch, R.W., Mason, K.O., Drew, J.E.: Hubble space telescope and R-band eclipse maps of the UX Ursae Majoris accretion disk. ApJ **448**, 395 (1995)
10. Baptista, R., Horne, K., Wade, R.A., et al.: HST spatially resolved spectra of the accretion disc and gas stream of the nova-like variable UX Ursae Majoris. MNRAS **298**, 1079 (1998)
11. Baptista, R., Morales-Rueda, L., Harlaftis, E.T., Marsh, T.R., Steeghs, D.: Tracing the spiral arms in IP Pegasi. A&A **444**, 201 (2005)
12. Baptista, R., Santos, R.F., Faúndez-Abans, M., Bortoletto, A.: A study of the evolution of the accretion disk of V2051 Ophiuchi through two Outburst cycles. AJ **134**, 867 (2007)
13. Baptista, R., Silveira, C., Steiner, J.E., Horne, K.: Spatially resolved spectra of the accretion disc of the nova-like variable UU Aquarii. MNRAS **314**, 713 (2000)
14. Baptista, R., Steiner, J.E.: Accretion disk image reconstructions in V4140 SGR (NSV 12615). A&A **249**, 284 (1991)
15. Baptista, R., Steiner, J.E.: Improving the eclipse mapping method. A&A **277**, 331 (1993)
16. Baptista, R., Steiner, J.E., Horne, K.: Multicolour eclipse studies of UU Aquarii – II. The accretion disc. MNRAS **282**, 99 (1996)
17. Bath, G.T.: Dynamical instabilities and mass exchange in binary systems. MNRAS **171**, 311 (1975)
18. Bennie, P.J., Hilditch, R., Horne, K.: Decomposing interacting binary light curves: the eclipses of the mean light, secular variability and flickering in RW Tri, in Evans, A., Wood, J. (eds.) Cataclysmic Variables and Related Objects. IAU Coll., vol. 158, p. 33. Kluwer, Dordrecht (1996)
19. Bobinger, A., Horne, K., Mantel, K.H., Wolf, S.: Eclipse maps of the dwarf nova IP Peg on the decline from outburst. A&A **327**, 1023 (1997)
20. Bobinger, A.: Genetic eclipse mapping and the advantage of Black Sheep. A&A **357**, 1170 (2000)
21. Bruch, A.: Flickering in cataclysmic variables – its properties and origins. A&A **266**, 237 (1992)
22. Bruch, A.: Studies of the flickering in cataclysmic variables. VI. The location of the flickering light source in HT Cassiopeiae, V2051 Ophiuchi, IP Pegasi and UX Ursae Majoris. A&A **359**, 998 (2000)
23. Bruch, A., Beele, D., Baptista, R.: Multicolor photometry and eclipse mapping of OY Carinae in a superoutburst. A&A **306**, 151 (1996)
24. Cannizzo, J.K.: The accretion disk limit cycle model: toward an understanding of the long-term behavior of SS Cygni. ApJ **419**, 318 (1993)
25. La Dous, C.: Synthetic optical and ultraviolet spectra of stationary accretion disks. A&A **211**, 131 (1989)
26. Feline, W.J., Dhillon, V.S., Marsh, T.R., Watson, C.A., Littlefair, S.P.: ULTRACAM photometry of the eclipsing cataclysmic variables GY Cnc, IR Com and HT Cas. MNRAS **364**, 1158 (2005)
27. Garcia, A., Sodré, L., Jablonski, F.J., Terlevich, R.J.: Optical monitoring of quasars – I. Variability. MNRAS **309**, 803 (1999)
28. Geertsema, G.T., Achterberg, A.: Turbulence in differentially rotating thin disks – a multi-component cascade model. A&A **255**, 427 (1992)

29. Groot, P.J.: Optical Variability in Compact Source, PhD Thesis, University of Amsterdam (1999)
30. Harlaftis, E.T., Baptista, B., Morales-Rueda, L., Marsh, T.R., Steeghs, D.: Spiral shock detection on eclipse maps: simulations and observations. A&A **417**, 1063 (2004)
31. Herbst, W., Shevchenko, K.S.: A photometric catalog of herbig AE/BE stars and discussion of the nature and cause of the variations of UX Orionis stars. AJ **118**, 1043 (1999)
32. Hirose, M., Osaki, Y.: Hydrodynamic simulations of accretion disks in cataclysmic variables – superhump phenomenon in SU UMa stars. PASJ **42**, 135 (1990)
33. Horne, K.: Images of accretion discs. I – The eclipse mapping method. MNRAS **213**, 129 (1985)
34. Horne, K., Marsh, T.R.: Indirect imaging of accretion disks in binaries. In: Mason, K., Charles, P. (eds.) The Physics of Accretion onto Compact Objects, p. 1. Springer, Berlin (1986)
35. Horne, K.: Eclipse mapping of accretion disks: the first decade. In: Wheeler, J.C. (ed.) Accretion Disks in Compact Stellar Systems, p. 117. World Scientific Publ. Co., Singapore (1993)
36. Ioannou, Z., Naylor, T., Welsh, W.F., Catalán, M.S., Worraker, W.J., James, N.D.: The 'outside-in' outburst of HT Cassiopeiae. MNRAS **310**, 398 (1999)
37. Kawaguchi, T., Mineshige, S., Machida, M., et al.: Temporal $1/f^{\alpha}$ fluctuations from fractal magnetic fields in black-hole accretion flow. PASJ **52**, L1 (2000)
38. Knigge, C., Drew, J.E.: Eclipse mapping of the accretion disk wind in the cataclysmic variable UX Ursae Majoris. ApJ **486**, 445 (1997)
39. Knigge, C., Long, K.S., Wade, R.A., et al.: Hubble space telescope eclipse observations of the nova-like cataclysmic variable UX Ursae Majoris. ApJ **499**, 414 (1998)
40. Lasota, J.P.: The disc instability model of dwarf novae and low-mass X-ray binary transients. New Astron. Rev. **45**, 449 (2001)
41. Long, K.S., Wade, R.A., Blair, W.P., Davidsen, A.F., Hubeny, I.: Observations of the bright novalike variable IX Velorum with the Hopkins ultraviolet telescope. ApJ **426**, 704 (1994)
42. Lubow, S.H.: Impact-driven eccentricity in accretion disks. ApJ **432**, 224 (1994)
43. Meyer, F., Meyer-Hofmeister, E.: Vertical structure of accretion disks. A&A **106**, 34 (1982)
44. O'Donoghue, D.: Observational evidence for tidal effects in cataclysmic variable accretion discs. MNRAS **246**, 29 (1990)
45. Robinson, E.L., Wood, J.E., Wade, R.A.: Application of realistic model atmospheres to eclipse maps of accretion disks: the effective temperature and flare of the disk in the dwarf nova Z Chamaeleontis. ApJ **514**, 952 (1999)
46. Rolfe, D., Haswell, C.A., Patterson, J.: Superhumps in V348 Pup. MNRAS **317**, 759 (2000)
47. Rutten, R.G.M., Dhillon, V.S., Horne, K., Kuulkers, E., van Paradijs, J.: Spectrally resolved eclipse maps of the accretion disk in UX Ursae Majoris. Nature **362**, 518 (1993)
48. Rutten, R.G.M., Dhillon, V.K.: Roche tomography: imaging the stars in interacting binaries. A&A **288**, 773 (1994)
49. Rutten, R.G.M., Dhillon, V.S., Horne, K., Kuulkers, E.: Spectral eclipse mapping of the accretion disk in the nova-like variable UX Ursae Majoris. A&A **283**, 441 (1994)
50. Rutten, R.G.M.: 3D eclipse mapping. A&AS **127**, 581 (1998)
51. Rutten, R.G.M., Kuulkers, E., Vogt, N., van Paradijs, J.: OY Carinae revisited – development of the accretion disk during a normal outburst. A&A **265**, 159 (1992)
52. Rutten, R.G.M., van Paradijs, J., Tinbergen, J.: Reconstruction of the accretion disk in six cataclysmic variable stars. A&A **260**, 213 (1992)
53. Saito, R.K., Baptista, R.: Hubble space telescope faint object spectrograph spectral mapping of V2051 Ophiuchi in a low state. AJ **131**, 2185 (2006)
54. Shakura, N.I., Sunyaev, R.A.: Black holes in binary systems. Observational appearance. A&A **24**, 337 (1973)
55. Shu, F.H.: Mass transfer in semi-detached binaries. In: Eggleton, P., Mitton, S., Whelan, J. (eds.) Structure and Evolution of Close Binary Systems, p. 253. D. Reidel Pub. Co., Dordrecht (1976)

56. Skilling, J., Bryan, R.K.: Maximum entropy image reconstruction – general algorithm. MNRAS **211**, 111 (1984)
57. Skilling, J.: Theory of maximum entropy image reconstruction. In: Justice, J.H. (ed.) Maximum Entropy and Bayesian Methods in Applied Statistics, p. 156. Cambridge University Press, Cambridge (1987)
58. Spruit, H.C.: Fast eclipse mapping. A&A **289**, 441 (1994)
59. Steeghs, D., Harlaftis, E.T., Horne, K.: Spiral structure in the accretion disc of the binary IP Pegasi. MNRAS **290**, L28 (1997)
60. Warner, B.: Cataclysmic Variable Stars. Cambridge University Press, Cambridge (1995)
61. Wade, R.A.: A test of synthetic accretion disk spectra using ultraviolet flux distributions of novalike variables. ApJ **335**, 394 (1988)
62. Warner, B., Nather, R.E.: Observations of rapid blue variables – II. U Geminorum. MNRAS **152**, 219 (1971)
63. Warner, B., O'Donoghue, D.: High-speed photometry of Z Chamaleontis during outbursts. MNRAS **233**, 705 (1988)
64. Whitehurst, R.: Numerical simulations of accretion disks. I – Superhumps – a tidal phenomenon of accretion disks. MNRAS **232**, 35 (1988)
65. Wood, J.H.: Eclipse mapping of accretion discs in cataclysmic variables. In: Shafter, A.W. (ed.) Interacting Binary Stars. ASP Conference Series, vol. 56, p. 48. ASP, San Francisco (1994)

Chapter 10
Stokes Imaging: Mapping the Accretion Region(s) in Magnetic Cataclysmic Variables

Stephen B. Potter

10.1 Introduction

Magnetic cataclysmic variables (mCVs) are interacting binaries in which material overflows the Roche lobe of a red-dwarf secondary star and is accreted onto a magnetic white dwarf star (Fig. 10.1). See [11] and [42] for a review of mCVs. A subclass of mCVs, known as polars, have a sufficiently strong white dwarf magnetic field that the system is locked into synchronous rotation and prevents an accretion disc from forming.

The accretion material initially leaves the inner Lagrange point and follows a ballistic trajectory. At some distance from the white dwarf, the stress of the white dwarf magnetic field overwhelms the ram pressure of the ballistic stream and its flow becomes directed by the magnetic-field lines. The magnetically confined supersonic flow eventually becomes subsonic at a shock front at some height above the white dwarf surface.

The shock, which typically has a temperature of $\sim$10–50 keV, heats and ionizes the accreting material. The shocked-heated material cools as it settles onto the white dwarf surface, forming a stratified (in density and temperature) relaxation region. Further details of the standard shock model can be found in the following review by Wu and references therein [48].

The main radiative-transport processes, in the post-shock region, are bremsstrahlung and cyclotron cooling. Bremsstrahlung radiation is emitted in the keV X-ray band, and the post-shock region is optically thin to the bremsstrahlung X-rays. Cyclotron radiation is emitted in the optical and IR bands, and it is strongly polarised (Fig. 10.2). The post-shock region is opaque to the optical/IR cyclotron emission at low frequencies but is transparent in the blue/UV frequencies.

S.B. Potter (✉)
South African Astronomical Observatory, Cape Town, South Africa
e-mail: sbp@saao.ac.za

H.M.J. Boffin et al. (eds.), *Astronomy at High Angular Resolution*, Astrophysics and Space Science Library 439, DOI 10.1007/978-3-319-39739-9_10

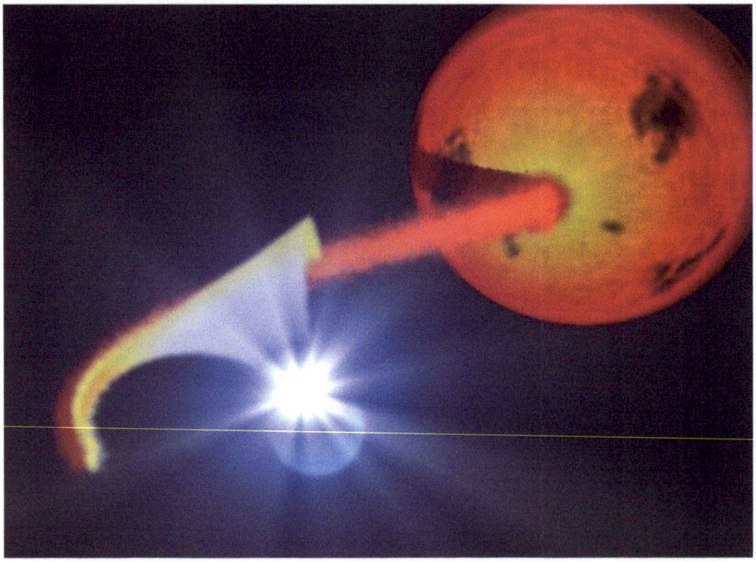

Fig. 10.1 A schematic of a polar, a sub-class of the magnetic Cataclysmic Variables. The *red dwarf* secondary star, the ballistic stream, the magnetic accretion curtain and the bright accretion region on the surface of the *white dwarf* are shown here

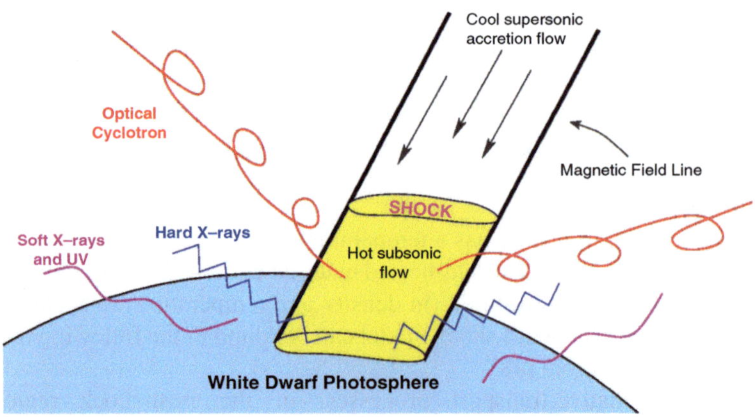

Fig. 10.2 A schematic of the accretion region on the surface of the *white dwarf* highlighting the emission mechanisms

10.2 Polarisation Modelling

Over the course of a binary orbit (typically a few hours) the polarised optical emission is observed to be highly variable. This is because the cyclotron emission from the shock is highly dependent on viewing angle. For example, when our viewing angle is such that we are looking down towards the top of the shock, effectively parallel to the magentic field lines, an increase in the amount of circularly polarised flux is observed. When the viewing angle changes to being perpendicular to the magnetic field lines, effectively viewing the side of the shock, an increase in linear polarised flux is observed. The actual morphology of the orbitally phase-resolved polarised variations will depend on the location, size and shape of the localised cyclotron emission region(s) on the surface of the white dwarf.

With appropriate modelling of the polarised cyclotron emission, the location, size and shape of the localised cyclotron emission region(s) can be reconstructed, (e.g., [2, 6–8, 10, 15, 25–28, 31, 34, 40, 44, 45, 53]). At the core of these studies are the cyclotron-opacity and radiative-transfer calculations. Two formulations are commonly used in the calculations: the decomposed normal-mode formulation (e.g., [7]), which is valid in the limit of large Faraday rotation; and the more general 4-Stoke formulation (e.g., [25, 26]).

Meggitt & Wickramashinghe provided tables of the results of their calculations assuming that the heated shock region can be represented as a simple hemisphere of constant temperature and density (Fig. 10.3 and [25, 26]). The tables provide the polarised Stokes parameters as a function of viewing angle from which one could then produce model phase-resolved polarised light curves. They calculated multiple tables, each representing shocks of different magnetic field strengths, temperatures, densities and a plasma parameter Λ (representing the electron density and characteristic size of the emission region).

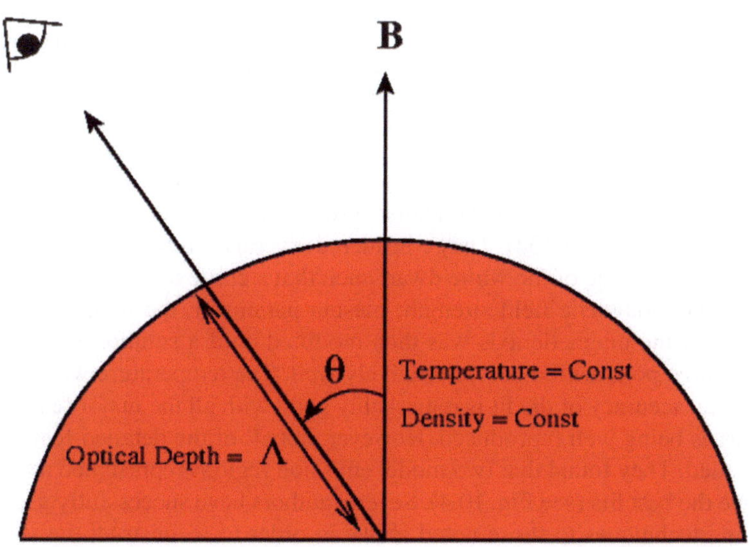

Fig. 10.3 A schematic of a *hemisphere* used in performing the radiative transfer calculations

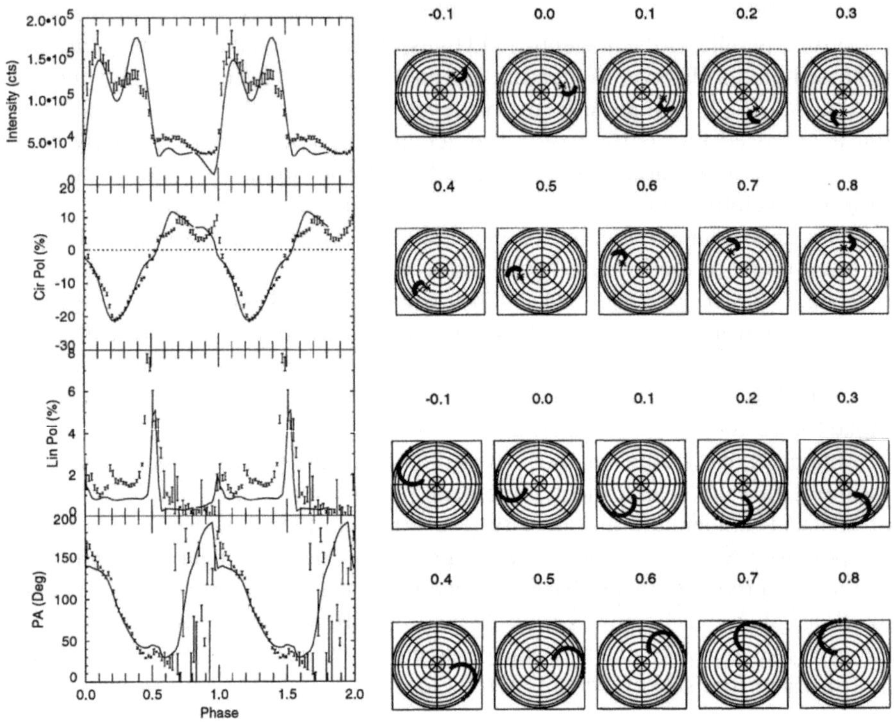

Fig. 10.4 *Left*: Orbitally phase-resolved optical photo-polarimetric observations of the polar REJ1844-74, with model over-plotted. Intensity, circular and linear polarisation and position angle can be seen from top to bottom. *Right*: Views of the upper and lower accretion arcs used to produce the model polarisation (Reproduced with permission from Ramsay et al. [34])

Cyclotron emission regions of arbitrary shapes, sizes and locations could then be constructed by using several such hemispheres. Figure 10.4 shows such an example taken from Ramsay et al. [34]. They considered several accretion regions of differing extents on the surface of the white dwarf such that a crude fit to the light curve was obtained. The magnetic field strength, plasma parameter, the inclination, and the colatitude of the magnetic axis was then modified until a solution was found that fitted all three polarisation curves relatively well. The temperature was taken to be 10 keV. The accuracy of the fit is remarkably good with all the major features in the light curves being well reproduced. However, not all of the detailed features could be modelled. They found that two model emission regions represented as long, thin arcs gave the best fits (see Fig. 10.4). Several authors have successfully used similar modelling techniques to those noted above in order to reconstruct the shape, size and location of the cyclotron emission regions and have generally found that long thin arcs best fit the observations.

10.3 Stokes Imaging

Despite the success of modelling the cyclotron emission regions, it was clear that the approach was largely by trial and error until the model gave a good fit to the observations. A more objective technique was required in order to take the modelling procedure to the next step and to eliminate any initial biases the astronomer may have.

Stokes imaging as described by Potter, Hakala & Cropper is such a technique [31]. The trial and error approach has been replaced with a Genetic Algorithm (GA) which optimises the model to the observations.

The GA works by first generating a set of random solutions. Each solution consists of a large number of hemispheres placed randomly over the surface of the white dwarf. The "fitness" of each solution is then calculated using:

$$F(p) = \chi^2 + \lambda \sum_i ||\nabla p_i||^2;$$

where χ^2 is the χ^2-fit to the observations, λ is a Lagrangian multiplier, p_i is a single emission element on the surface of the white dwarf, $||\nabla p_i||$ is the mean gradient of the number of emission points at point i.

Minimising this with a suitable choice of the Lagrangian multiplier, λ, will produce the maximum entropy solution of the distribution of emission points. The GA then breeds all the solutions by using a type of natural selection procedure; the "fittest" solutions have a higher probability of being selected for breeding. Eventually the improvement in fitness of the GA solutions starts to level out and a more analytical approach is used to improve the fit further. The technique is best demonstrated through the example test case displayed in Fig. 10.5. Figure 10.5a shows a model white dwarf with two arc-like emission regions. The corresponding model light curves are shown in Fig. 10.5b. After several generations the solutions start to converge. One such solution is displayed in Fig. 10.5c which has clearly started to resemble the input image. At this stage the GA approach becomes inefficient at progressing further. A more analytical technique now takes over which effectively "cleans up" the image to give the final solution in Fig. 10.5d. The model prediction is overplotted on the input light curve in Fig. 10.5b. See [31] for details and more test cases.

Stokes imaging has been applied to several mCVs, each with differing morphologies and quality of observations. These include CP Tuc [35], ST Lmi [28], V347 Pav [30], RXJ2115 [36], QS Tel [37], V834 Cen [32], HU Aqr [16] and CTCVJ1928-5001 [29].

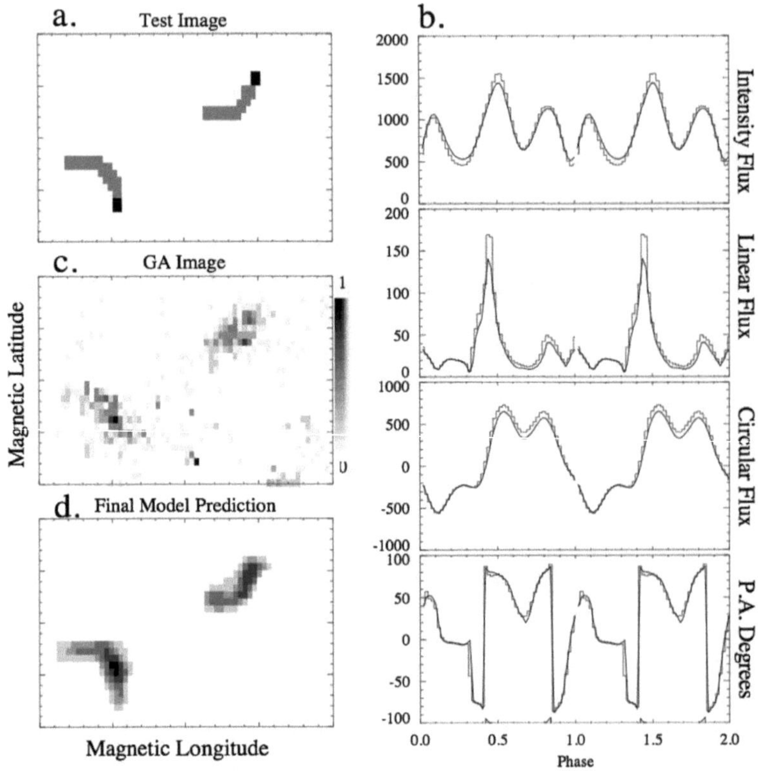

Fig. 10.5 Model white dwarf with two arc-like emission regions: (**a**) the input test image; (**b**) the input test data and final model solution; (**c**) the GA optimised image; (**d**) the final optimised image

10.4 Photo-polarimetric Observations of the Eclipsing Polar CTCV J1928-5001

Figure 10.6 shows polarimetric observations of the eclipsing polar CTCV J1928-5001 taken in August 2003 on the 1.9 m of the South African Astronomical Observatory using the UCT polarimeter [10]. These results were extracted from work published in [29]. The photometry and polarimetry are modulated on a period of ~101 min. Circular polarisation variations are seen from approximately −8 to 12 % in the unfiltered observations. Two linear polarised pulses are detected at orbital phases coinciding with the reversals in the circular polarisation.

Figure 10.6 shows the model fit from Stokes imaging that has reproduced the main polarisation morphology and predicted two emission regions. The predicted upper emission region is also displayed in Fig. 10.6 for two different assumed system inclinations. Simple, single particle ballistic and magnetic trajectories were also calculated and over-plotted. As can be seen from Fig. 10.6, the magnetic footprints were found to have the best overlap with the emission region for

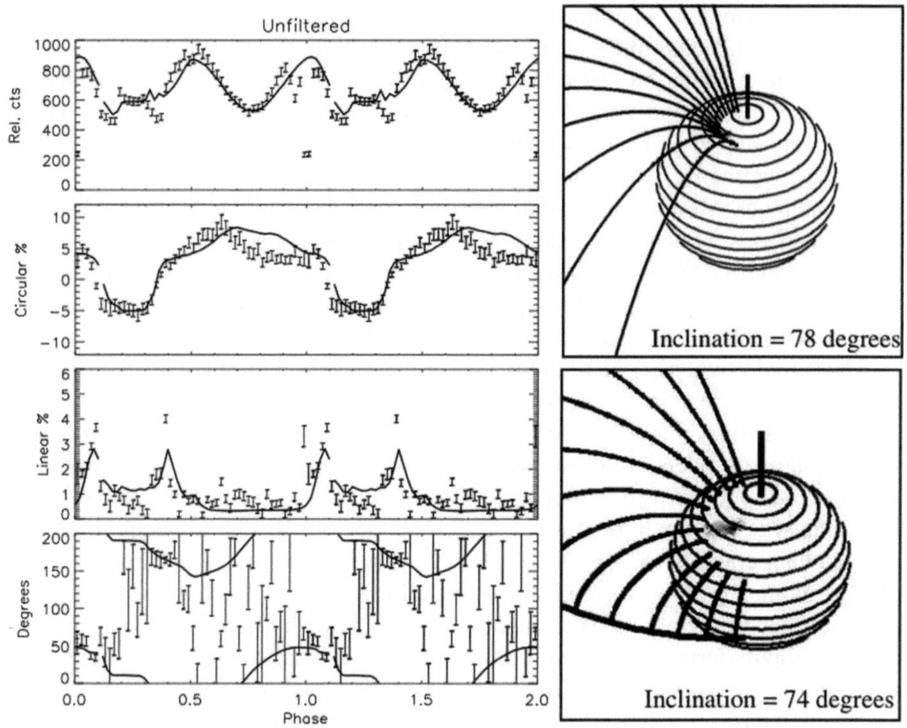

Fig. 10.6 *Left*: folded, unfiltered photo-polarimetric observations and model fit of the eclipsing polar CTCVJ1928-5001. *Right*: the location of the accretion region on the surface of the white dwarf for different model solutions. Magnetic trajectories also shown (Reproduced with permission from Potter, Augesteijn & Tappert [29])

inclinations $\sim 78°$. Reassuringly such an inclination is in agreement with the implied mass ratios and primary masses for CVs with similar orbital periods [41] and is consistent with the range of inclinations estimated from using the eclipse length [17]. The location of the lower accretion region (not shown) does not overlap with the magnetic footprints so well. The authors concluded that this is an indication that the magnetic field of the white dwarf is not a simple dipole but is at least an offset dipole field (see [29] for further details).

10.5 Future Work: Stratified Accretion Shocks

The work discussed so far relied on simple emission regions with constant density, temperature and magnetic field. The geometry was assumed to be either an infinitely small hemisphere with finite opacity (e.g., [25, 26]) or a semi-infinite slab (e.g., [2, 3, 7]). However, it was first suggested by Ferrario, Bailey & Wickramasinghe

[13], that the single temperature models are insufficient to reproduce the cyclotron spectrum and indeed multi-waveband polarimetric light curves.

Some later calculations included the spatial variations in the temperature, density and magnetic field and considered more complicated geometries, e.g., cylinders, for the emission region. These inhomogeneous models (e.g., [45, 49–51]) were generally parametric, with the temperature and density variations treated in an ad hoc manner. More realistic settings were considered in the next generation of inhomogeneous models. For example, 3-D emission regions confined by a dipole magnetic field [4–6], or ridge-like emission regions [53]. Eventually, models were used in which the density and temperature structures of the emission regions were determined by hydrodynamic calculations [14, 46, 53, 54].

These hydrodynamic calculations, performed in parallel to the radiative-transfer calculations, were carried out to model the temperature and density structures of the emission regions properly. The models evolved from one-temperature hydrodynamic models with bremsstrahlung (or a single) cooling [1, 9, 18]; one-temperature models with multiple cooling processes (e.g., [12, 47, 52]); to the full-fledged two-temperature models which include effects due to unequal electron and ion temperature [14, 20, 21, 38, 39, 46]. Other effects, such as variation of the gravity along the flow, were also considered (e.g., [12, 19]). Cropper et al. [12] demonstrated that gravity has substantial effects on the hydrodynamics of the flow. When a gravity term is included in the hydrodynamic calculations, it increases the thickness of the shock-heated region but reduces the shock temperature. The X-rays from the resulting post-shock region are therefore softer than when the gravity term is omitted. Thus, if gravity effects are neglected, white dwarf masses will be overestimated when using X-ray spectra to determine the white dwarf masses of mCVs [33].

Figure 10.7 shows an example of hydrodynamic calculations for three different white dwarf masses. In each case (white dwarf masses of 0.5, 0.85 and 1.2 $M_{\odot}$) the vertical temperature structure is plotted as a function of height for different accretion rates. As can be seen, the temperature-height relation is strongly dependent on the white dwarf mass. Furthermore, there is also a strong dependence on magnetic field strength (not shown).

We performed the radiative transfer calculations in order to derive the intensities and polarisations from structured emission regions, constructed using the hydrodynamic model. The aim being to produce more realistic grids of spectra and polarisations for modelling the polarised optical/IR radiation from mCVs such as Stokes imaging.

We considered ray-tracing calculations along the lines-of-sight starting from designated emitting points to an observer at infinity. The emission region is divided into strata, each with homogeneous temperatures and densities. The temperature and density of each stratum are obtained by discretising the temperature and density structures of the emission region obtained from the hydrodynamic calculations.

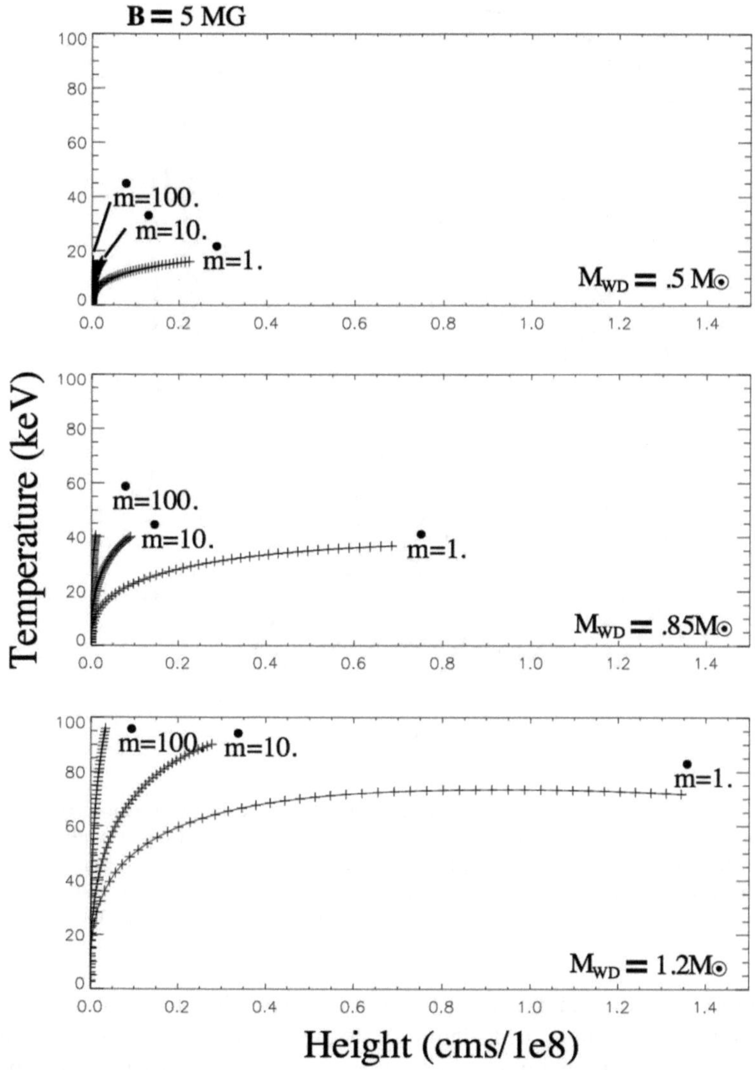

Fig. 10.7 Temperature profiles in the post-shock regions for white dwarf mass of 0.5, 0.85 and 1.2 $M_{\odot}$. In each panel, three cases with the specific accretion rate of 1, 10 and 100 g cm^{-2} s^{-1} are shown. The magnetic field strength of the white dwarf is fixed at 5 MG

We lay a rectangular grid at the bottom of each stratum and consider rays starting from emitting points located on the grid (Fig. 10.8). Each ray is independent but its intensity and polarisations are modified by the line-of-sight materials in the strata above.

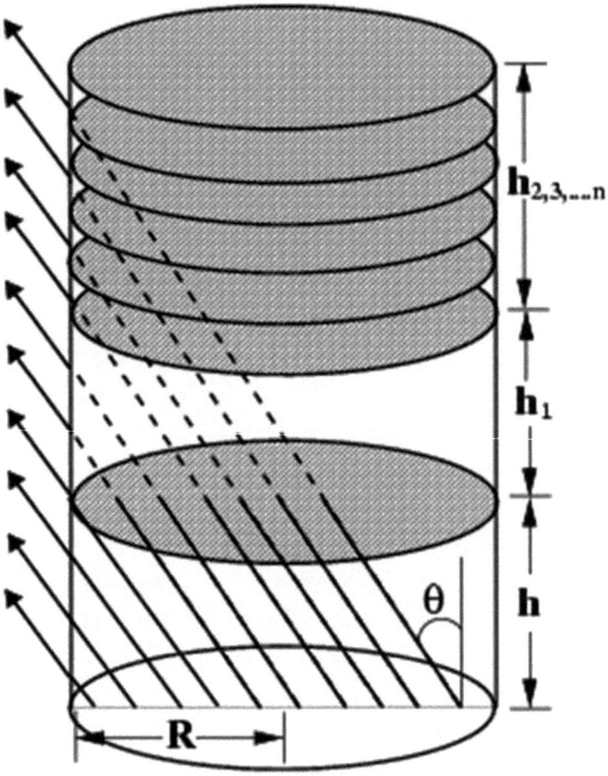

Fig. 10.8 An illustration of the ray-tracing algorithm used in the polarisation radiative-transfer calculations. The *cylindrical* emission region is divided into strata of equal thickness and cross-section radius. The cylinder is embedded in a magnetic field parallel to the symmetry axis. Shown are the light paths of emission from one of the strata for a particular viewing angle. The Stokes parameters are calculated for each ray and the resultant Stokes parameters are the sum of all rays

Figure 10.9 shows an extract of the grid corresponding to the structured shocks shown in Fig. 10.7. These grids demonstrate the future potential of Stokes imaging to derive fundamental parameters of mCVs, such as white dwarf mass, magnetic field strength and the temperature and density profile of the shock(s). For example, simply comparing the intensity grids only (top row, from left to right) one can see that there is not only a peak intensity wavelength dependence on the white dwarf mass (decreasing from $\sim$0.61 to $\sim$0.38 microns for 0.5 to 1.2 $M_{\odot}$) but also a significant spectral morphology change, i.e., the lower mass white dwarf shows significant cyclotron humps in the spectra which are not so apparent in the higher mass white dwarf. The circular and linear polarisation show significantly more morphological dependence on white dwarf mass as a function of viewing angle. Indeed, Fig. 10.9 is only an extract from the grid of calculations. The other dimensions of the grid include magnetic field strength, specific accretion rate and radius of the accretion column. Combined with orbitally phase-resolved spectro-

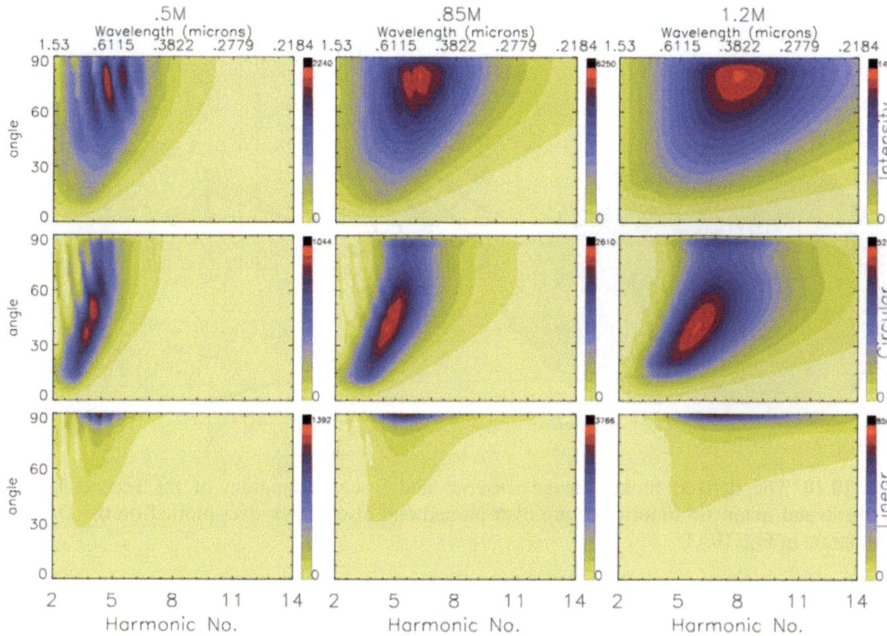

Fig. 10.9 The cyclotron intensity and polarisations from a structured emission region for white dwarf masses of 0.5, 0.85 and 1.2 $M_{\odot}$ (left to right columns respectively). The accretion rate is $\dot{m} = 1.0\,\mathrm{g\,cm^{-2}\,s^{-1}}$, the magnetic field, $B = 35\,\mathrm{MG}$, and the radius of the accretion column, $R = 10^4\,\mathrm{cm}$

polarimetry, we anticipate these grids will be a powerful diagnostic for deriving the fundamental parameters of the white dwarfs in mCVs.

10.6 Future Work: Multi-tomography

The true benefit of Stokes imaging comes when combined with other tomographic techniques (e.g., [32]). Stokes imaging was combined with the analysis of spectroscopic observations using the Doppler tomography and Roche tomography techniques of Marsh & Horne and Watson & Dhillon [23, 43]. Up to then, all three techniques had been used separately to investigate the geometry and accretion dynamics in cataclysmic variables. For the first time, Potter et al. [32] applied all three techniques to simultaneous polarimetric and spectroscopic observations for a single system (V834 Cen). This allowed them to compare and test each of the techniques against each other and hence to derive a better understanding of the geometry, dynamics and system parameters. Figures 10.10 and 10.11 show the results.

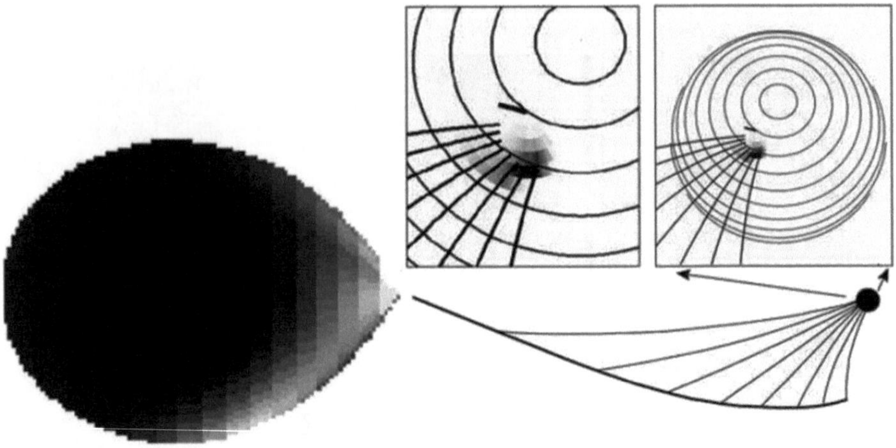

Fig. 10.10 The derived Stokes image (*inserts*) and Roche tomogram of the secondary star. Ballistic and magnetic trajectories are over-plotted and also shown over-plotted on the Doppler tomogram in Fig. 10.11

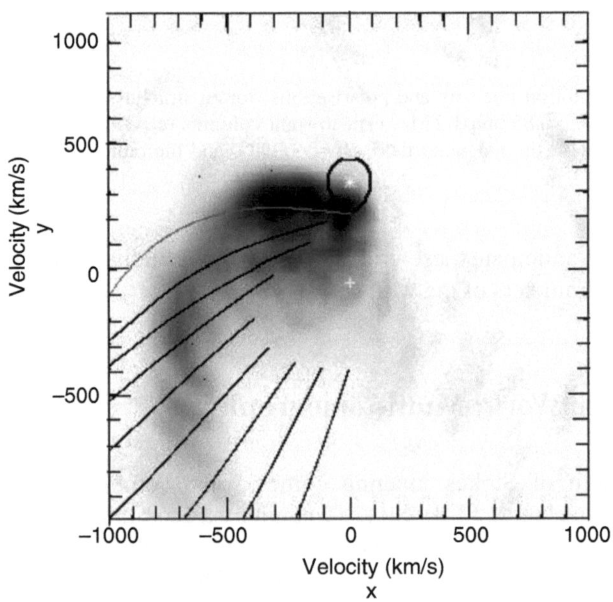

Fig. 10.11 Doppler tomogram of the HeII emission line form V834 Cen. The same ballistic and magnetic trajectories have been over-plotted as in Fig. 10.10

Stokes imaging predicted a single emission region, visible for the whole orbit, to be responsible for the observed polarised variations. The location of the region on the white dwarf was found to be consistent with EUV observations [24]. Both analyses placed the main emission region approximately more than 40° and up to

70° in azimuth from the line of centres of the two stars. The Doppler tomographic analysis showed several components associated with the secondary star, the ballistic stream and parts of the magnetically confined stream. Magnetic Doppler trajectories were also calculated and shown to be generally coincident with the magnetic signatures found in the Doppler maps and, furthermore, the footprints of the magnetic field lines on the surface of the white dwarf were shown to be coincident with the prediction for the location of the cyclotron emission region from Stokes imaging. Roche tomography of the narrow component seen in the He II 4686 Å emission line revealed two asymmetries in the emission distribution on the Roche surface of the secondary. Namely, there was generally more emission on the inner surface and also on the leading face of the secondary. They argued that this was consistent with irradiation of the secondary and not only supported by the Doppler tomogram, but also by the Stokes imaging results, which showed that most of the accreting material comes along field lines that intersect the later parts of the ballistic stream.

We next plan to exploit recent advances made in Doppler tomography, namely the inside-out variant as described by Kotze, Potter & McBride [22]. An example of this technique is shown in Fig. 10.12, which displays a He II Doppler tomogram of the polar UZ For together with its inside-out Doppler tomogram.

The inside-out technique effectively reverses the velocity axis to create an inside-out Doppler coordinate framework with the intent to expose/enhance emission details that are overly compacted/washed-out in the standard Doppler framework. The technique appears to be especially effective at separating the velocity components in polars. Figure 10.12 clearly shows how emission from the secondary, the

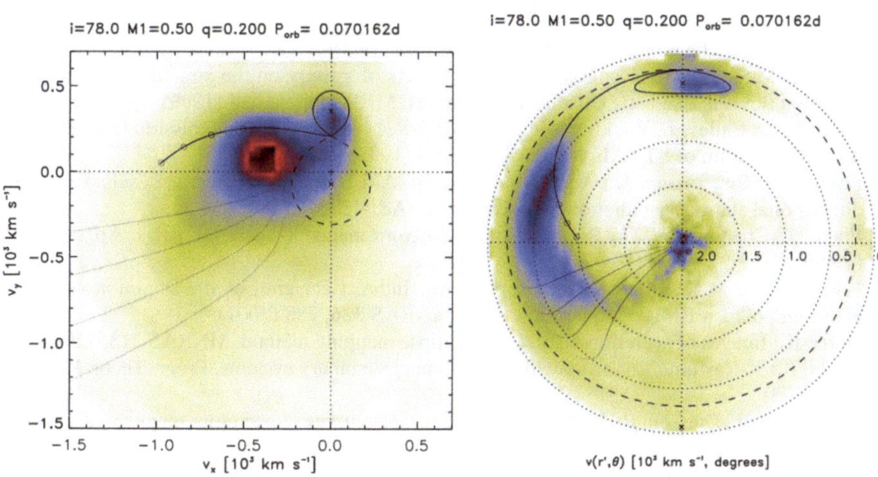

Fig. 10.12 *Left*: A Doppler tomogram of the HeII emission from the polar UZ For. *Right*: An inside-out Doppler tomogram of the same data set. In both cases the same model ballistic and magnetic accretion curtains are over-plotted as well as the roche lobes of the secondary and primary stars

ballistic and magnetic streams are easier to distinguish from each other compared to the normal tomogram. In particular the high-velocity components of the emission line, which is washed out towards the lower left quadrant of the normal Doppler map, becomes more compact and enhanced in the inside-out framework. Combining this technique with the other tomographic techniques, described above, will be our next step.

References

1. Aizu, K.: X-Ray emission region of a white dwarf with accretion. Progr. Theor. Phys. **49**, 1184 (1973)
2. Barrett, P.E., Chanmugam, G.: Polarized radiation from hot plasmas and applications to AM Herculis binaries. II – effect of collisions and Thomson scattering. ApJ **278**, 298 (1984)
3. Barrett, P.E., Chanmugam, G.: Cyclotron lines in accreting magnetic white dwarfs with an application to VV Puppis. ApJ **298**, 743 (1985)
4. Canalle, J.B.G., Opher, R.: The cyclotron spectrum of VV Puppis. A&A **189**, 325 (1988)
5. Canalle, J.B.G., Opher, R.: Cyclotron spectrum from a dipole magnetic field accretion column. A&A **219**, 334 (1989)
6. Canalle, J.B.G., Opher, R.: The accretion column of DQ Herculis binaries. A&A **251**, 474 (1991)
7. Chanmugam, G., Dulk, G.A.: Polarized radiation from hot plasmas and applications to AM Herculis binaries. ApJ **244**, 569 (1981)
8. Chanmugam, G., Frank, J., King, A.R., et al.: The magnetic field of the intermediate polar BG Canis Minoris. ApJ **350**, L13 (1990)
9. Chevalier, R.A., Imamura, J.N.: Linear analysis of an oscillatory instability of radiative shock waves. ApJ **261**, 543 (1982)
10. Cropper, M.S.: Simultaneous linear and circular polarimetry of EF ERI. MNRAS **212**, 709 (1985)
11. Cropper, M.S.: The polars. Space Sci. Rev. **54**, 195 (1990)
12. Cropper, M.S., Wu, K., Ramsay, G., et al.: Effects of gravity on the structure of post-shock accretion flows in magnetic cataclysmic variables. MNRAS **306**, 684 (1999)
13. Ferrario, L., Bailey, J., Wickramasinghe, D.T.: Detection of cyclotron emission features in the infrared spectrum of ST LMi. MNRAS **262**, 285 (1993)
14. Fischer, A., Beuermann, K.: Accretion physics of AM Herculis binaries. I. Results from one-dimensional stationary radiation hydrodynamics. A&A **373**, 211 (2001)
15. Frank, J., Chanmugam, G.: Polarized radiation from magnetic accretion rings. ApJ **365**, 660 (1990)
16. Harrop-Allin, M.K., Potter, S.B., Cropper, M.: Indirect imaging of the accretion stream in eclipsing polars – III. HU Aquarii low state. MNRAS **326**, 788 (2001)
17. Horne K.: Images of accretion discs. I – the eclipse mapping method. MNRAS **213**, 129 (1985)
18. Hoshi, R.: X-Ray emission from white dwarfs in close binary systems. Progr. Theor. Phys. **49**, 776 (1973)
19. Imamura, J.N., Durisen, R.H.: X-ray spectra and light curves of accreting magnetic degenerate dwarfs. ApJ **268**, 291 (1983)
20. Imamura, J.N., Aboasha, A., Wolff, M.T., et al.: The stability properties of two-temperature white dwarf radiative shock waves. ApJ **458**, 327 (1996)
21. Imamura, J.N., Durisen, R.H., Lamb, D.Q., et al.: X-ray and ultraviolet radiation from accreting white dwarfs. IV – two-temperature treatment with electron thermal conduction. ApJ **378**, 665 (1987)

22. Kotze, E., Potter, S., McBride, V.: Exploring inside-out Doppler tomography: non-magnetic cataclysmic variables. A&A **579**, A77 (2015)
23. Marsh, T.R., Horne, K.: Images of accretion discs. II – doppler tomography. MNRAS **235**, 269 (1988)
24. Mauche, C.W.: Extreme ultraviolet explorer phase-resolved spectroscopy of V834 Centauri. ApJ **578**, 439 (2002)
25. Meggitt, S.M.A., Wickramasinghe, D.T.: The polarization properties of magnetic accretion columns. MNRAS **198**, 71 (1982)
26. Meggitt, S.M.A., Wickramasinghe, D.T.: An analysis of the cyclotron spectrum and polarization properties of VV puppis. MNRAS **198**, 975 (1982)
27. Piirola, V., Coyne, G.V., Takalo, S.J.L., et al.: UBVRI polarimetry of AM Herculis-type binaries. 5: the asynchronous (?) polar BY Camelopardalis (H0538+608). A&A **283**, 163 (1994)
28. Potter, S.B.: Stokes imaging of the AM Her system ST LMi. MNRAS **314**, 672 (2000)
29. Potter, S.B., Augusteijn, T., Tappert, C.: Photopolarimetric observations of the new eclipsing polar CTCVJ1928 – 5001. MNRAS **364**, 565 (2005)
30. Potter, S.B., Cropper, M., Hakala, P.J.: Stokes imaging of the accretion region in magnetic cataclysmic variables – II. V347 Pav. MNRAS **315**, 423 (2000)
31. Potter, S.B., Hakala, P.J., Cropper, M.: 'Stokes imaging' of the accretion region in magnetic cataclysmic variables – I. conception and realization. MNRAS **297**, 1261 (1998)
32. Potter, S.B., Romero-Colmenero, E., Watson, C., et al.: Stokes imaging, doppler mapping and Roche tomography of the AM Herculis system V834 Cen. MNRAS **348**, 316 (2004)
33. Ramsay, G.: Determining the mass of the accreting white dwarf in magnetic cataclysmic variables using RXTE data. MNRAS **314**, 403 (2000)
34. Ramsay, G., Cropper, M., Wu, K., et al.: Optical polarization and X-ray data on the AM HER star RE J1844-74. MNRAS **282**, 726 (1996)
35. Ramsay, G., Potter, S.B., Buckley, D.A.H., et al.: Simultaneous optical polarimetry and X-ray observations of the magnetic CV CP TUC (AX J2315-592). MNRAS **306**, 809 (1999)
36. Ramsay, G., Potter, S., Cropper, M., et al.: Simultaneous optical polarimetry and X-ray data of the near-synchronous polar RX J2115-5840. MNRAS **316**, 225 (2000)
37. Romero-Colmenero, E., Potter, S.B., Buckley, D.A.H.: An eclipsing geometry for the polar QS Tel. ASPC **330**, 429, (2005)
38. Saxton, C.J., Wu, K.: Perturbative analysis of two-temperature radiative shocks with multiple cooling processes. MNRAS **310**, 677 (1999)
39. Saxton, C.J., Wu, K.: Stability analyses of two-temperature radiative shocks: formulation, eigenfunctions, luminosity response and boundary conditions. MNRAS **324**, 659 (2001)
40. Schwope, A.D., Beuermann, K., Jordan, S.: One-pole and two-pole X-ray emission in the AM Herculis binary BL HYI. A&A **223**, 179 (1989)
41. Smith, D.A., Dhillon, V.S.: The secondary stars in cataclysmic variables and low-mass X-ray binaries. MNRAS **301**, 767 (1998)
42. Warner, B.: Cataclysmic Variable Stars. Cambridge Astrophysics Series 28. Cambridge University Press, Cambridge (1995)
43. Watson, C.A., Dhillon, V.S.: Roche tomography of cataclysmic variables – I. artefacts and techniques. MNRAS **326**, 67 (2001)
44. Wickramasinghe, D.T., Meggitt, S.M.A.: The polarization properties of magnetic accretion columns. III – a grid of uniform temperature and shock front models. MNRAS **214**, 605 (1985)
45. Wickramasinghe, D.T., Ferrario, L.: Cyclotron emission from inhomogeneous shocks in AM Herculis-type systems. ApJ **334**, 412 (1988)
46. Woelk, U., Beuermann, K.: Stationary radiation hydrodynamics of accreting magnetic white dwarfs. A&A **306**, 232 (1996)
47. Wu, K.: Structure of inhomogeneous accretion shocks. Proc. Astron. Soc. Aust. **11**, 61 (1994)
48. Wu K.: Accretion onto magnetic white dwarfs. Space Sci. Rev **93**, 611 (2000)
49. Wu, K., Chanmugam, G.: Cyclotron spectra from inhomogeneous accretion columns. ApJ **331**, 861 (1988)

50. Wu, K., Chanmugam, G.: Cyclotron spectra from inhomogeneous accretion columns. II – polarization. ApJ **344**, 889 (1989)
51. Wu, K., Chanmugam, G.: Polarized cyclotron emission from three-dimensional accretion shocks. ApJ **354**, 625 (1990)
52. Wu, K., Chanmugam, G., Shaviv, G.: Structure of steady state accretion shocks with several cooling functions: closed integral-form solution. ApJ **426**, 664 (1994)
53. Wu, K., Wickramasinghe, D.T.: Three-dimensional structured shocks in Am-Herculis type systems – part one – the angular dependence of intensity and polarization from axisymmetric shocks. MNRAS **246**, 686 (1990)
54. Wu, K., Wickramasinghe, D.T.: Three-dimensional structured shocks in AM Herculis-type systems. II – cyclotron emission from ridge-shaped emission regions. MNRAS **256**, 329 (1992)

Chapter 11
Doppler Tomography

Thomas R. Marsh and Axel D. Schwope

11.1 Introduction

The spectra of binary stars continually vary as the two stars orbit, but underlying this are two invariant spectra which simply shift in wavelength according to the changing radial velocities of their respective stars. In this case we know, or at least we can greatly restrict, the possible ways in which the radial velocities vary, and from a large set of observed spectra it becomes possible to deduce the individual stellar spectra, as is accomplished with the technique known as "spectral disentangling" ([46]; see Chap. 7). A different but related situation arises in accreting binary stars. Again the spectra continually change, and again, to a first approximation there is an underlying fixed pattern. However, in this case it is not so much the spectrum that we are interested in (it is usual to focus on individual atomic lines with presumed near delta-function intrinsic profiles), but the distribution of velocities of emission sites within the binary. It was this problem in the context of the emission line profiles from cataclysmic variable stars that led to the development of the method of Doppler tomography [30]. The complex variations of these stars have long been recognised, a good example being the star WZ Sge which shows strong line emission from the impact of the mass transfer stream with the accretion disc which traces an "S-wave" in the spectra trailed across photographic plates taken by Kraft et al. in the early 1960s [26]. Since its application to cataclysmic variable stars, Doppler tomography

T.R. Marsh (✉)
Department of Physics, University of Warwick, Coventry CV4 7AL, UK
e-mail: T.R.Marsh@warwick.ac.uk

A.D. Schwope
Leibniz-Institute for Astrophysics Potsdam (AIP), An der Sternwarte 16, 14482 Potsdam, Germany
e-mail: ASchwope@aip.de

H.M.J. Boffin et al. (eds.), *Astronomy at High Angular Resolution*, Astrophysics and Space Science Library 439, DOI 10.1007/978-3-319-39739-9_11

has been applied to many other types of stars, but the same principle of a fixed rotating structure links them all.

In this chapter we describe the method, and recent variants of it, and then discuss significant results that have come from its application.

11.2 Principles of Doppler Tomography

In this section we present the principles underlying Doppler tomography. The reader is also referred to past reviews containing a pedagogical element for different perspectives on this topic which can be confusing on first encounter [9, 28, 29]. We begin by defining the coordinate system generally used.

11.2.1 Coordinates

In Doppler tomography, we usually work with images as a function of velocity coordinates, which introduces some subtleties as to the precise coordinate system being used. For instance, while one is used to spatial images of binary stars in "the rotating frame", the velocity space coordinate frame used in Doppler images is *not* its direct equivalent. If it were, all emission from either star would be placed at zero velocity since the stars are fixed in the rotating frame. Therefore this section is devoted to a detailed explanation of the coordinates used to present Doppler maps and the reasoning behind the choices made.

For some target (usually an accreting binary star) under consideration, assume that there exists a uniformly rotating frame of reference F' in which the target's brightness distribution remains constant. That is, the brightness at every point in this reference frame is unchanging, even if there is movement of material within it. Ideally we would like to know the distribution of emission sources in this frame, but, unfortunately the systems under study are far too small to be spatially-resolved and we have to fall back on more indirect methods. For binary stars in circular orbits, F' rotates with the two stars, and we will refer to it as "the" rotating frame, although other effects such as precession or spin of a magnetic accretor could serve to define it differently in some cases. Each point in F' has an associated position (x', y', z') and velocity (v'_x, v'_y, v'_z), although note that material with very different velocities can exist at closely-spaced spatial positions, and there is not necessarily a simple translation between position and velocity in this or any other frame. Without loss of generality, we will take the rotation axis to be parallel to the z'-axis, the zero-point of time and phase to be the time when the x-axis points as close as it can to Earth (so that Earth at this time lies in the positive quadrant of the x'–z' plane), and the y'-axis to be defined in accordance with a right-handed set of axes, i.e. $\hat{\mathbf{y}}' = \hat{\mathbf{z}}' \wedge \hat{\mathbf{x}}'$. In the case of binary stars we will also define the origin to be at the centre of one of the two stars and the x-axis to point towards its companion, usually from the accreting star to

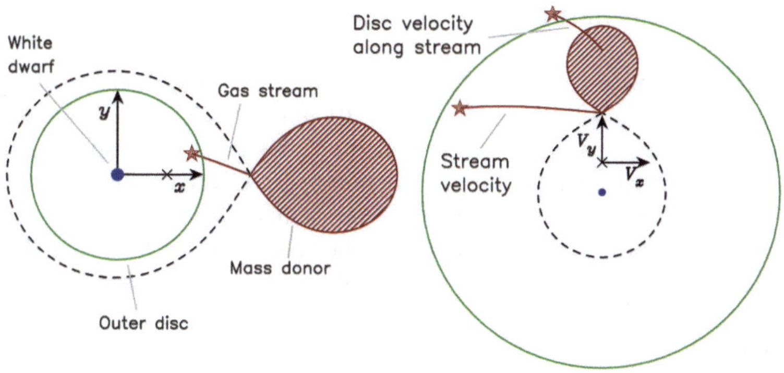

Fig. 11.1 *Left:* a schematic representation of an accreting binary in spatial coordinates, showing the definition of the x and y axes described in the text. The binary rotates anti-clockwise in this representation. *Right:* the equivalent in terms of the inertial V_x, V_y coordinates discussed in the text. Any features in "solid body" rotation, such as the two stars and the Roche lobes, preserve their shapes but are rotated 90° anti-clockwise. The mass transfer stream is plotted in two ways, (**a**) the actual velocity of the stream (lower track), and (**b**) the Keplerian velocity of the disc along the track of the stream (upper). The solid circle marking the outer edge of the disc in the spatial representation on the left transforms to the outermost circle on the right, an inversion of radial ordering characteristic of velocity space Doppler maps. The rest of the disc lies even further out than the outermost circle. The binary rotates around the centre of mass, indicated by a 'X' on the x-axis of the left-hand figure, which is therefore the origin of the velocity space plot on the right. If the accretor spins faster than the binary, its emission would continue to be centred at the point indicated on the right-hand plot, but it would be more broadly distributed

the mass donor (see Fig. 11.1 for a visual representation). This means that the centre of the accreting star (star 1) will be located at $(x', y', z') = (0, 0, 0)$, the centre of the donor star (star 2) is at $(a, 0, 0)$ where a is the binary separation, while the centre of mass is at $\mathbf{C} = (\mu a, 0, 0)$, where $\mu = M_2/(M_1 + M_2)$, with M_1 and M_2 the two stellar masses. By definition in this case, the two stars are fixed within the rotating frame and have velocity $(0, 0, 0)$, but we view the system from the (approximately) inertial frame of Earth, and need to transform the velocity to see what we observe. It is simplest to translate to the Earth's frame via an intermediate inertial frame F defined as the inertial frame that is instantaneously aligned with the rotating frame F' at any particular time. In other words, at any given orbital phase F is defined by the orientation of the F' axes. This leads to the following transformations between the position $\mathbf{r}'$ and velocity $\mathbf{v}'$ in F to the position $\mathbf{r}$ and velocity $\mathbf{v}$ in F:

$$\mathbf{r} = \mathbf{r}', \tag{11.1}$$

$$\mathbf{v} = \mathbf{v}' + \boldsymbol{\Omega} \wedge \left(\mathbf{r}' - \mathbf{C}\right), \tag{11.2}$$

where $\boldsymbol{\Omega} = \Omega\,\hat{\mathbf{z}}$ is the angular velocity characterising the rotating frame. By assumption, a given point in the rotating frame F' has a fixed value of $\mathbf{r}'$ and $\mathbf{v}'$. Since Ω is also constant then so too are $\mathbf{r}$ and $\mathbf{v}$. Doppler maps as usually presented are maps of emission line intensity as a function of $\mathbf{v}$ as defined by Eq. 11.2, i.e. $I(\mathbf{v})$.[1]

Applying the above relation to the centre of mass of the donor star which has $\mathbf{r}' = (a, 0, 0)$ and $\mathbf{v}' = (0, 0, 0)$, gives $\mathbf{v} = (0, a(1-\mu)\Omega, 0)$, so if the donor star emits, it contributes at a point on the positive y axis in velocity-space, which is a characteristic and important feature of Doppler maps. Given the definition of F, it is not one inertial frame but a set of inertial frames that maintains alignment with the rotating frame of the system, and therefore the "constant" vector $\mathbf{v}$ steadily rotates with respect to us. Spectroscopy via the Doppler effect leaves us sensitive only to its component along our line of sight or "radial velocity" V_R which is straightforwardly shown to be

$$V_R = \gamma - v_x \sin(i) \cos(2\pi\phi) + v_y \sin(i) \sin(2\pi\phi) - v_z \cos(i),$$

with the various signs the result of the standard choice of a positive sign for the radial velocity for objects receding from Earth. Here, γ is the "systemic velocity" or radial velocity of the centre of mass of the system (assumed fixed), ϕ is the orbital phase (from 0 to 1) and i is the inclination of the orbital axis with respect to the line of sight (assumed fixed). An edge-on, eclipsing system is characterised by $i = 90°$. Except for eclipsing systems, we usually have little idea of i, which leaves Doppler maps degenerate to changes in i which stretch and compress the three axes through the $\sin i$, $\cos i$ factors if we use v_x, v_y, and v_z as our coordinates. Therefore we instead subsume the $\sin i$ and $\cos i$ factors into the coordinates to give instead

$$V_R = \gamma - V_x \cos(2\pi\phi) + V_y \sin(2\pi\phi) - V_z, \tag{11.3}$$

where $(V_X, V_y, V_z) = (v_x \sin i, v_y \sin i, v_z \cos i)$. Equation 11.3 is the origin of Kraft's "S-waves" [26] and is fundamental to Doppler imaging. As has been shown, a fixed point in the rotating frame F' corresponds to a fixed point in F which has a unique set of velocity coordinates (v_x, v_y, v_z). The Doppler effect introduces an extra uncertainty due to projection effects, but Eq. 11.3 shows that if we measure V_R as a function of orbital phase and fit a sinusoid plus a constant, we can deduce accurate values for the closely-related triad, (V_x, V_y, V_z) (assuming we know γ). It is worth remarking at this point that although the final coordinate system is related to the instantaneous frame F by a modest scaling transformation, if we want to get back to the rotating frame F', we need to invert Eq. 11.2 which requires that we know i, μ (or equivalently $q = M_2/M_1$) and $\mathbf{r}$ as well. We can in some cases learn i and q, but in most cases we have very little purchase on the spatial position which does not enter Eq. 11.3 at all. There is thus little reason to try to effect this transformation. The same information is needed to obtain emissivity as a function of spatial position $\mathbf{r}$, and so it is similarly rare to compute spatial Doppler maps.

[1] With a slight scaling due to projection effects as will be discussed.

In the final (V_x, V_y, V_z) coordinates the centre of donor star is placed at $(0, a(1-\mu)\Omega \sin(i), 0)$, which is on the V_y axis with ordinate $(M_1/(M_1 + M_2))\Omega a \sin(i) = +K_2$, the standard semi-amplitude one would measure from a radial velocity measurement of the donor star. Similarly, emission from the accretor is centred on $(0, -K_1, 0)$. The shapes of the components plotted in Fig. 11.1 are entirely defined by K_1 and K_2 alone, which fix the scale and relative sizes of the components. This is the advantage of subsuming the projection factors $\sin(i)$ and $\cos(i)$ into the coordinates as then, if one is lucky, K_1 and K_2 can be directly read from the map just by measuring how far corresponding features are from the origin.

11.2.2 3D Profile Formation

From the previous section it is clear that a spot of emission leads to a sharp emission line which varies sinusoidally in radial velocity with time. If we saw such a feature we could associate the strength of the component with a particular point (V_x, V_y, V_z) in velocity space, a very crude "image". To go further we need to consider a continuous distribution of emission, so suppose that we have a distribution of emissivity so that

$$I(V_x, V_y, V_z)\, dV_x\, dV_y\, dV_z,$$

is the flux at Earth (units of power per unit area) from an infinitesimal cuboid in velocity space defined by V_x to $V_x + dV_x$, V_y to $V_y + dV_y$, and V_z to $V_z + dV_z$. This flux will appear centred at radial velocity V_R given by Eq. 11.3, but it will be somewhat broadened by processes intrinsic to the source, such as thermal broadening, and also by instrumental broadening. For simplicity we assume that these can be represented by a single broadening profile $g(V)$ that satisfies

$$\int_{-\infty}^{+\infty} g(V)\, dV = 1.$$

In this case the line profile that will be observed is given by

$$f(V, \phi) = \int_{-\infty}^{+\infty} \int_{-\infty}^{+\infty} \int_{-\infty}^{+\infty} I(V_x, V_y, V_z) g(V - V_R)\, dV_x\, dV_y\, dV_z, \tag{11.4}$$

with V_R given by Eq. 11.3. The profile f as defined here has units of power per unit area per unit velocity interval. Doppler tomography is only effective if the broadening function g is relatively narrow compared to the width of the emission region. In the ideal case, $g(V) = \delta(V)$, a delta-function, and Eq. 11.4 can be viewed as selecting a nested set of surfaces in three-dimensional velocity space defined by the condition $V = V_R$ with the surfaces labelled by the particular velocity V in the profile to which they contribute. Since Eq. 11.3 can be written as

$$aV_x + bV_y + cV_z = d, \tag{11.5}$$

where $a = -\cos(2\pi\phi)$, $b = +\sin(2\pi\phi)$, $c = -1$ and $d = V - \gamma$ all constants for a particular phase ϕ, we can see that the surfaces are in fact a set of parallel planes.

The profile formation encapsulated in Eq. 11.4 converts a 3D distribution of emission into a 2D function and information is certainly lost. Unique inversion from 2D line profiles to a fully 3D Doppler image is not possible, even though as we saw with the discussion of a single spot in the previous section, some 3D information is encoded in the profiles through the mean velocity of S-waves [2, 29]. This has been used in some studies of Algol systems [1, 33, 34] and its potential for polars has been explored [25], but not so far applied. Since this review is focussed on cataclysmic variable stars we do not discuss it further but refer the interested reader to the cited studies. With the extra condition that the image is two-dimensional in form, there results the conventional 2D Doppler tomography used in the majority of of published Doppler maps, and we turn to this now.

11.2.3 2D Profile Formation

The solid-body rotation term of Eq. 11.2, $\boldsymbol{\Omega} \wedge (\mathbf{r} - \mathbf{C})$, only leads to velocities parallel to the *x*–*y* orbital plane of the system. As the result of strong symmetry between $+z$ and $-z$, it is often the case that bulk motions within the rotating frame are also largely parallel to the orbital plane. This applies to the commonly-seen mass transfer stream and accretion disc components for instance. This means that in many instances one can assume that there is only significant emission close to $V_z = 0$, allowing us to integrate out the *z*-component and leaving

$$f(V,\phi) = \int_{-\infty}^{+\infty} \int_{-\infty}^{+\infty} I(V_x, V_y) g(V - V_R)\, dV_x\, dV_y, \tag{11.6}$$

(see Eq. 5 in [28]). The set of planes in 3D defined by Eq. 11.5 then becomes a set of lines in 2D:

$$aV_x + bV_y = d. \tag{11.7}$$

Given the values of the coefficients a and b, the lines at orbital phase ϕ run parallel to the direction $\mathbf{P} = (\sin 2\pi\phi, \cos 2\pi\phi)$. The profile of Eq. 11.6 can thus be thought of as a "projection" (or collapse) of the 2D image along the direction defined by $\mathbf{P}$, a direction that rotates clockwise as the orbital phase advances, which can be thought of as the direction that the Earth rotates as seen from the rotating frame of the binary.

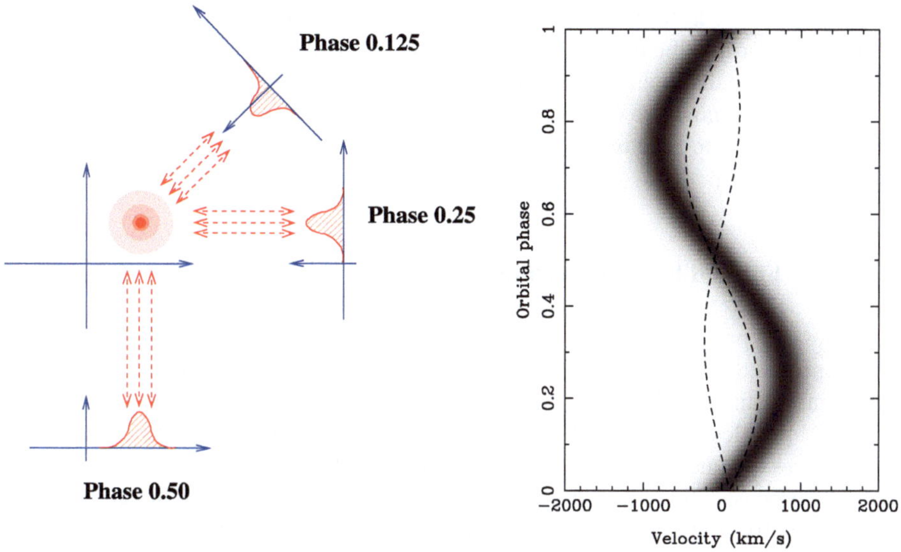

Fig. 11.2 *Left:* a spot in velocity-space with corresponding line profiles at three orbital phases which are the projections of the image in the directions shown. In reverse if the profiles are smeared along the same directions, a process called "back-projection", they will add together at the position of the spot. *Right:* a "trailed" spectrum with profiles plotted as a 2D image with phase ascending corresponding to the image on the *left*. Back-projection in this dataspace corresponds to line integrals along multiple paths (Re dashed lines show two examples); paths that coincide with the S-wave lead to a large intensity at the corresponding (V_x, V_y) position (Figure from [28])

To emphasise these points here is Eq. 11.6 written with the V_R term expanded and the broadening function set to a δ-function:

$$f(V, \phi) = \int_{-\infty}^{+\infty} \int_{-\infty}^{+\infty} I(V_x, V_y)\delta\left(V - \gamma + V_x \cos 2\pi\phi - V_y \sin 2\pi\phi\right) dV_x dV_y. \tag{11.8}$$

Figure 11.2 is an attempt to visualise this relation. On the left, the profiles equivalent to a single spot of emission are displayed at the correct projection direction for three different phases. On the right, the result of measurements at many phases are plotted as a greyscale running up the page, a plot known as a "trailed spectrum" in deference to the process of trailing objects along the slits of spectrograph that was employed when observing such objects in the days of photographic plates.

11.2.4 Inversion

The profile formation encapsulated in Eqs. 11.6 and 11.8 means that a series of line profiles taken at different phases around a binary orbit is the equivalent of a series of projections of a 2D image along straight lines into 1D. This is mathematically identical in form to the use in medicine of a series of X-ray images taken at different angles to image the interior of the human body. Such X-ray photographs are imaged with a method known as computerised tomography, where the word "tomography" is connected to the reconstruction of images from projections. This led to Doppler tomography [30].

11.2.4.1 Filtered Back-projection

A series of projections can be inverted to produce an image in a two-stage linear process called filtered back-projection [28, 30]. In the context of Doppler tomography, first of all a filter is applied in velocity to each line profile to produce modified profiles $\tilde{f}(V, \phi)$. In Fourier space the filter boosts the amplitudes of high frequencies (frequencies in the sense of cycles per step in velocity). This is a simple and fast process, although some care is needed to avoid over-amplification of noise which usually dominates at the very highest frequencies. The next step is back-projection:

$$I(V_x, V_y) = \int_0^{0.5} \tilde{f}(\gamma - V_x \sin 2\pi\phi + V_y \sin 2\pi\phi, \phi)\, d\phi, \tag{11.9}$$

which can also be written in a way that resembles Eq. 11.8 as

$$I(V_x, V_y) = \int_0^{0.5} \tilde{f}(V, \phi)\delta\left(V - \gamma + V_x \sin 2\pi\phi - V_y \sin 2\pi\phi\right) d\phi.$$

Figure 11.2 contains two means of visualising this relation. On the left, the approximate inverse of profile formation by projection is instead to smear each profile back along the original projection direction, as indicated by the double-direction arrows. Where multiple streaks reinforce, a spot will form. This is why the process is called back-projection. The right-hand panel shows a more literal interpretation of the equation as a series of integrals along sinusoidal paths in a trailed spectrum over the data. When the path of such an integral runs along an S-wave, a strong contribution will emerge resulting in a spot.

Although not often used in practice, the concept of back-projection is a very helpful one for interpreting Doppler images. For instance it is clear that bad pixels, perhaps the result of cosmic rays, will lead to the appearance of linear features in maps, and caution should always be exercised if such features are apparent.

11.2.4.2 Regularised Fitting

It is more common to regard Doppler imaging as a model fitting problem. One represents the map by (usually) a square grid of pixels covering the necessary range

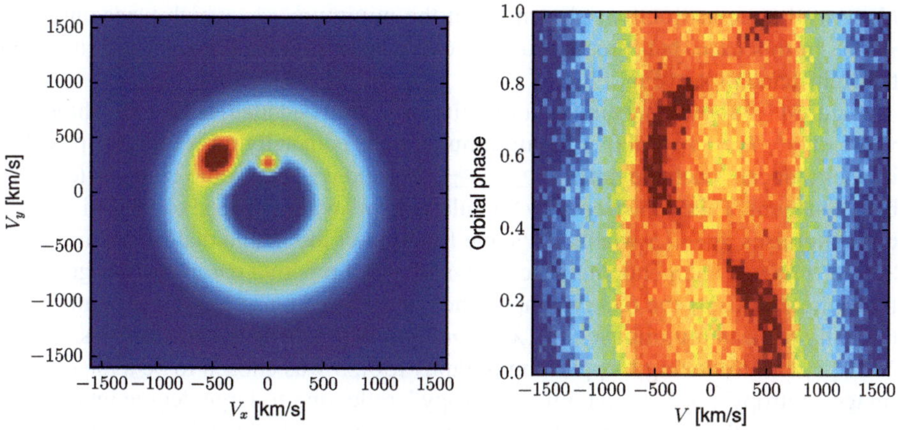

Fig. 11.3 *Left:* an artificial generated Doppler map including a component to represent a disc (ring-like structure), a gas stream/disc impact spot (left-hand bright spot), and a weak spot from the donor star. *Right:* data generated from the model with added pseudo-random gaussian white noise

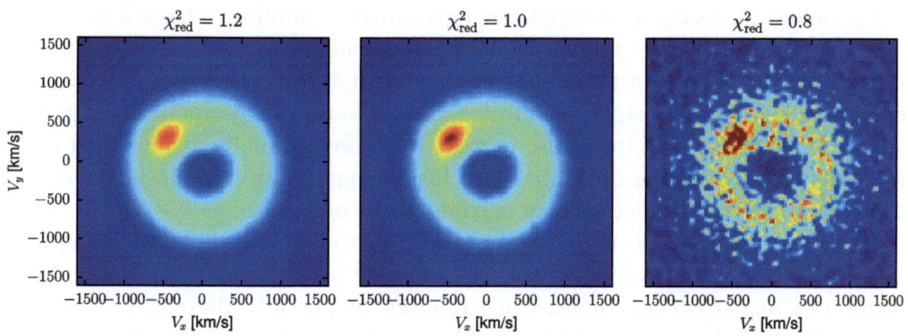

Fig. 11.4 Three image reconstructions from the data of Fig. 11.3, differentiated by the χ^2 per data point reached in each case. All are plotted to the same maximum intensity used to display the model in Fig. 11.3 to which they should be compared. Note the marginal signs of emission from the donor star, which is in accord with its level in the data of Fig. 11.3

of velocities with enough resolution to match the data, and then asks the question "what set of pixel intensities best matches the data?". The immediate problem one faces is that if "best" is regarded as the minimum χ^2, the usual answer is a useless mass of high and low values dominated by noise. To show this, in Fig. 11.3 we show an artificial image and some data computed from it with added pseudo-random gaussian white noise. Two spots representing the gas stream/disc impact region and emission from the donor star have been added as features commonly seen in real data. In Fig. 11.4 we show three reconstructions from the data of Fig. 11.3 for different values of $\chi^2_{\rm red} = \chi^2/N$, with N set to the number of data points rather than the number of degrees of freedom, the latter being difficult to define in this case.

None of the reconstructions shown are at the minimum χ^2_{red} which turns out to be $\sim$0.68 in this case, but already noise is overwhelming the reconstructed image in the right-hand plot of Fig. 11.4.

Figure 11.4 shows that aiming for minimum χ^2 is a bad idea, but for any threshold value of χ^2, $T > \chi^2_{min}$ above the minimum, there is usually an infinity of possible images that can reach the range $T \geq \chi^2 \geq \chi^2_{min}$, and we need some criterion to select between them. This is answered by "regularisation" where instead of minimising χ^2, one minimises $\chi^2 - \lambda S$ where S is some function of the image that is maximum for "good" images, independent of the data, and λ is a positive parameter (Lagrangian multiplier) that sets the relative importance of minimising χ^2 versus maximising S. (Equally one can miniminise $\chi^2 + \lambda S'$ if *minimum* S' selects the best images.) In the 1988 paper introducing Doppler tomography [30], S was chosen to be a measure of image "entropy", hence "maximum entropy", reflecting to large degree inheritance from its use in the method of eclipse mapping developed by [17], but other measures can easily be imagined, for instance measures of maximum local smoothness, and may indeed have advantages in practice [57].

Figure 11.4 shows that there is no unique inversion, as there is a trade-off between noise and goodness of fit. In this case the particular solution selected is labelled by the χ^2 achieved. As this is lowered, one moves from a solution of large λ where S is more influential to one of small λ where χ^2 dominates. The choice of where to stop is really one of taste: most would agree that the right-hand image of Fig. 11.4 has been pushed too far, whereas the left-hand image looks overly smooth. The precise values of χ^2 and S vary from case to case, but there is usually a significant range of values over which the qualitative conclusions one draws from an image do not change, even if the detailed level of apparent noise may vary.

The noise–smoothness trade-off is not specific to regularised fitting, but applies to filtered back-projection too where one must set a parameter to limit the amplification of high-frequency noise when filtering. A possible alternative method to approach the problem would be to parameterise the problem by describing the disc through a small number of components, and then one could simply fit for the unique solution of minimum χ^2. This is a method that has achieved some success in the case of fitting the light curves of eclipsing dwarf novae for example [18, 39]. Many observed Doppler maps can be described in terms of a disc plus one or two spots, so such a method may be possible, but would run the risk of missing unexpected structures, and is probably not a good avenue to pursue.

11.2.5 Doppler Tomography Extras

There are a number of additional effects that one can allow for when using Doppler tomography which we briefly discuss here.

11.2.5.1 Systemic Velocity

We have already referred to 3D imaging that can cope with a range of V_z values. This has the advantage of automatically allowing for the systemic velocity γ which simply displaces the image in the V_z direction (see Eq. 11.3). In 2D imaging we assume $V_z = 0$ and therefore it is important to use the correct value for the systemic velocity γ. If not, the resultant image will be blurred, as discussed in [30]. If no independent value for γ exists, the standard approach is to optimise γ through minimisation of χ^2 during the imaging cycle. This is generally a fast operation.

11.2.5.2 Orbital Phase Uncertainty

In non-eclipsing systems, without clear donor stars it can be difficult to know the absolute orbital phase. This leads to a rotational uncertainty in the resultant image that must be considered when interpreting it. In the case of cataclysmic variables, the emission line radial velocities are often used as a proxy for the white dwarf although it has been known for many years that they are very often not reliable, and that the orbital phase zero point that one deduces from emission lines tends to lag the true zero phase [56]. This causes an anti-clockwise rotation of a Doppler map based upon an emission-line phase. As a consequence, a bright-spot component for instance will appear lower in the map than expected. Caution is needed if such an effect is seen; the problem is most severe in short-period systems where the white dwarf's radial velocity semi-amplitude is small.

11.2.5.3 Finite Exposures

For faint and very short period systems, it can be difficult to acquire spectra in a time much shorter than the orbital period. Consider for instance the 21-st magnitude 5-min binary HM Cnc where it is a challenge even to take exposure in less time than the binary takes to complete an entire orbit [37]. This leads to azimuthal smearing of the Doppler map by an amount $2\pi \Delta t/P$ radians where Δt is the exposure time and P the orbital period. Whether this is significant depends upon the intrinsic width of the line profile and the instrumental resolution. A feature K km/s from the origin of the map will be smeared in azimuth by $2\pi K \Delta t/P$, and particularly when planning observations, it is good to consider how this compares to the spectrograph velocity resolution, $c\Delta\lambda/\lambda$, for instance. Blurring can be combatted by de-convolution, and in the code used by [30] (see below), one can compute models allowing for finite exposures, which can be helpful in sharpening bright-spot features in difficult cases.

11.2.5.4 Blended Lines

It is fairly common to encounter blended (overlapping) lines. In the case of AM CVn stars, HeII 4686 is only separated by $\sim$1700 km s^{-1} from HeI 4713, and similarly the Ca II triplet lines in the I-band each have a nearby line of the Paschen series that can be significant. In such cases it is straightforward within the regularised fit approach to allow for multiple lines, perhaps with independent images or as scaled versions of the same image. In practice this rarely works perfectly, and particularly in the weaker member of a blend, it is common to see ring-like features resulting from the nearby stronger line. Still, the results are improved over simply ignoring the effect.

11.2.5.5 Modulation Mapping

Standard Doppler tomography produces a map that can be regarded as an orbitally-averaged image, although strictly speaking one of the "axioms" of standard Doppler tomography is that the image does not change [28]. In reality there are almost always geometric effects which lead to a break-down of this assumption. Most obviously for instance, emission on the heated face of the donor star will move into and out of our sight as the binary rotates. In the modulation mapping method introduced by [51], one regards the image as a Fourier series in orbital phase:

$$I(V_x, V_y) = I_0(V_x, V_y) + A(V_x, V_y)\cos 2\pi\phi + B(V_x, V_y)\sin 2\pi\phi.$$

The reconstruction then finds I_0, as well as the cosine and sine components A and B. This has some value in interpreting the sharp emission components and can deliver greatly improved fits to the data [4].

11.2.6 Codes for Doppler Tomography

We finish this section with a brief description of three publicly-available codes that we are aware of for implementing Doppler tomography. The original maximum entropy code using the MEMSYS algorithm [47] is available as an add-on[2] to the STARLINK software suite. Its chief disadvantage is probably the learning curve that must be surmounted if one is unfamiliar with STARLINK. Henk Spruit distributes a simple maximum entropy-based code[3] that is straightforward to get going, although it misses some of the features of the original code, such as dealing with blended lines. Finally, a recent variant using a different regularisation measuring

[2]http://www.astro.warwick.ac.uk/people/marsh/software/

[3]http://www.mpa-garching.mpg.de/~henk/

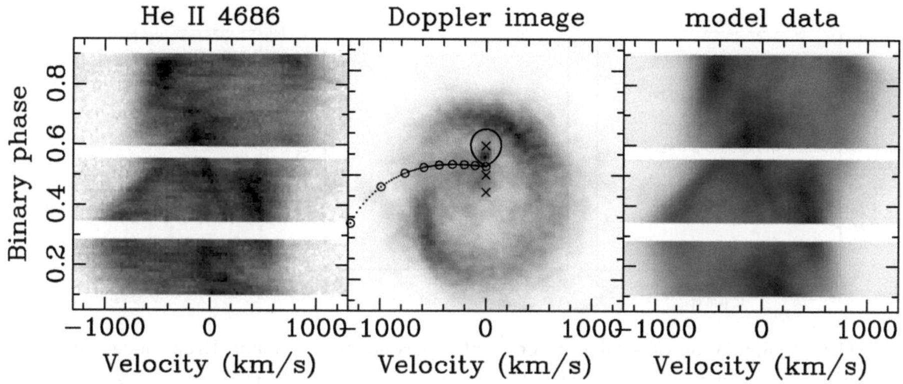

Fig. 11.5 Spiral shock structures seen in the dwarf nova IP Peg during outburst [14]

the variation across an image has been developed [57] and is available for general use.[4] An advantage of the regularisation function chosen is that it copes naturally with negative image values. Although these should not occur in theory, in practice absorption effects, particularly in highly-inclined systems, often lead to a need for negative image values which are tricky to accommodate with entropy-based codes.

11.3 Doppler Tomography in Practice

Doppler tomography has achieved the status of a standard tool of the field and, with over 400 citations to the original paper [30] and many others besides, it is not possible to cover all of the results that have come from its application. In this section therefore we concentrate upon some highlights, particularly those where it has led to insights that were not obvious from the data alone.

11.3.1 Spiral Shocks

Coming only a decade after the development of the method, the discovery of global spiral structures in the accretion disc of the dwarf nova IP Peg [53] is still a stand-out result from the application of Doppler tomography. Very complex profile variations (Fig. 11.5) in which the classic double-peaks from the accretion disc appear to show discontinuities, were only convincingly interpreted after they were mapped. Similar variations have been seen in a number of other systems [3, 12, 20] and are thought to be induced by the tidal potential of the binaries.

[4]http://home.hiroshima-u.ac.jp/~uemuram/dttvm/

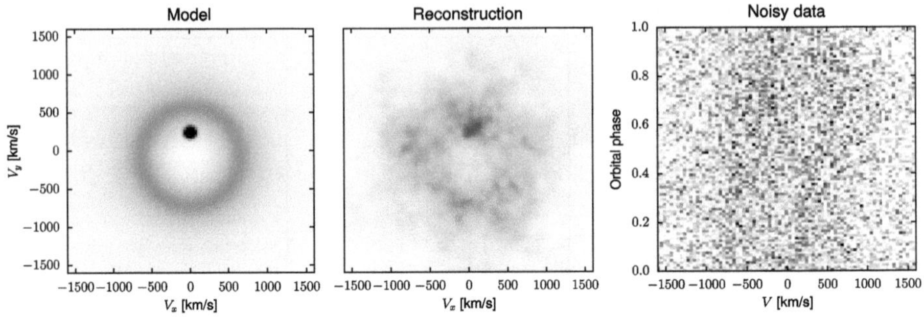

Fig. 11.6 A simulation of a faint system with emission from a donor. Even though the donor is barely visible in the data on the *right*, a significant excess appears at its location in the reconstruction

11.3.2 Donor Star Emission

In many short period cataclysmic variable stars and low-mass X-ray binaries it is practically impossible to obtain reliable radial velocity measurements of the compact object because it has a low amplitude and one has to use very broad emission lines as a proxy for the object itself. In these cases, detection of the donor can be very useful indeed for the precise phase constraint that it can return as well as the constraint it places upon the mass of the compact object. Doppler tomography can be a useful detection method here. Figure 11.6 shows a simulation where significant emission from the donor can be recovered in a map even though it is impossible to see in the trailed spectrum. This has been put to good use in the case of X-ray binaries using Bowen fluorescence emission (C III/N III 4640) from the donor [6, 52], and using Balmer and Ca II emission to reveal brown dwarf-mass donors in cataclysmic variable stars [42, 55, 58]. The last cited work reveals a further interesting feature (Fig. 11.7), namely that Ca II emission reveals the donor with particular clarity in some cases, and is also distinctly sharper when it comes from the disc. There is a distinct impression that the Balmer line map (Hβ) is limited by significant intrinsic line broadening. For other examples of donor stars in Doppler maps see Sect. 11.4.3.

11.3.3 AM CVn Stars

Over the past decade there have been many new discoveries of the hydrogen-deficient accreting binaries known as AM CVn stars. The line profiles of these systems are unusual for sometimes featuring emission from the white dwarf [27, 31]. The position of this emission in Doppler maps along with the position of the stream/disc impact spot has been used to constrain the system parameters in some of these systems [36] in which no donor star has ever been directly detected. Doppler

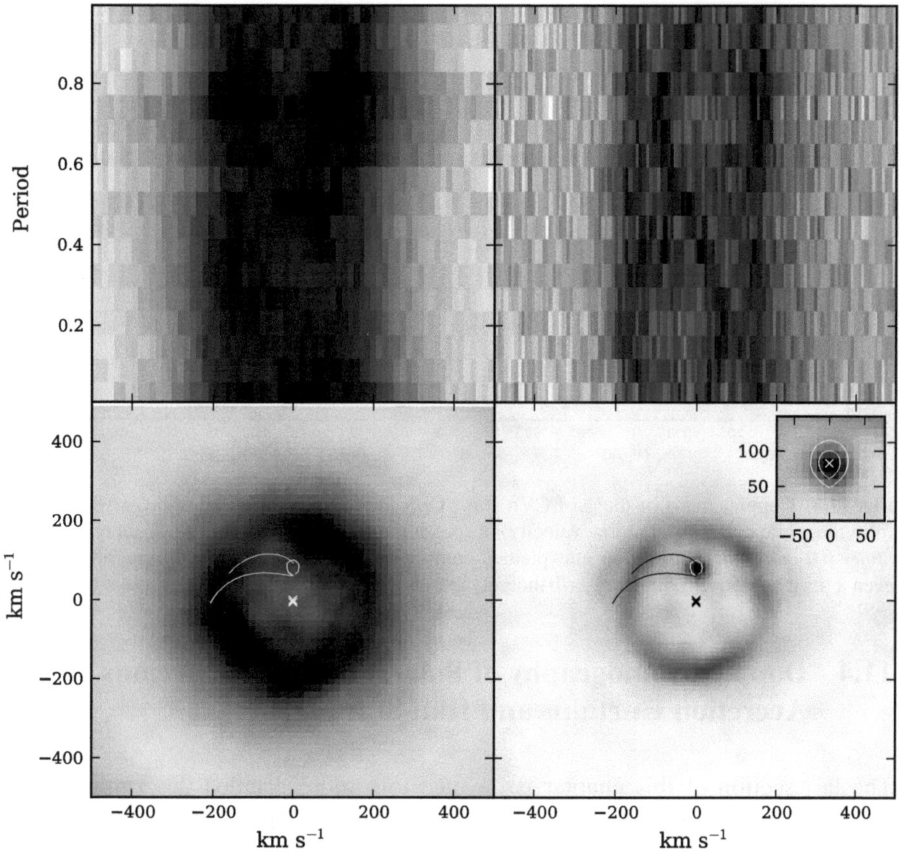

Fig. 11.7 Hβ (*left*) and Ca II 8662 (*right*) trailed spectra and Doppler images of the low-inclination dwarf nova GW Lib [58]. Emission from the donor star is particularly clear in Ca II emission

maps of AM CVn stars often also feature emission spots in unusual locations (see Fig. 11.8) which have yet to have had a convincing explanation. AM CVn systems have extreme mass ratios leading to low radial velocity amplitudes for the white dwarf. This makes the bright-spots ("S-waves") particularly significant in measurement of their orbital periods. This has led to the use of Doppler maps for period measurement by through alignment of bright-spots over multiple nights of data [35].

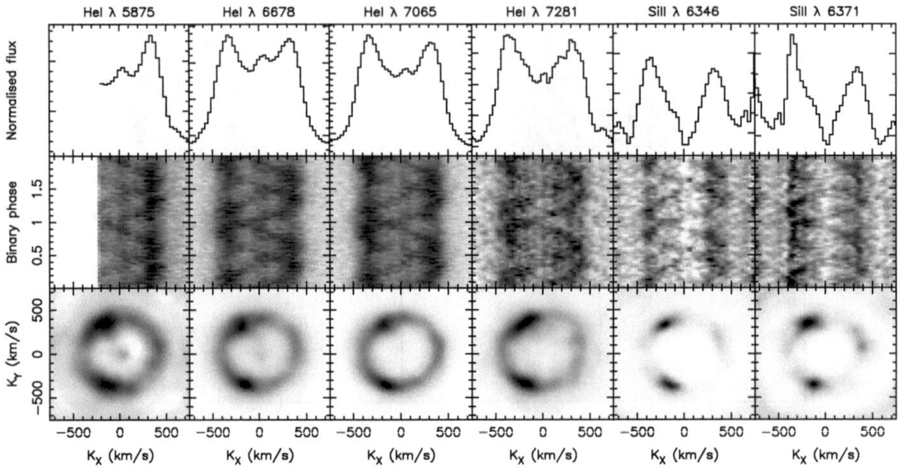

Fig. 11.8 Doppler maps of the AM CVn star SDSS J124058.03-015919.2 show weak emission from the white dwarf (spot at low velocity), a spot at the standard stream/disc impact region in the *upper-left quadrant*, but also an unexplained component in the *lower-left quadrant*, and possibly even a third component on the right-hand side [35]

11.4 Doppler Tomography of Polars: Accretion Streams, Accretion Curtains and Half Stars

The last section of this chapter is devoted to a more detailed description of the achievements, main results but also the limitations of Doppler tomography applied to a specific class of close binaries, the polars.

Polars are cataclysmic binaries where accretion is dominated by strong magnetic fields of the white dwarf stars. The magnetic moment and the Alfven radius of the accreting white dwarf are so large that both stars of the binary are kept in synchronous rotation; no accretion disc is formed. Instead matter is transferred to the white dwarfs via accretion streams and similar structures. The gravitational energy of the instantaneously accreted matter is released in small regions (spots, arcs) via X-ray bremsstrahlung, optical cyclotron radiation and an intensive radiation component in the UV/EUV spectral regime. While the former may deposit its energy in the irradiated hemisphere of the mass-donating late-type star (and distort its radial velocity curves), the latter is responsible for photoionisation of gaseous matter in the binary system. Recombination radiation from streaming and orbiting matter in the binary gives rise to the very complex and ever-changing line profiles that were impossible to disentangle with 'classical' methods like line-profile and radial-velocity fitting (see [19] for a photographic trailed spectrogram of the 81 min binary EF Eri).

The first ever tomogram of a magnetic CV, an Hα map of the high-inclination system VV Pup, was published by [8] who noticed the fundamental difference of their Doppler map to those of non-magnetic CVs.

11.4.1 Accretion Streams and Curtains

The main features that are typically found in trailed spectrograms and Doppler maps of polars are highlighted in Fig. 11.9. The original data that were obtained with integration times between 20 and 60 s and covering several orbital cycles of the $P_{\text{orb}} = 125$ min eclipsing binary were phase-averaged and arranged as pseudo-trailed spectrograms in the left column of the Figure. The emission lines belong to ionised Helium (HeIIλ4685). Their existence immediately point to the presence of an ionising source (the ionisation potential of Helium is 54 eV) on the white dwarf. The recombination lines of the H-Balmer series and of neutral Helium show very similar features, the H-Balmer lines, however, typically show a larger optical thickness and their line width is broader. Their Doppler maps are less structured and therefore more difficult to interpret.

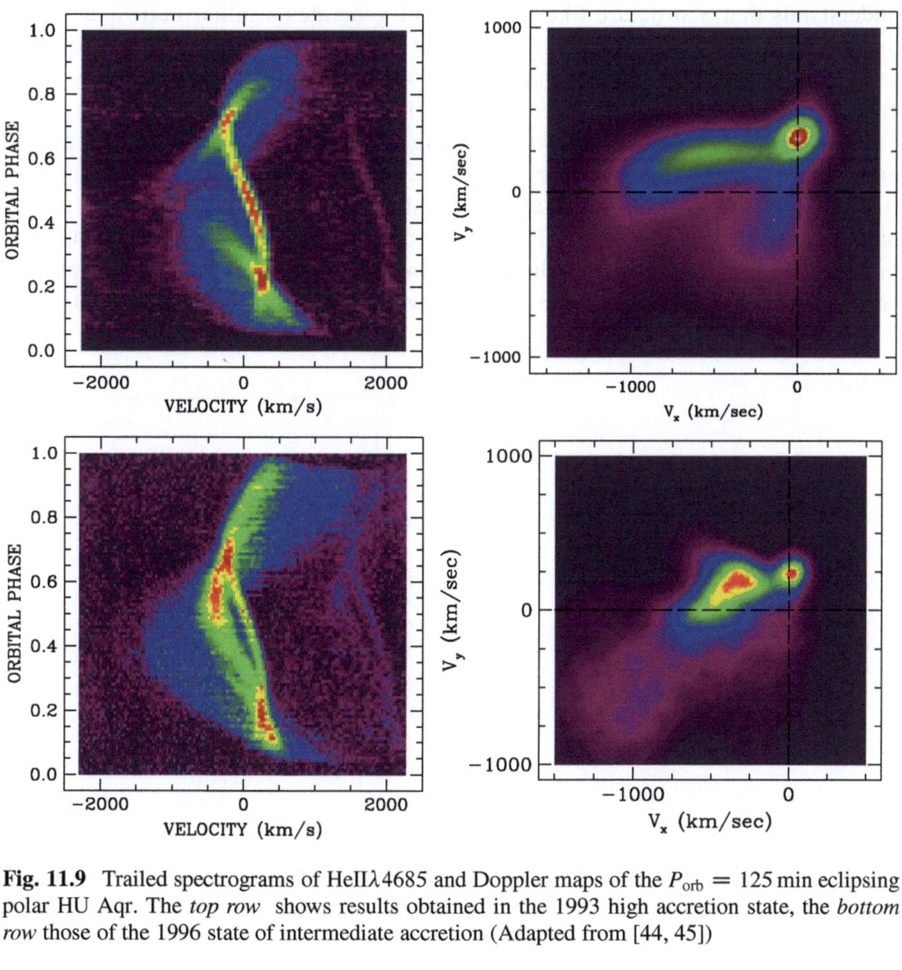

Fig. 11.9 Trailed spectrograms of HeIIλ4685 and Doppler maps of the $P_{\text{orb}} = 125$ min eclipsing polar HU Aqr. The *top row* shows results obtained in the 1993 high accretion state, the *bottom row* those of the 1996 state of intermediate accretion (Adapted from [44, 45])

The trailed spectrograms show an S-wave best visible around phase 0.5. At this phase it has zero velocity and its radial velocity changes from being redshifted to become blueshifted. This narrow emission line (NEL) traces the mass-donating star (see Sect. 11.4.3 for a more detailed description of its properties).

Without tomography researchers were disentangling the line profiles into three or four Gaussians that were traced throughout the orbital cycle or only parts of it. The relative phasing, velocity amplitudes and widths of those components were used to locate their origin in different parts of the accretion stream and were used to imply considerable structure of the line-emitting gaseous matter (see, e.g., [38]).

The Doppler maps of the data reveal such divisions to be artificial. The high-state map (upper right) shows three main structures, a point-like spot of emission at $(v_x, v_y) \simeq (0, 300)\,\mathrm{km\,s^{-1}}$, a comet-like tail attached to it that extends to large negative velocities in the x-direction, and a more symmetric structure in the lower left quadrant, i.e. mainly at negative x- and y-velocities. If a trailed spectrogram is regarded as the projection of a velocity map onto the observers frame one understands that a structure like the long-tail generates a rather narrow emission line at phases 0.25 and 0.75 and a very broad line of width $1000\,\mathrm{km\,s^{-1}}$ at phases 0 and 0.5. This insight was only possible by Doppler tomography and it clearly limits any interpretation of Gaussian-deconvolved line profiles.

The structures in the Doppler map of Fig. 11.9 (top right) are interpreted quite naturally as originating from the donor star, the ballistic accretion stream falling freely in the gravitational potential of the white dwarf and the magnetically guided part of the stream. Figure 11.10 helps to reach a basic understanding of structures in polars. The choice of the coordinate system etc. follows the convention explained above (see Fig. 11.1). It is important to emphasise that the diagram to the right shows

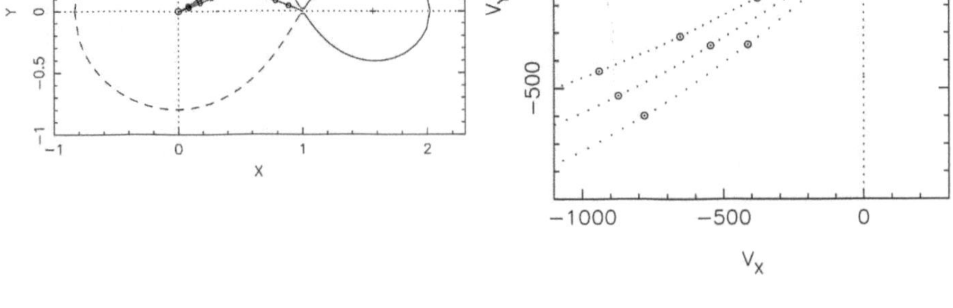

Fig. 11.10 Accretion geometry of a polar in spatial (*left*) and Doppler coordinates (*right*). The inclination of the dipole axis with respect to the rotation axis is 15° and points towards the donor star

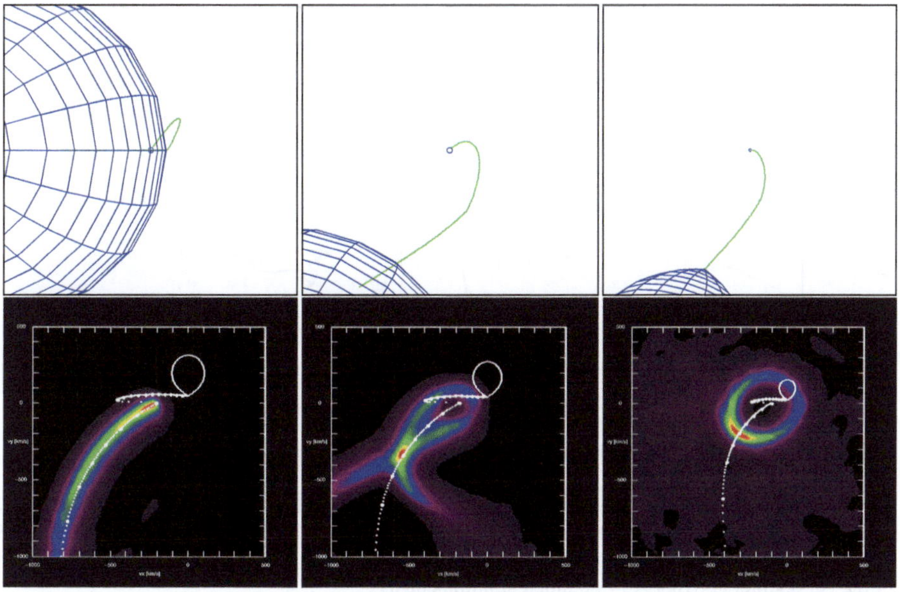

Fig. 11.11 Effect of finite v_z component on Doppler maps for $i = 30°, 60°$, and $90°$ [49]

velocity projections onto the orbital plane because the streaming matter may have large velocities perpendicular to it. Non-zero streaming velocities out of the plane may have significant impact on the Doppler maps of low-inclination systems.

The adverse effect of non-zero V_z-velocities is illustrated in Fig. 11.11. A numerical experiment was set up in the following way. Matter freely falling along a dipolar field line was assumed to emit optically thin radiation in a non-specified atomic line. Each element along the trajectory from a coupling region towards the white dwarf was assumed to radiate away the same amount of energy. The components of the 3D velocity vectors were projected onto the observers frame as a function of the phase angle. The resulting trailed spectrogram was used to generate the Doppler maps shown in the bottom row of the figure for orbital inclinations of 90°, 60°, and 30° (from left to right). While the emission lines for the 90° case are mapped exactly on the expected trajectory, fish- and ring-like maps emerge if the inclination angle is smaller. The largest brightness is still observed somewhere on the single-particle trajectory but the constant presence of blue- and red-shifted emission line components from the rising and falling parts along the field line leads to a map with emission in all four quadrants.

This example just illustrates the possible consequences of violating one of the basic assumptions ("axioms") of Doppler tomography that all motion is parallel to the orbital plane.

Let's come back to the Doppler map of HU Aqr obtained in the 1993 high accretion state (Fig. 11.9, top right). It seems as if there were two features whose locations in the map are influenced (to say the least) or are even fixed by the masses

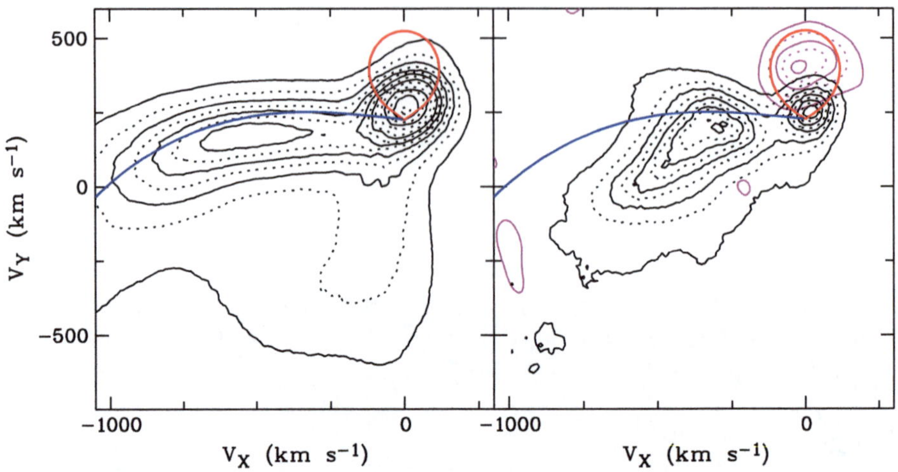

Fig. 11.12 Shape of the Roche-lobe and location of the ballistic stream in Doppler coordinates for the best-known binary parameters of HU Aqr compared with Doppler maps of the HeII λ 4685 obtained in high (*left*) and intermediate states of accretion. Shown in *magenta* is the Doppler map of the Na-doublet at 8183/9194 obtained simultaneously

of the two stars: the position of the donor star and of the ballistic stream. Can those be used to measure the masses of the stars in the binary?

Perhaps this is possible in principle, but in practice it turned out to be very difficult. The radius of the white dwarf could be measured with high signal-to-noise ratio with ULTRACAM at the VLT and the mass estimated using a mass-radius relationship. The radial velocity of the two hemispheres of the donor star were measured using Na absorption lines originating from the non-irradiated and Ca II emission lines from the irradiated hemispheres, respectively. This set of measurements constrains the mass ratio to $Q = M_{\rm wd}/M_2 = 4.6 \pm 0.2$, the mass of the white dwarf to about $0.8\,{\rm M}_\odot$, and the inclination to 87°. The shape of the Roche lobe and the location of the ballistic stream are overlaid over the Doppler maps in Fig. 11.12.

In the high state the bulk of emission from the 'ballistic' accretion stream does not follow the single-particle trajectory but is displaced by 60–70 km s^{-1} towards negative $v_{\rm y}$. A mass ratio of about 2.5 would be needed to give a best match between the model and the observation, which is impossible to achieve given the other observational constraints. In both maps there is a considerable amount of emission also at higher positive $v_{\rm y}$ velocities which is difficult to understand given the well-known accretion geometry of HU Aqr as sketched in Fig. 11.10. The high inclination of the object rules out $v_{\rm z}$-effects as a possible interpretation. Also, the free-falling accretion stream is thought to be incompressible, hence one expects no velocity broadening perpendicular to the streaming direction (see the model maps for this object by [16]).

While the use of the He II Doppler maps cannot be used directly for a determination of the binary star parameters, they offer deep insight into the flow pattern in the binary. While the high state ballistic stream stretches to $V_x < -1000\,\mathrm{km\,s^{-1}}$, it is stopped at $V_x \simeq -600\,\mathrm{km\,s^{-1}}$ at reduced accretion rate. Hence, the stream does not reach the same high-infall velocities along the ballistic part before it is overcome by the magnetic pressure. Hence the reduction in the mass accretion rate leads to a reduction in the mass density. The stream then seems to be re-directed so that it appears in the lower left quadrant in Doppler coordinates. There it occupies rather different regions in the two accretion states. This reflects that accretion happens along different field lines in different accretion states. Extended phases of absorption in the soft X-ray light curves reveal the existence of an extended accretion curtain, which is spanned between the inner Lagrangian point and the magnetically guided accretion streams that transports the bulk of the accreting matter. Ionisation and re-emission will happen all along the accretion curtain and reveals broad structures in the region of the Doppler maps to the lower-left of the ballistic trajectory. Attempts were made to map the brightness distribution along the accretion stream with and without velocity information [13, 15, 59].

UZ For is in many respects a twin system of HU Aqr as far as the orbital period and the X-ray properties are concerned. Its trailed spectrogram and the resulting Doppler map for He II 4685 are shown in Fig. 11.13. As in HU Aqr, there is an apparently clear indication of a ballistic stream reaching a maximum free-fall velocity of $v_x \simeq= -650\,\mathrm{km\,s^{-1}}$. The stream is then re-directed into the lower left quadrant where it still can be found as a rather well-focused structure. The overlaid streams in the right panel of the Figure were computed for matter streaming along dipolar field lines and allowed to determine the orientation of the magnetic axis (azimuth $\varphi = 45°$, co-latitude $\beta = 15°$).

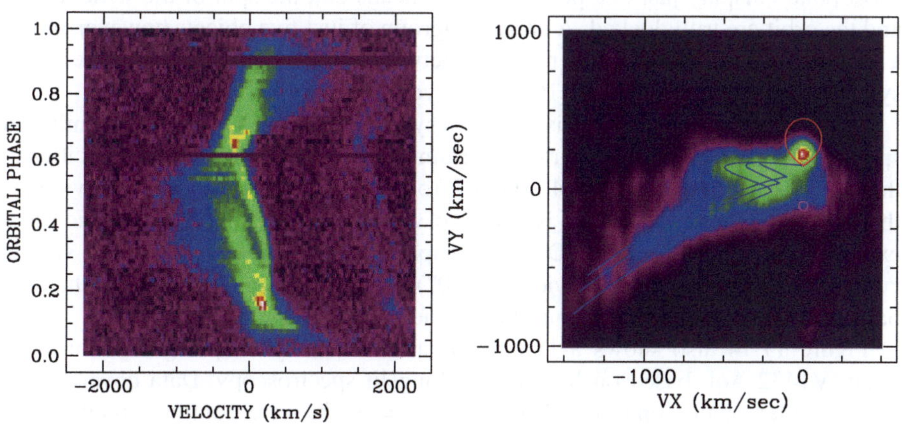

Fig. 11.13 Trailed spectrogram (HeIIλ4685, folded into 70 phase bins) and Doppler map of the $P_{orb} = 126.5$ min eclipsing polar UZ For

The trajectories in Fig. 11.13 were computed for a mass ratio $Q = M_{\rm wd}/M_2 = 3$. If the donor star followed Knigge's sequence, it would have a mass $M_2 = 0.19\,{\rm M}_\odot$ [23, 24]. The two numbers imply an undermassive white dwarf (see the recent compilation by [10]). As discussed in [32] a high-speed, high S/N low-state light curve, which would allow to measure the radius of the white dwarf directly, is still missing but current high-state photometry already gives strong evidence for a more massive white dwarf with ${\rm M_{wd}} \simeq 0.74\,{\rm M}_\odot$. Hence, the two apparently cleanest examples for Doppler maps with matter freely falling along single-particle trajectories are in disagreement with mass ratio measurements and thus challenging our understanding of both the ionisation processes and the dynamics of the accretion stream.

The long-period polar QQ Vul shows even more complexities of those emission patterns by displaying a time-dependent shift of the v_y velocity of the putative "ballistic" trajectory [41]. There is no obvious reason for such a displacement. However, the observation of an almost stationary N V line in AM Her which was interpreted as being due to matter in a magnetic slingshot prominence emanating from the secondary star [11] may be stimulating thoughts about models with other magnetic structures, perhaps even those with open field lines that may exist in the vicinity of L_1 and that eventually acts as venting system for matter to be released towards the white dwarf. Currently it seems difficult if not impossible to uniquely locate the single-particle trajectory of freely falling matter from L_1 into the Roche lobe of the white dwarf.

11.4.2 Accretion Curtains in Asynchronous Polars

Most polars display just one period, which means that the spin of the white dwarf and the orbit are truly locked. A small subgroup of just five objects however shows a small asynchronism of a percent or less between those two quantities, the reported synchronisation time scales are $10^2 \ldots 10^3$ years.

The first tomographic analysis of such a system, BY Cam, revealed line emission spread out over a large velocity range forming a crescent at negative v_y velocities in the Doppler maps (see Fig. 11.14, adapted from [40]). This was regarded as evidence that the majority of the matter is accreted via an extended curtain. Location and extent of the structure in the Doppler maps could be reproduced with a simple curtain model raised over a wide ($\sim$180°) range in azimuth implying that the ballistic stream stretches to a point far behind the white dwarf.

Figure 11.14 also shows a Doppler map of the only eclipsing asynchronous polar, V1432 Aql, based on VLT/FORS longslit spectroscopy. Data covering 2.3 binary cycles of the long-period polar, $P_{\rm orb} = 3.4$ h, were continuum-subtracted and phase-averaged for the creation of the Doppler map. The orbital inclination of BY Cam is moderate, the map thus might be affected by the v_z effects described above. They do not play a role in V1432 Aql. The spread of emission predominantly into the two lower quadrants in the Doppler map of this system therefore is real.

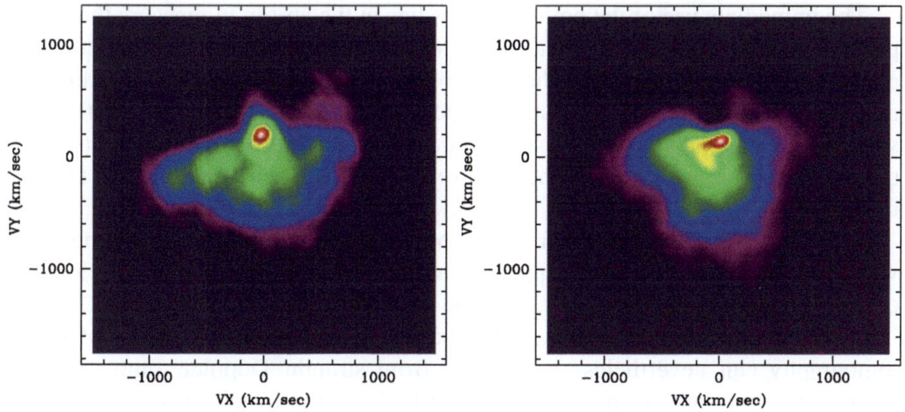

Fig. 11.14 Doppler maps of the asynchronous polars BY Cam (Hβ, *left*) and V1432 Aql (He II 4685, *right*)

One observes large velocities in negative y-direction over a wide range of positive and negative v_x-velocities. Similarly to BY Cam, the map may be interpreted by an azimuthally extended accretion curtain. Emission from matter up to an azimuth of 180° is required to explain the negative v_y-velocity component at all velocities v_x. Emission in the upper right quadrant, i.e. positive velocities in both projected velocity components, is suggestive of matter that has made it around the white dwarf to an azimuth of up to 270° and being accreted at the far side of the white dwarf. However, a proper interpretation of the maps requires support from hydrodynamical models of the accretion flow in asynchronous polars.

11.4.3 The Donor Stars

The donor stars in polars could be traced and measured by the narrow emission lines originating in a quasi-chromosphere from the EUV-irradiated part of those stars (examples are shown in Figs. 11.9 and 11.13) and through photospheric absorption lines. Long-period polars are harbouring larger and hotter, thus brighter donors than their short-period brothers. Objects above the CV period gap thus may show emission and absorption lines even in their high accretion states. Below the period gap one typically needs a low state to uncover photospheric radiation from the donors while below $P_{\rm orb} \sim 110$ min the donors become too faint and are always outshone by accretion-induced radiation or even by the white dwarfs. Examples for long-period objects with Doppler maps based on photospheric absorption lines are V1309 Ori, AM Her, and QQ Vul [7, 49, 50]. HU Aqr is a very good example for a short-period object with both, emission and absorption lines [43] and EF Eridani for a polar showing only emission lines from its irradiated substellar secondary [42].

The immediate use of the emission lines lies in the ability to measure the orbital period unequivocally and to locate all other observed features with respect to true phase zero, i.e. with respect to the line joining both stars in the binary. While this may be achieved easily in low-state objects via Gaussian fits to line profiles and sine-fits to the radial velocity curves of the NEL, high-state objects and in particular the asynchronous polars reveal the location of the donor stars only in their Doppler maps.

When looking at the donor stars not all points are equally visible at all times. This violation of another axiom of Doppler tomography was motivating the development of Roche tomography (see [60], and the following chapter in this book). Some important conclusions as well as complexities and potential limitations of Roche tomography can nevertheless be derived from straight Doppler tomography, in particular if the narrow emission lines (NEL) from the irradiated hemisphere are concerned. There is no doubt that the NEL in various objects is a tracer of the donor, but its exact location is a variable. Lines of different atomic species may show different radial velocity amplitudes at a given epoch, hence appear at different v_{y} (Fig. 11.15). Lines of the same atomic species may show different radial velocity amplitudes at different epochs (Fig. 11.9). The size of these effects hasn't been investigated systematically but will depend on the irradiation spectrum and the distribution of gaseous matter around the L_1-point. This becomes obvious from, e.g., the He II maps in HU Aqr (Figs. 11.9 and 11.12), which always show their peak emission at the trailing side of the donor star due to shielding of the leading side by an accretion curtain. This also means that the radial velocity curves of those lines have a small but non-negligible phase-shift with respect to true phase zero.

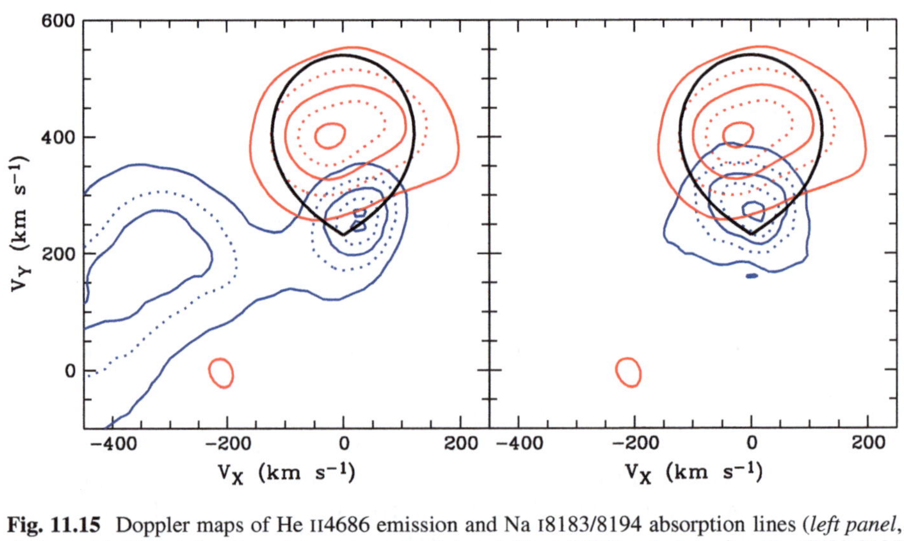

Fig. 11.15 Doppler maps of He II 4686 emission and Na I 8183/8194 absorption lines (*left panel*, shown with *blue* and *red* contours, respectively) and Ca II 8498 emission plus Na I 8183/8194 absorption lines (*right panel*) obtained in a high state of HU Aqr. The size of the Roche lobe corresponds to $Q = M_{\mathrm{wd}}/M_2 = 4.58$ [43]

Thus, if it comes to mass estimates based on the emission lines (with an appropriate K-correction) it is recommended to use preferentially the Ca II triplet both in magnetic and non-magnetic cataclysmic variables [5, 58]. The low Ca I ionisation potential of 6.1 eV, lower than the ionisation potential of hydrogen, seems to ensure full and mainly unshielded irradiation of the secondary, while the ionising photons causing H and He emission have to pass an accretion curtain or other structures which may lead to a very complex irradiation pattern.

Kafka et al. [21, 22] reported the occurrence of satellite lines to Hα in the low states of five polars. Their location in the Doppler maps at high positive v_y-velocities and close to the Roche lobes of the donor star were interpreted as being of magnetic origin in the form of large loops being fed with prominence-like material released from the active donor star (for an illustration of a similar structure in AM Her see [49], while for further reports of sling-shot prominences see [11, 48, 54, 61]). It appears that those structures might stay in place for times significantly longer than the lifetimes of prominences in single fast rotating stars. [22] argue that the magnetic interaction of the two stellar components acts as a stabilising element.

11.4.4 Summary on Polars

Doppler tomography of magnetic CVs has uncovered structures as small as $0.1\,R_\odot$ at distances as large as a few 100 pc. The resolution thus achieved is below the micro-arcsec scale. It has made visible the accretion streams in the strongly magnetic CVs called polars, it has uncovered the existence of accretion curtains and structures on or associated with the donor stars in this type of CVs. The interpretation of the Doppler maps is still not fully developed and needs theoretical support.

References

1. Agafonov, M., Richards, M., Sharova, O.: Three-dimensional Doppler tomogram of gas flows in the Algol-type binary U Coronae Borealis. ApJ **652**, 1547 (2006)
2. Agafonov, M.I., Sharova, O.I.: Few projections astrotomography: radio astronomical approach to 3D reconstruction. Astronomische Nachrichten **326**, 143 (2005)
3. Baba, H., Sadakane, K., Norimoto, Y., Ayani, K., Ioroi, M., Matsumoto, K., Nogami, D., Makita, M., Taichi, K.: Spiral structure in WZ Sagittae around the 2001 outburst maximum. PASJ **54**, L7 (2002)
4. Bassa, C.G., Jonker, P.G., Steeghs, D., Torres, M.A.P.: Optical spectroscopy of the quiescent counterpart to EXO0748-676. MNRAS **399**, 2055 (2009)
5. Beuermann, K., Reinsch, K.: High-resolution spectroscopy of the intermediate polar EX Hydrae. I. Kinematic study and Roche tomography. A&A **480**, 199 (2008)
6. Casares, J., Steeghs, D., Hynes, R.I., Charles, P.A., O'Brien, K.: Bowen fluorescence from the companion star in X1822-371. ApJ **590**, 1041 (2003)

7. Catalán, M.S., Schwope, A.D., Smith, R.C.: Mapping the secondary star in QQ Vulpeculae. MNRAS **310**, 123 (1999)
8. Diaz, M.P., Steiner, J.E.: Locating the emission line regions in polars: Doppler imaging of VV Puppis. A&A **283**, 508 (1994)
9. Echevarría, J.: Doppler tomography in cataclysmic variables: an historical perspective . Mem. Soc. Astron. Ital. **83**, 570 (2012)
10. Ferrario, L., de Martino, D., Gänsicke, B.T.: Magnetic white dwarfs. Space Sci. Rev. (2015)
11. Gänsicke, B.T., Hoard, D.W., Beuermann, K., Sion, E.M., Szkody, P.: HST/GHRS observations of AM Herculis. A&A **338**, 933 (1998)
12. Groot, P.J.: Evolution of spiral shocks in U Geminorum during outburst. ApJ **551**, L89 (2001)
13. Hakala, P., Cropper, M., Ramsay, G.: 3D eclipse mapping in AM Herculis systems – 'genetically modified fireflies'. MNRAS **334**, 990 (2002)
14. Harlaftis, E.T., Steeghs, D., Horne, K., Martín, E., Magazzú, A.: Spiral shocks in the accretion disc of IP Peg during outburst maximum. MNRAS **306**, 348 (1999)
15. Harrop-Allin, M.K., Hakala, P.J., Cropper, M.: Indirect imaging of the accretion stream in eclipsing polars – I. Method and tests. MNRAS **302**, 362 (1999)
16. Heerlein, C., Horne, K., Schwope, A.D.: Modelling of the magnetic accretion flow in HU Aquarii. MNRAS **304**, 145 (1999)
17. Horne, K.: Images of accretion discs. I – the eclipse mapping method. MNRAS **213**, 129 (1985)
18. Horne, K., Marsh, T.R., Cheng, F.H., Hubeny, I., Lanz, T.: HST eclipse mapping of dwarf nova OY Carinae in quiescence: an 'Fe II curtain' with Mach approx. = 6 velocity dispersion veils the white dwarf, ApJ **426**, 294 (1994)
19. Hutchings, J.B., Fisher, W.A., Cowley, A.P., Crampton, D., Liller, M.H.: The complex emission-line structure in the magnetic white dwarf binary 2A 0311-227 /EF Eridani/. ApJ **252**, 690 (1982)
20. Joergens, V., Spruit, H.C., Rutten, R.G.M.: Spirals and the size of the disk in EX Dra. A&A **356**, L33 (2000)
21. Kafka, S., Ribeiro, T., Baptista, R., Honeycutt, R.K., Robertson, J.W.: New complexities in the low-state line profiles of AM Herculis. ApJ **688**, 1302 (2008)
22. Kafka, S., Tappert, C., Ribeiro, T., Honeycutt, R.K., Hoard, D.W., Saar, S.: Low-state magnetic structures in polars: nature or nurture? ApJ **721**, 1714 (2010)
23. Knigge, C.: The donor stars of cataclysmic variables. MNRAS **373**, 484 (2006)
24. Knigge, C.: Erratum: the donor stars of cataclysmic variables. MNRAS **382**, 1982 (2007)
25. Kononov, D.A., Agafonov, M.I., Sharova, O.I., Bisikalo, D.V., Zhilkin, A.G., Sidorov, M.Y.: The applicability of 3D Doppler tomography to studies of polars. Astron. Rep **58**, 881 (2014)
26. Kraft, R.P., Mathews, J., Greenstein, J.L.: Binary stars among cataclysmic variables. II. Nova WZ Sagittae: a possible radiator of gravitational waves. ApJ **136**, 312 (1962)
27. Marsh, T.R.: Kinematics of the helium accretor GP COM. MNRAS **304**, 443 (1999)
28. Marsh, T.R.: Doppler tomography. In: Boffin, H.M.J., Steeghs, D., Cuypers, J. (eds.) Astrotomography, Indirect Imaging Methods in Observational Astronomy. Lecture Notes in Physics, vol. 573, p. 1. Springer, Berlin (2001)
29. Marsh, T.R.: Doppler tomography. Ap&SS **296**, 403 (2005)
30. Marsh, T.R., Horne, K.: Images of accretion discs. II – Doppler tomography. MNRAS **235**, 269 (1988)
31. Morales-Rueda, L., Marsh, T.R., Steeghs, D., Unda-Sanzana, E., Wood, J.H., North, R.C.: New results on GP Com. A&A **405**, 249 (2003)
32. Perryman, M.A.C., Cropper, M., Ramsay, G., Favata, F., Peacock, A., Rando, N., Reynolds, A.: High-speed energy-resolved STJ photometry of the eclipsing binary UZ For. MNRAS **324**, 899 (2001)
33. Richards, M.T., Agafonov, M.I., Sharova, O.I.: New evidence of magnetic interactions between stars from three-dimensional Doppler tomography of Algol binaries: β PER and RS VUL. ApJ **760**, 8 (2012)

34. Richards, M.T., Sharova, O.I., Agafonov, M.I.: Three-dimensional Doppler tomography of the RS Vulpeculae interacting binary. ApJ **720**, 996 (2010)
35. Roelofs, G.H.A., Groot, P.J., Marsh, T.R., Steeghs, D., Barros, S.C.C., Nelemans, G.: SDSS J124058.03-015919.2: a new AM CVn star with a 37-min orbital period. MNRAS **361**, 487 (2005)
36. Roelofs, G.H.A., Groot, P.J., Nelemans, G., Marsh, T.R., Steeghs, D.: Kinematics of the ultracompact helium accretor AM canum venaticorum, MNRAS **371**, 1231 (2006)
37. Roelofs, G.H.A., Rau, A., Marsh, T.R., Steeghs, D., Groot, P.J., Nelemans, G.: Spectroscopic evidence for a 5.4 min orbital period in HM cancri. ApJ **711**, L138 (2010)
38. Rosen, S.R., Mason, K.O., Cordova, F.A.: Phase-resolved optical spectroscopy of the AM HER system E1405-451. MNRAS **224**, 987 (1987)
39. Savoury, C.D.J., Littlefair, S.P., Dhillon, V.S., Marsh, T.R., Gänsicke, B.T., Copperwheat, C.M., Kerry, P., Hickman, R.D.G., Parsons, S.G.: Cataclysmic variables below the period gap: mass determinations of 14 eclipsing systems. MNRAS **415**, 2025 (2011)
40. Schwarz, R., Schwope, A.D., Staude, A., Remillard, R.A.: Doppler tomography of the asynchronous polar BY Camelopardalis, A&A **444**, 213 (2005)
41. Schwope, A.D., Catalán, M.S., Beuermann, K., Metzner, A., Smith, R.C., Steeghs, D.: Multi-epoch Doppler tomography and polarimetry of QQ Vul. MNRAS **313**, 533 (2000)
42. Schwope, A.D., Christensen, L.: X-shooting EF Eridani: further evidence for a massive white dwarf and a sub-stellar secondary. A&A **514**, A89 (2010)
43. Schwope, A.D., Horne, K., Steeghs, D., Still, M.: Dissecting the donor star in the eclipsing polar HU Aquarii. A&A **531**, A34 (2011)
44. Schwope, A.D., Mantel, K.H., Horne, K.: Phase-resolved high-resolution spectrophotometry of the eclipsing polar HU Aquarii. A&A **319**, 894 (1997)
45. Schwope, A.D., Staude, A., Vogel, J., Schwarz, R.: Indirect imaging of polars. Astronomische Nachrichten **325**, 197 (2004)
46. Simon, K.P., Sturm, E.: Disentangling of composite spectra. A&A **281**, 286 (1994)
47. Skilling, J., Bryan, R.K.: Maximum entropy image reconstruction – general algorithm. MNRAS **211**, 111 (1984)
48. Southworth, J., Marsh, T.R., Gänsicke, B.T., Steeghs, D., Copperwheat, C.M.: Orbital periods of cataclysmic variables identified by the SDSS. VIII. A slingshot prominence in SDSS J003941.06+005427.5? A&A **524**, A86 (2010)
49. Staude, A., Schwope, A.D., Hedelt, P., Rau, A., Schwarz, R.: Tomography of AM Her and QQ Vul. In: Vrielmann, S., Cropper, M. (eds.) IAU Colloq. 190: Magnetic Cataclysmic Variables. Astronomical Society of the Pacific Conference Series, vol. 315, p. 251. Astronomical Society of the Pacific, San Francisco (2004)
50. Staude, A., Schwope, A.D., Schwarz, R.: System parameters of the long-period polar V1309 Ori. A&A **374**, 588 (2001)
51. Steeghs, D.: Extending emission-line Doppler tomography: mapping-modulated line flux. MNRAS **344**, 448 (2003)
52. Steeghs, D., Casares, J.: The mass donor of scorpius X-1 revealed. ApJ **568**, 273 (2002)
53. Steeghs, D., Harlaftis, E.T., Horne, K.: Spiral structure in the accretion disc of the binary IP Pegasi. MNRAS **290**, L28 (1997)
54. Steeghs, D., Horne, K., Marsh, T.R., Donati, J.F.: Slingshot prominences during dwarf nova outbursts? MNRAS **281**, 626 (1996)
55. Steeghs, D., Marsh, T., Knigge, C., Maxted, P.F.L., Kuulkers, E., Skidmore, W.: Emission from the secondary star in the old cataclysmic variable WZ Sagittae. ApJ **562**, L145 (2001)
56. Stover, R.J.: A radial-velocity study of the dwarf nova RU Pegasi. ApJ **249**, 673 (1981)
57. Uemura, M., Kato, T., Nogami, D., Mennickent, R.: Doppler tomography by total variation minimization. PASJ **67**, 22 (2015)

58. van Spaandonk, L., Steeghs, D., Marsh, T.R., Torres, M.A.P.: Time-resolved spectroscopy of the pulsating CV GW Lib. MNRAS **401**, 1857(2010)
59. Vrielmann, S., Schwope, A.D.: Accretion stream mapping of HU Aquarii. MNRAS **322**, 269 (2001)
60. Watson, C.A., Dhillon, V.S.: Roche tomography of cataclysmic variables – I. Artefacts and techniques. MNRAS **326**, 67 (2001)
61. Watson, C.A., Steeghs, D., Shahbaz, T., Dhillon, V.S.: Roche tomography of cataclysmic variables – IV. Star-spots and slingshot prominences on BV Cen. MNRAS **382**, 1105 (2007)

Chapter 12
Tomographic Imaging of Stellar Surfaces and Interacting Binary Systems

Julien Morin, Colin Alastair Hill, and Christopher Allan Watson

12.1 Doppler and Zeeman-Doppler Imaging of Stellar Surfaces

Doppler imaging (hereafter DI) and related methods provide the highest angular resolution available to stellar physicists. Relying on the relationship between the spatial position of inhomogeneities on stellar surfaces and the radial velocity of the corresponding features in spectral lines, these inverse methods are indeed the only techniques able to resolve features at the photospheres of main sequence dwarf stars and even brown dwarfs [18]. This section first presents the main scientific application of DI techniques: to study stellar magnetism, followed by an overview of the principles of "classical" DI (aimed at mapping temperature inhomogeneities) and then Zeeman-Doppler imaging (hereafter ZDI, aimed at mapping large scale magnetic fields). We conclude with a brief summary of scientific highlights enabled by these techniques.

12.1.1 Doppler Imaging as a Tool to Study Stellar Magnetism

DI techniques mainly aim at studying stellar magnetism. They allow astrophysicists to map either large scale surface magnetic fields (with ZDI based on the inversion

J. Morin (✉)
LUPM, Université de Montpellier & CNRS, Place Eugène Bataillon, 34095 Montpellier Cedex 05, France
e-mail: julien.morin@umontpellier.fr

C.A. Hill • C.A. Watson
Astrophysics Research Centre, Queen's University Belfast, BT7 1NN, Belfast, UK
e-mail: chill17@qub.ac.uk; c.a.watson@qub.ac.uk

H.M.J. Boffin et al. (eds.), *Astronomy at High Angular Resolution*, Astrophysics and Space Science Library 439, DOI 10.1007/978-3-319-39739-9_12

of polarised spectra), or spots with different temperatures or chemical compositions than the surrounding "quiet" photosphere. These surface inhomogeneities are themselves in most cases thought to be intimately connected with stellar magnetism. As such, these techniques play an important role in modern stellar physics: to better constrain the role of magnetic fields in stellar evolution.

During the last two decades, surveys aimed at detecting and measuring stellar magnetic fields have demonstrated that magnetism is ubiquitous throughout the Hertzsprung-Russell diagram – surface magnetic fields have been detected across a wide range of masses and evolutionary stages – and that the detected fields exhibit a wide variety of properties in terms of modulus, geometry and temporal variability. The reader is referred to [23] for a review. Moreover, magnetic fields are thought to play an important role in a number of key physical processes in stellar physics; a few representative examples are listed below.

Star formation Star formation is often thought to be the phase of stellar evolution in which magnetic fields play the most important role. Magnetic effects are indeed among the main factors that drive the very collapse of a molecular cloud and therefore star formation efficiency (see, e.g., [41] for a review). Once a protostar is actually formed, magnetic fields play a key role in the processes that determine the exchange of mass and angular momentum between the star and its environment (accretion, outflows, star-disk coupling, see [11] for the case of T Tauri stars).

Transport phenomena and evolution of stellar rotation In addition to the evolution of the stellar radius, which results in a change in stellar spin if angular momentum is conserved, two main types of processes have an impact on surface rotation rates: firstly a net gain or loss of angular momentum associated with accretion and mass-loss processes; and secondly, the redistribution of angular momentum inside the star due to internal transport mechanisms. Some of these processes can also transport chemical species and result in inhomogeneous surface distributions of specific elements. Most mechanisms of both types can be strongly influenced by – or are even powered by – magnetic fields. The reader is referred to [12] for an account of the angular momentum evolution of low-mass stars, and to [67] for a review of transport processes in stellar interiors.

Activity and space weather The broad notion of "stellar activity" encompasses all the time-dependent phenomena resulting from the action of magnetic fields on the atmospheres of cool stars. Stellar activity is also associated with significant amounts of high-energy particles (stellar wind) and electromagnetic radiation (UV and X-rays), which are potentially harmful for putative life developing on planets orbiting these cool active stars [60].

Origin of stellar magnetic fields The origin of stellar magnetic fields is, in itself, a very rich field of research. The basic picture is that there are two main origins: stars with an outer radiative zone (spectral types hotter than mid-F) can host simple and steady fossil magnetic fields, which are the remnants of a previous evolutionary stage (although alternatives have been proposed, e.g., [13]), whereas cool stars with an outer convective envelope generate through dynamo action time-dependent magnetic fields which can be very complex. The reader is referred to [16] for an in-depth view of the issue of the origin of stellar magnetic fields.

12.1.2 Principles of Doppler Imaging (DI)

With tomographic imaging it is potentially possible to map any physical quantity which affects the formation of spectral lines, at the surface of a rapidly rotating star. For a spectral line with significant rotational broadening, because of the Doppler effect, the positions of heterogeneities on the visible stellar disc are directly connected to the wavelength of the corresponding distortions inside the spectral line. In the limiting case of a spectral line dominated by rotational broadening, i.e.:

$$v \sin i \gg W, \tag{12.1}$$

where $v \sin i$ is the equatorial rotation velocity of the star projected on the line-of-sight, and W is the typical width of the local line profile, a stellar spectral line can be essentially seen as a 1-D map of the visible stellar disc [21, 52]. This "map" would indeed be only resolved in the direction perpendicular to the rotation axis of the star and completely blurred in the orthogonal direction. This is illustrated in Fig. 12.1, which represents the *direct (or forward) problem* computationally: from the distribution of surface inhomogeneities and a model of line formation it is possible to compute the corresponding integrated line profile.

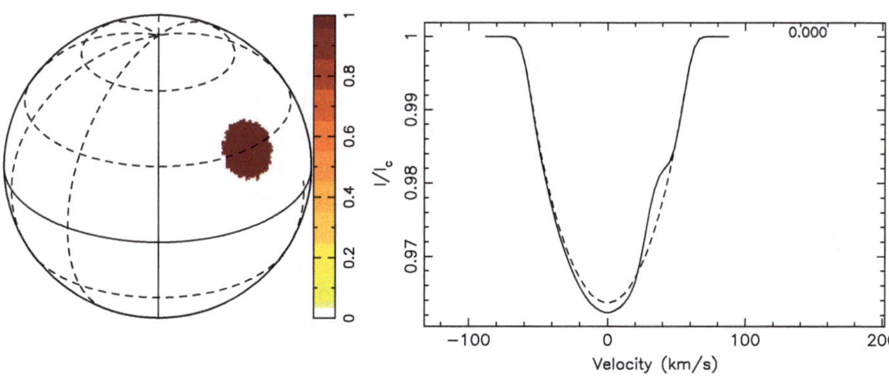

Fig. 12.1 Basic principles of Doppler Imaging. *Left:* schematic view of a stellar surface with one single infinitely dark spot covering 1 % of the surface. *Right:* corresponding spectral line profile (*solid line*) and the same line profile if there were no spot (*dashed line*). An inclination angle of $i = 60°$ and a projected equatorial velocity of $v \sin i = 60\,\mathrm{km\,s^{-1}}$ have been assumed in this forward modelling. The position of the bump in the line profile is directly linked to the position of the spot on the visible stellar disc

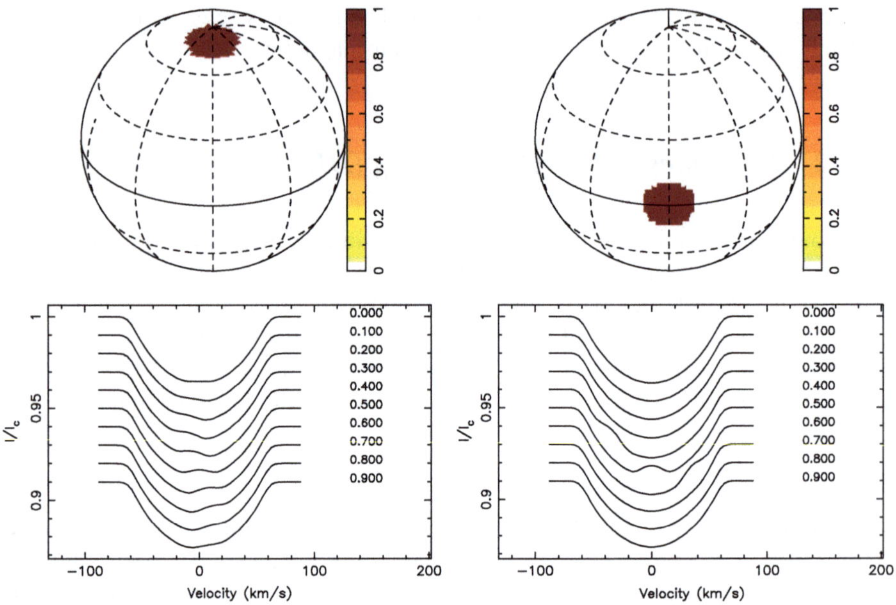

Fig. 12.2 For each column the *top panel* represents the stellar surface model with one spot covering 1 % of the surface and crossing the central meridian at phase 0.5; the *bottom panel* displays the corresponding time-series of ten spectra evenly spread over one stellar rotation period. Stellar parameters are the same as in Fig. 12.1. *Left:* Spot latitude $l = 80°$: the corresponding spectral line distortion is confined at the centre of the line (low radial velocities) and is always visible. *Right:* Spot latitude $l = 0°$: the corresponding spectral line distortion is crosses the whole line profile and is only visible during a fraction of the stellar rotation cycle

The basic idea of DI consists of recording a number of spectra that densely sample the stellar rotation cycle in order to solve the *inverse problem*, i.e. to reconstruct a 2-D map of the studied physical quantity at the stellar surface [36, 86, 104]. The information on the latitude of surface inhomogeneities is encoded in the temporal evolution of the spectral line, see Fig. 12.2. Practically, the model of the stellar surface is discretised in the form of a grid with a few thousands cells or pixels. Imaging codes start from an initial assumption (usually a homogeneous photosphere) and iteratively add spots to the reconstructed map until the corresponding line profiles match the observed ones within the noise level. Because the DI problem is generally underdetermined, due to the presence of noise and of the limited spatial resolution of the reconstruction, multiple maps can produce line profiles that are in agreement with the observations. A unique image can be obtained by choosing the solution that satisfies an additional criterion; this can be gradient minimisation of the reconstructed quantity with Tikhonov regularisation [101] or maximisation of the Shannon entropy [97], as is often done in radio interferometric image reconstruction.

This method has been successfully applied to map temperature inhomogeneities at the surfaces of many cool stars (see reviews [8, 100]) as well as to map chemical abundance patches on chemically peculiar stars (see, e.g., [1, 111]).

To conclude this description of DI, it is worth mentioning the main limitations of DI techniques and discussing the properties of the most appropriate targets. Firstly, DI is obviously limited to map the portion of the star which is visible at a given point during the rotation cycle, this ranges from half the stellar surface for a pole-on star, to the whole stellar surface for an equator-on star. Note moreover that neither of these inclination angles are particularly favourable for DI: a star seen close to pole-on tends to have a low $v \sin i$ and can therefore not be resolved (see below), while for a star seen close to equator-on a degeneracy appears: spots located symmetrically with respect to the equator result in the same series of spectroscopic signatures, producing a mirroring effect in the maps. Secondly, the best achievable resolution of a DI map is directly set by the ratio of its $v \sin i$ to the width of its spectral lines W: the larger the ratio, the higher the number of resolved elements at the stellar surface. A simple rule is that for spectral lines dominated by rotational broadening the maximum number of resolved elements, n, is roughly set by:

$$n = \frac{2\pi\ v \sin i}{W}, \tag{12.2}$$

while for slower rotators – as different parts of the stellar photosphere are visible at different rotational phases – about four elements can still potentially be resolved. Stars with very high $v \sin i$ are therefore natural targets for DI. However for very fast rotators the very low amplitude of line profile distortions to be modelled becomes a limiting factor. Furthermore, in order to limit phase smearing it is important to keep the exposure time a small fraction of the stellar rotation period. For very fast rotators it can become a challenge to obtain sufficient S/N with integration times that are limited to a few percent of the rotation period. As a result of this resolution vs signal-to-noise ratio (S/N) trade-off, stars with $v \sin i$ ranging from 20 to 100 km s^{-1} are generally considered the best targets for DI. For a more in-depth discussion of the limitations of DI, of the effect non-ideal datasets and of uncertainties on stellar parameters ($v \sin i$, inclination angle of the rotation axis on the line-of-sight, etc.) see [17, 85]. Finally, it is important to stress that only relatively large-scale structures can be mapped with DI and that both theoretical and observational studies indicate that small-scale spots are present on most active stars but cannot be resolved with DI [49, 76, 99].

12.1.3 Measurements of Stellar Magnetic Fields and Zeeman-Doppler Imaging

Zeeman-Doppler Imaging is an extension of DI aimed at mapping stellar magnetic fields from spectropolarimetric time-series [91]. In addition to the effects pointed

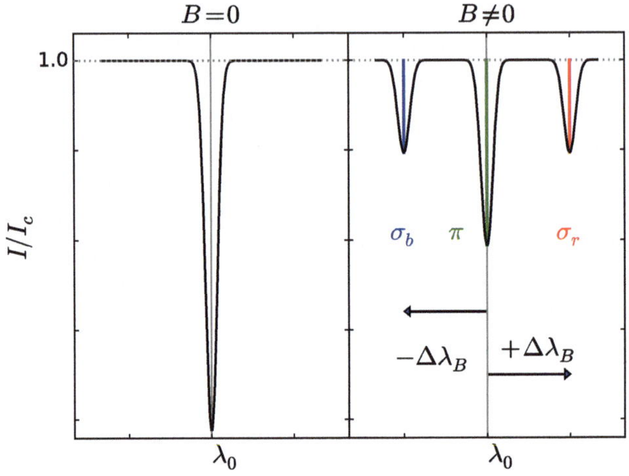

Fig. 12.3 Schematic view of the Zeeman effect for a normal triplet and a large splitting. *Left:* A magnetically sensitive spectral line formed without any magnetic field. *Right:* The same spectral line formed with a magnetic field. The π, σ_b and σ_r components are respectively marked with a *green*, *blue* and *red vertical line*. The x-axis ticks are in units of Zeeman splitting $\Delta\lambda_B$ (Eq. 12.3), and λ_0 is the central wavelength of the line without a magnetic field

out in Sect. 12.1.2 (Doppler effect, rotational modulation), it relies on the properties of the Zeeman effect to disentangle the different field orientations and actually map the magnetic field vector.

Zeeman measurements in unpolarised light Direct measurements of stellar photospheric magnetic fields rely on the Zeeman effect [112], which is the splitting of spectral lines formed in the presence of a magnetic field. In the basic case of the normal Zeeman triplet a spectral line with a central wavelength, λ_0, is split into three components: one π component at λ_0 and two σ components located at $\lambda_0 \pm \Delta\lambda_B$, see Fig. 12.3. The Zeeman splitting, $\Delta\lambda_B$, is proportional to the magnetic field modulus, B, such that (in CGS units):

$$\Delta\lambda_B = \frac{\lambda_0^2 e \,\overline{g} B}{4\pi m_e c^2} = 4.67 \times 10^{-12} \lambda_0^2 \overline{g} B, \tag{12.3}$$

where the effective Landé factor, $\overline{g}$, is a dimensionless factor ranging from 0 to 3 (depending on the spectral line considered) and the numerical constant is computed for wavelength in nm, and B in G. The reader is referred to [62] for a more complete description of the Zeeman effect in the context of stellar physics. According to this equation, for a given magnetically-sensitive spectral line (i.e. with $\overline{g} \neq 0$ and preferably $\overline{g} > 1$) the magnetic field modulus is directly inferred from the measured splitting, this is the method used by G.E. Hale in 1908 to perform the first magnetic field measurement in a sunspot [38]. This method requires high-resolution spectra and high S/N data. Indeed, in the favourable case of the Fe I line at $\lambda_0 = 630.25$ nm

($\overline{g} = 2.5$), and a magnetic field modulus, $B = 2\,\mathrm{kG}$ (typical of a sunspot), Eq. 12.3 yields $\Delta\lambda_B = 9.2 \times 10^{-3}$ nm (equivalent radial velocity shift, $\Delta v_B = 4.4\,\mathrm{km\,s^{-1}}$).

In the case of stars other than the Sun the stellar surface is not spatially resolved and the measured Zeeman effect corresponds to an average over the visible stellar disc. For cool, active stars featuring a distribution of surface magnetic field strengths and convective velocities, combined with projected rotational velocities of at least a few $\mathrm{km\,s^{-1}}$, the individual Zeeman components are generally not resolved, and the presence of a magnetic field results in a spectral line broadening rather than an actual splitting.

For the most active cool stars it is possible to derive the information on the range of surface magnetic field strengths from high-quality red optical and near-infrared spectra [51, 96], but in most cases only the magnetic field modulus averaged over the visible stellar disc $\langle||\mathbf{B}||\rangle$ can be inferred. This quantity is often termed the magnetic flux to stress that it is not a local magnetic field strength measurement, although it has the dimension of a magnetic flux density expressed in Gauss (CGS units) or Tesla (SI units), also see the discussion in [73]. Average magnetic field measurements based on the Zeeman effect in unpolarised light can detect and quantify strong magnetic fields such as those observed intermediate mass Ap/Bp stars (e.g., [68]) or cool active stars (see review [83]) very efficiently. They are however not well-suited to stars hosting weak magnetic fields such as the Sun – magnetic fluxes weaker than a few hundreds Gauss cannot be securely disentangled from other potential sources of spectral line broadening [2] – or to stars with $v \sin i \gg v_B$ (about $20\,\mathrm{km\,s^{-1}}$ for cool active stars) for which the Zeeman broadening is completely blurred by rotational broadening and cannot be recovered. Finally, Zeeman broadening measurements are almost insensitive to the magnetic field orientation: they can probe efficiently simple ordered fields as well as complex highly-tangled ones, but provide little information on the magnetic field orientation.

Zeeman effect measurements in polarised light A complementary approach relies on another property of the Zeeman effect: the polarisation of the π- and σ-components, which also encodes information on the vector properties of the magnetic field: circular polarisation is sensitive to the line-of-sight projection of the vector magnetic field, while linear polarisation is sensitive to the component orthogonal to the line-of-sight (see Fig. 12.4). A brief account of spectropolarimetric techniques is given below; for a comprehensive introduction to astrophysical spectropolarimetry the reader is referred to [19]. H.W. Babcock was the first to successfully use this effect to measure stellar and solar magnetic fields [3, 4] It essentially consists of measuring the shift between two spectra of the same region seen through a right- and a left-handed circular polariser, corresponding to $2\Delta\lambda_B$ and hence directly proportional to the magnetic field strength. The same basic idea was also used to detect and monitor magnetic fields of strongly magnetic stars across the Hertzsprung-Russell diagram [5, 9, 10]. A modern extension of this method is the so-called "regression method" applied to low-resolution spectropolarimetric data [6]. The main advantage of the magnetograph method is that it is differential and does not suffer modelling errors, thus allowing it to be very sensitive (B_ℓ down

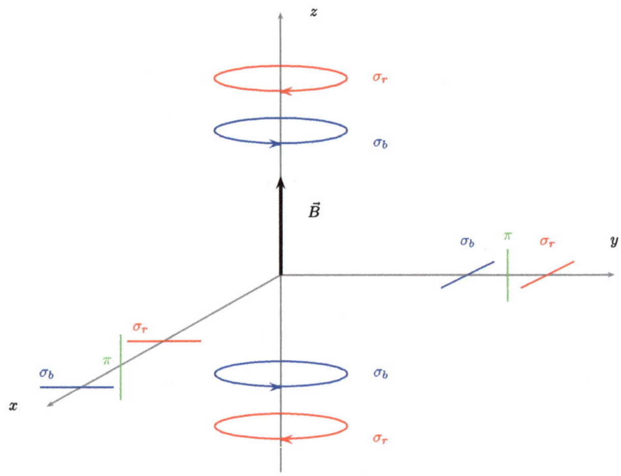

Fig. 12.4 Schematic representation of the polarisation of the components of the normal Zeeman triplet as a function of the relative orientation of **B** with respect to the line-of-sight. Radiation is circularly (linearly) polarised when the line-of-sight is aligned with (perpendicular to) the magnetic field vector

to a few tens of Gauss for stars). Indeed the null-field model is perfectly known: it corresponds to right- and left-hand circularly polarised spectra being exactly identical, while any wavelength shift between the two spectra is indicative of a magnetic field.

It is however not optimal in the sense that it assesses the so-called "longitudinal magnetic field", often called B_ℓ or B_z, which is the average over the visible stellar disc of the line-of-sight component of the magnetic field vector. A star hosting a strong magnetic field is composed of spots of opposite polarities visible simultaneously and therefore may have a longitudinal magnetic field close to zero, which would not be detectable by a magnetograph.

For stars hosting strong and simple magnetic fields – such as chemically peculiar Ap/Bp stars – time-series of such longitudinal field measurements (generally combined with Zeeman splitting measurements in unpolarised light and transverse field measurements from linear polarisation data) can be used to model the surface magnetic field as a low-order multipole expansion [63]. An improvement of the technique consists in performing high-resolution spectropolarimetric measurements, i.e. measuring directly the net circular (Stokes V parameter) and linear (Stokes Q and U parameters) polarisation in spectral lines as a function of wavelength. The development of this technique has largely relied on the advent of high-resolution high-efficiency stellar spectropolarimeters, such as the twin instruments ESPaDOnS at CFHT and NARVAL at TBL [26], and more recently, the polarimetric module for HARPS on the ESO 3.6 m telescope [98]. Due to the aforementioned relation between the position of a feature at the surface of a rotating star and the wavelength of the corresponding contribution to the spectral line profile, polarimetric signatures of spots of opposite polarities do not necessarily cancel each other and can be detected at the surface of rapidly rotating stars, even if the corresponding B_ℓ is almost zero. More complex magnetic fields can be detected for stars with larger $v \sin i$ (Equation 12.2) but magnetic fields occurring at the spatial scale of surface convection in cool main sequence stars always cancel out. In order to take full advantage of the information encoded in high-resolution spectropolarimetric data M. Semel proposed to extend Doppler Imaging to spectropolarimetric data in order to map the magnetic field vector at the surface of cool stars, and termed this technique Zeeman-Doppler Imaging or ZDI [91].

Zeeman-Doppler Imaging The implementation of ZDI is very similar to that in DI codes: starting from an initial assumption (usually a stellar surface without any magnetic field) magnetic regions are iteratively added until the synthetic polarised spectra computed from the map match the observed ones within noise. Uniqueness of the solution is also ensured by a regularisation criterion, in particular the maximum entropy algorithm of Skilling and Bryan [97] was adapted to minimize a quantity based on the magnetic energy of the reconstructed map [14, 24]. Several variants of ZDI have been implemented and successfully used, all relying largely on the same basic principles [15, 45, 80]. A few specificities of ZDI are pointed out below.

Due to the low-amplitude of Zeeman-induced linear polarisation in spectral lines of most main sequence stars, only circularly polarised spectra are available in most cases – linear polarisation is indeed a second-order effect in the weak-field regime whereas circular polarisation is a first-order effect, see [19]. Reliable ZDI inversions can however be performed from circular polarisation data only [22]. In particular, the ability of ZDI to disentangle between different magnetic field orientations is mainly due to the fact that they result in characteristic evolution of circularly polarised signatures during a stellar rotation (Fig. 12.5). It was, however, shown that mapping abundance patterns and magnetic fields simultaneously at the surface of Ap/Bp stars benefits from the use of the full information from Stokes I, Q, U and V spectra [22, 54]. The latest investigations of ZDI inversions of a cool very active star also reveals that using the full-Stokes data can improve the spatial resolution of the resulting large scale magnetic maps [88].

ZDI relies on the analysis of polarisation in spectral lines, which is very small for most stars. For instance circular polarisation peaks at $\sim 10^{-3}$ of the level of the unpolarised continuum in very active cool stars, hence detecting this signal in an individual spectral line therefore requires a signal to noise ratio of several thousands. In order to apply spectropolarimetric techniques and in particular ZDI to a wider range of stellar systems, it was identified from the start that one should take advantage of the fact that the many absorption lines present in stellar spectra essentially repeat the same information and can therefore be combined together [10, 91]. The Least-Squares Deconvolution technique implements this idea using a method similar to cross-correlation used to derive radial velocities [22]. Although it was originally designed for circular polarisation in the weak-field regime – when the Zeeman splitting is smaller than the linewidth, i.e. fields below ~ 2 kG for cool stars – it was later adapted to linear polarisation, and shown to be of use for stronger fields with some limitations [56, 106]. Improvements of LSD as well as alternative multi-line methods have been proposed, which can solve some of its limitations [15, 66, 92, 93].

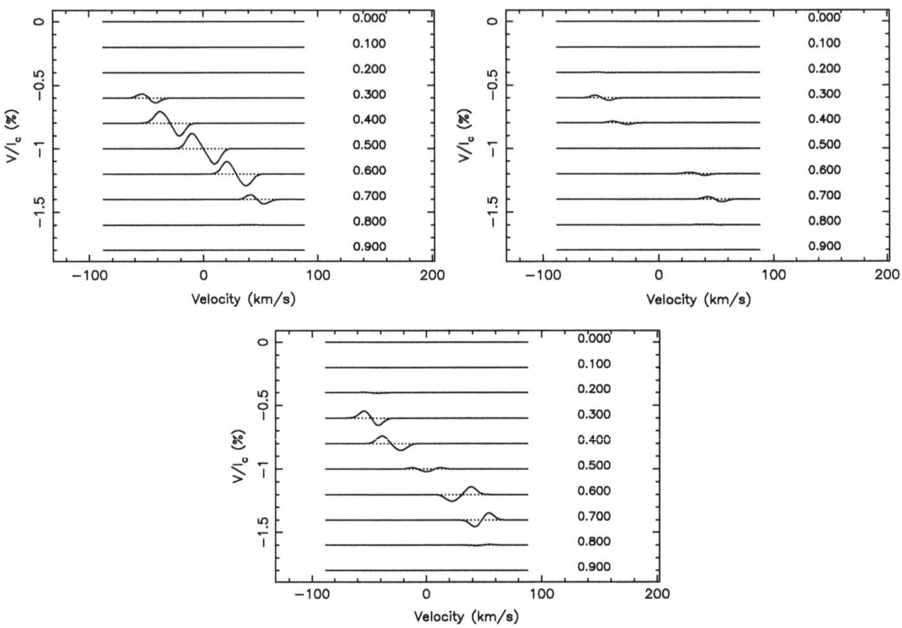

Fig. 12.5 Circularly polarised (Stokes V) line profiles for the same geometry and stellar parameters as Fig. 12.1, except that the dark spot has been replaced by a magnetic spot with $B = 2\,\mathrm{kG}$. The field orientation is radial (*top left*), meridional (*top right*) and azimuthal (*bottom*). The three different field orientations give rise to Stokes V time-series that have distinct differences from each other. In particular the azimuthal field behaves quite differently from the other orientations: the sign of the signature reverses during the stellar rotation cycle

In classical DI spots are reconstructed independently on each cell of the discretised map. In ZDI, the mapped magnetic field can be made divergence-free by construction if the reconstructed quantities are the coefficients of a poloidal-toroidal decomposition projected onto a spherical harmonics basis [25, 46]. This method not only introduces physical information in the reconstruction but also allows to provide directly useful diagnostics (e.g., fraction of magnetic energy residing in poloidal, toroidal, axisymmetric fields) and was shown to be more efficient at recovering simple multipolar geometries.

12.1.4 Zeeman-Doppler Imaging Science Highlights

To conclude this section on tomographic imaging of stellar surfaces, we present a brief selection of results highlighting the contribution of the Zeeman-Doppler Imaging technique to modern stellar physics.

12.1.4.1 Chemically Peculiar Stars

Among intermediate-mass main sequence stars (masses ranging from 1.5 to 8 $M_{\odot}$), chemically peculiar (CP) stars classified Ap and Bp have long been known to exhibit specific abundance patterns along with strong magnetic fields, while no magnetic fields were detected on for other classes of CP stars (e.g., Am and HgMn stars) [82]. Spectroscopic and spectropolarimetric monitoring of Ap/Bp stars have led to the conclusion that the magnetic fields of these objects are in most case simple dipoles tilted with respect to the rotation axis of the stars, steady (no evolution observed on timescales of decades), and that they exhibit elemental abundance spots which are connected to the surface magnetic field [61].

Recent detailed studies of Ap/Bp stars relying on tomographic imaging techniques have brought important precisions and corrections to these first ideas. Studies based on high-quality datasets (in particular including linear spectropolarimetry) and detailed modelling have concluded that although the radial component of the surface magnetic field has a dominant dipolar component, it also features important small-scale patches and horizontal fields [55, 57]. The relation between abundance spots and surface magnetic fields has been shown to be more complex than initially thought: for instance on one Ap star only a few elements exhibit ring-like structures correlating with the dipolar component (e.g., Oxygen), while heavy elements such as Ba, Nd and Eu display a variety of surface distributions [58]. It was also shown that previously noticed discrepancies between abundances derived from different lines of the same element can be reconciled by considering steep vertical abundance gradients in the atmosphere of these stars [53, 95]. On HgMn peculiar stars, time-evolving abundance spots have been discovered and the absence of significant magnetic field confirmed [59, 64]. This result questions the previously accepted idea that abundance patches are always connected to strong magnetic fields. All these results shed a new light on the relative importance of magnetic fields with respect to other physical processes causing or erasing chemical peculiarities (atomic diffusion, hydrodynamic instabilities, . . .).

12.1.4.2 Cool Main Sequence Stars

As opposed to Ap/Bp stars, almost no spectropolarimetric measurements of magnetic field at the surface of a cool main sequence star other than the Sun existed before the advent of second generation spectropolarimeters coupled with multi-line techniques.[1] Previous knowledge relied mainly on measurements of activity proxies, such as diagnostics of chromospheric (especially in CaII H&K lines) and coronal (at X-ray wavelengths) heating. Studies of G-K dwarfs demonstrated that the chromospheric and coronal activity levels depend on the stellar rotation rate (especially when measured as a Rossby number) until a saturation level is reached

[1]The only exception recorded here being the chromospherically active star ξ Boo [10].

[65, 75, 81]; and that cool main sequence stars also show activity cycles similar to those observed on the Sun [7, 39]. The first spectropolarimetric survey of G-K dwarfs has revealed the overall dependence of the magnetic topology on rotation rate – while solar-like slow rotators have predominantly poloidal fields, faster rotators generate strong toroidal fields [77] – as well as long-term changes in the large scale field topology, with some stars even showing regular magnetic cycles with polarity inversions [33, 50, 69, 78].

With the first spectropolarimetric survey of M dwarfs it was possible to show that a dramatic change occurs in the properties of the large-scale component of the magnetic field of M dwarfs close to the mass at which models predict that stars are fully-convective [29, 70], see Figs. 12.6 and 12.7. No such change had previously been identified from either chromospheric (or coronal) activity measurements or Zeeman broadening measurements (which are sensitive to the small scale magnetic fields, not only the large spatial scales as spectropolarimetry does). The ZDI studies show that the fraction of magnetic energy in the large-scale component dramatically increases for fully convective stars with respect to partly convective ones [84]. For the lowest mass M dwarfs the puzzling coexistence of two very different types of magnetic topologies for stars with very similar stellar parameters [71] has been interpreted as possible evidence for dynamo bistability in these objects [35, 72].

12.1.4.3 Classical T Tauri Stars

Stellar magnetic fields are known to be a key parameter for the evolution of accreting T Tauri stars (TTS). The main physical processes acting at this evolutionary stage – star-disc interaction, magnetospheric accretion, high-velocity outflows – are indeed largely controlled by the magnetic field of the star [11]. Zeeman-Doppler Imaging of TTS takes advantage of the information contained in both photospheric absorption lines, and emission lines formed mainly in the accretion shock, in order to map the surface brightness and magnetic field vector simultaneously as well as the accretion hotspots [28]. Recent ZDI analysis of a sample of TTS has revealed a wide variety of magnetic topologies which depend on the location of the star in a mass-age-rotation parameter space, and evolve on timescales of less than a year [27, 30, 48]. These results imply that the magnetic fields of these young stars (a few Myr) are not of fossil origin, but rather generated by dynamo action – with a similar dependence on stellar internal structure as M dwarfs [37]. They have also prompted theorists to develop more elaborate models of magnetospheric accretion taking into account the complexity of the stellar magnetic field [87].

12.1.4.4 Ongoing and Future Developments

Ongoing observing programs of cool stars aim to fill the gaps in mass-rotation-age parameter space in order to build a consistent picture of dynamo action in cool stars and of the effects of magnetic fields on their evolution, with first global trends

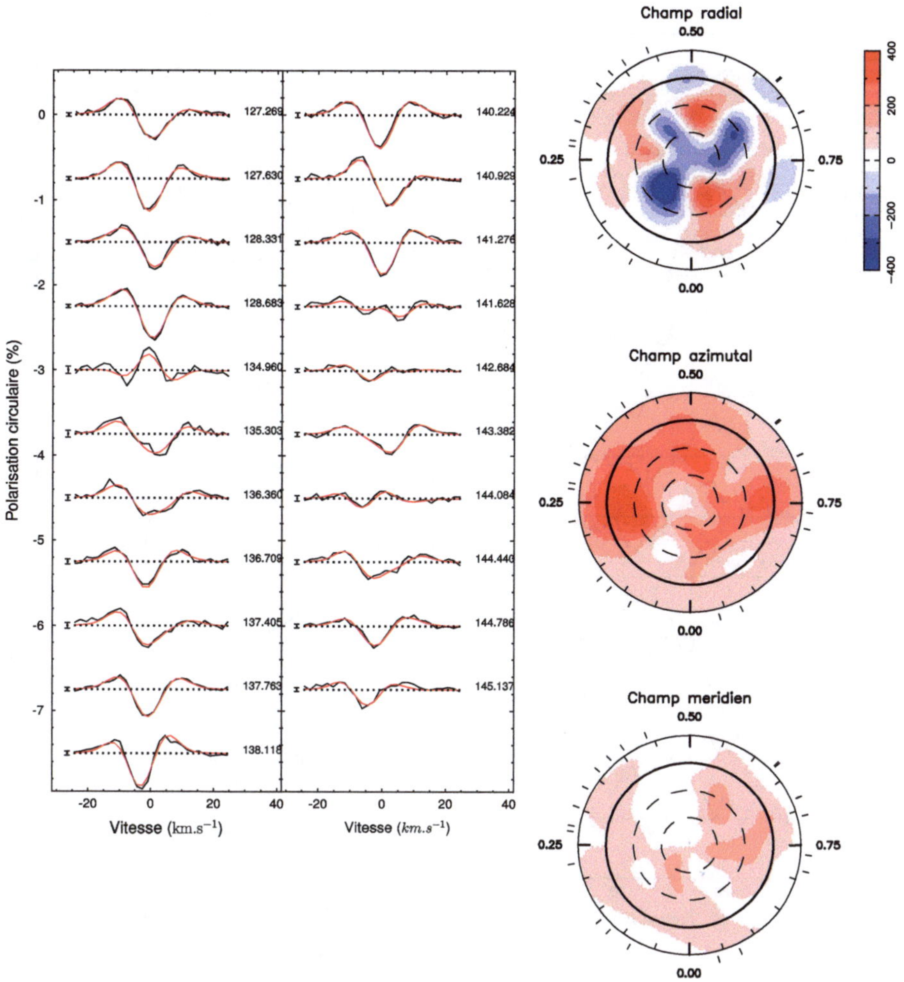

Fig. 12.6 ZDI reconstruction of the surface magnetic field of DT Vir = GJ 494A, $M_\star = 0.59\,M_\odot$, $P_{\rm rot} = 2.85\,$d. *Left:* Time-series of Stokes V LSD line profiles (*black*) and corresponding fit obtained with ZDI. The profiles are vertically shifted for display purposes, a *dotted line* shows the null-polarisation level for each LSD profile. A $\pm 1\sigma$ error-bar is displayed on the *left* of each profile. *Right:* Magnetic map reconstructed with ZDI in flattened polar projection for the three component of the magnetic field: radial (*top*), azimuthal (*middle*) and meridional (*bottom*). On each plot the visible is at the center and the equator is depicted by a *bold circle* (Adapted from [29])

already emerging (e.g., [102]). The first ZDI surveys are now being complemented to assess more specific points such as the effect binarity has on magnetic field generation and activity [74], or the study of planet-hosting stars [34, 103]. In the next few years, the synergy between between planet-search programs and ZDI

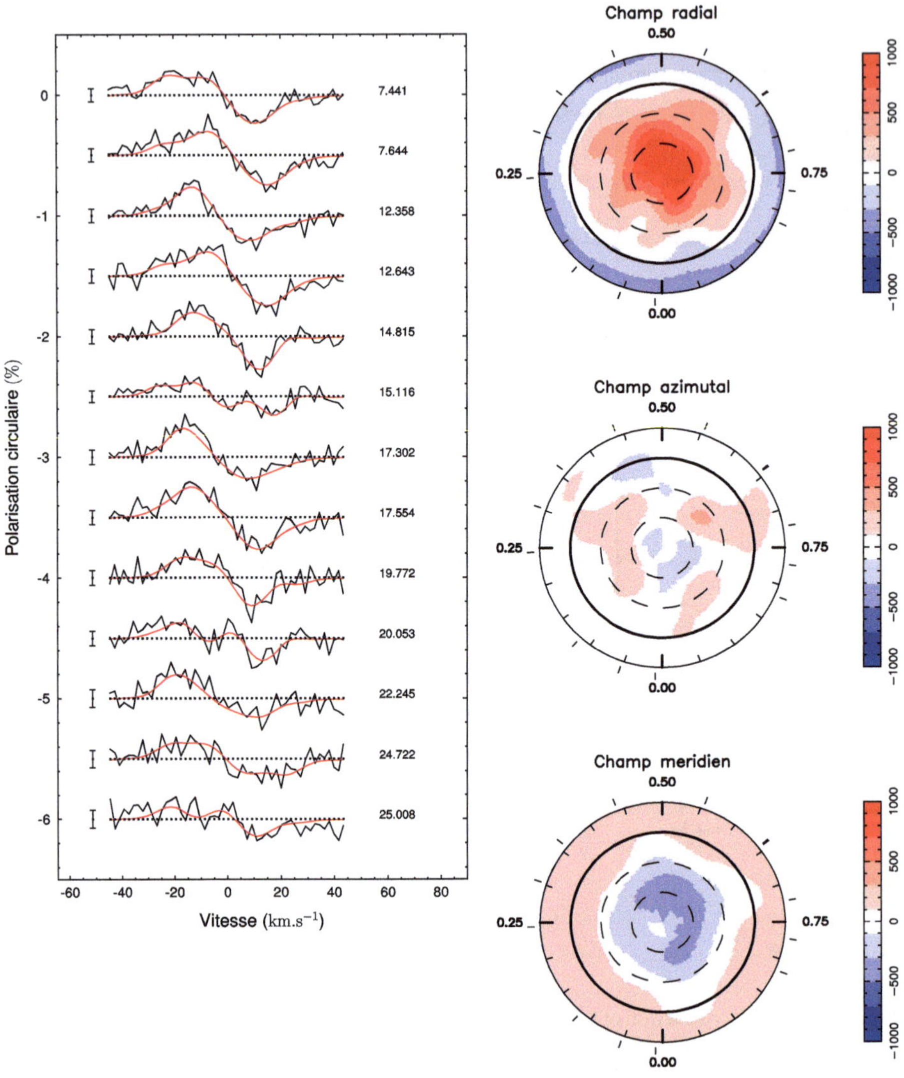

Fig. 12.7 Same as Fig. 12.6 for EQ Peg B = GJ 896B, $M_\star = 0.25\,M_\odot$, $P_{rot} = 0.405$ d. The partly convective star DT Vir hosts a complex multipolar magnetic field of moderate intensity, whereas the fully-convective star EQ Peg B displays a strong almost dipolar large-scale magnetic field (Adapted from [70])

studies is going to develop with new methods being developed [31, 40, 79] and new instruments – nIR spectropolarimeters [20, 32] – coming online soon.

12.2 Roche Tomography

Roche tomography was originally conceived by Rutten et al., specifically to provide surface maps of the Roche-lobe filling donor stars in cataclysmic variables (CVs) – semi-detached binaries (typically) consisting of a lower main-sequence star which is transferring mass to a white dwarf primary via Roche-lobe overflow [89]. However, the technique is equally applicable to Roche-lobe-filling stars of any system (for example low-mass X-ray binaries). Given the vast distances to these systems, direct imaging of the components of CVs is currently impossible and therefore indirect imaging techniques must be applied. In 1994 Roche tomography was computationally expensive and, arguably, better surface maps of single stars could be produced using Doppler imaging [89]. So why image the donor stars in CVs?

12.2.1 The Motivation for Roche Tomography

The initial motivation for developing Roche tomography was to better determine the component masses in interacting binaries, as these are crucial if one is to test binary evolutionary theory. It was well known that irradiation from the white dwarf (WD) primary and accretion regions could appreciably distort the measured radial velocity (RV) curve of the secondary star, resulting in systematic errors in the derived masses. This arises because irradiation typically suppresses the line absorption on the hemisphere of the secondary star facing the WD, causing the apparent centre-of-light of the absorption line to be shifted away from the centre-of-mass of the secondary. R. Wade and K. Horne [105] developed the 'K correction' to try and account for these effects, but while this can account for symmetric irradiation effects, it cannot account for any asymmetries or other features on the stellar surface that may also affect the measured RVs. By mapping the surface line intensity pattern across the secondary, it was envisaged that Roche tomography would simultaneously map the stellar surface while producing component masses that were corrected for the (now known) irradiation pattern.

As well as mapping the effects of irradiation, Roche tomography was developed to map stellar magnetic activity, which, at the time, was largely an aspirational goal – but one that has been achieved with great success. The magnetic field of the donor is thought to be responsible for sustaining the mass transfer that causes CVs to evolve to shorter orbital periods, by draining angular momentum from the system through a process known as magnetic braking – a standard ingredient of CV evolutionary theory for several decades. Furthermore, the rapidly-spinning ($P_{orb} \sim 10\,\mathrm{h}$) and tidally-locked nature of the donor star provides a unique parameter regime for tests of stellar dynamo theories. Indeed magnetic activity cycles have been invoked to explain variations in orbital periods, mean brightnesses, mean outburst frequencies, mean outburst durations and outburst shapes of interacting binaries, while starspots (a manifestation of magnetic activity) are thought to be able to regulate mass transfer on more rapid timescales. Despite its importance, there was no direct observational evidence of stellar activity other than a TiO study of SS Cyg by [110]. Roche tomography was set to change that by providing a crucial insight into the magnetic activity of these unique systems.

12.2.2 The Principles of Roche Tomography

Analogous to Doppler imaging of single stars, Roche tomography uses phase-resolved spectra to reconstruct the line-intensity distribution on the secondary star in CVs. However, Roche tomography differs in two important ways. First, the secondary stars are tidally distorted into a Roche-lobe, and are in synchronous rotation around the centre-of-mass of the binary. Secondly, the systems often show rapid variability due to accretion. This means that one usually requires simultaneous photometry with which to slit-loss correct the spectra before continuum subtraction, whereas spectra used for Doppler imaging can simply be normalised. To date, the principles and axioms of Roche tomography have not changed, other than taking into consideration that the secondary star is not necessarily tidally locked (see Sect. 12.2.5). The axioms of Roche tomography are:

1. The secondary star is Roche-lobe filling, synchronously rotating and in a circularised orbit.
2. The observed surface of the star coincides with the surface defined by the critical Roche potential.
3. The shape of the intrinsic line profiles remains unchanged.
4. The line profiles are due to the secondary star light only.
5. The secondary star exhibits no intrinsic variation during the observation.
6. The orbital period, inclination, stellar masses, limb darkening, intrinsic line width and systemic velocity are known.
7. The final map is the one of maximum entropy (relative to an assumed default image) which is consistent with the data.

By assuming the binary parameters, we can define the surface of secondary to lie on the surface of gravitational equipotential, as given by the Roche approximation at the point where Roche-lobe overflow occurs. We model the stellar surface using a grid of tiles lying on this critical surface. Each tile is assigned a copy of the local specific intensity profile – either Gaussian or taken from a spectral-type template – and then convolved with the instrumental profile. An intensity is assigned to each tile to scale its contribution to the total profile, which in turn is found by integrating over the visible surface of the star, taking into account limb-darkening, foreshortening, and the radial velocity of each element. In essence, Roche tomography carries out the reverse of this process by iteratively varying the contribution of each tile until a map is obtained that is consistent with the observed data. The goodness of fit for any given map is measured using the reduced χ^2 statistics, however, as there are many maps which can fit the observed data equally well, an additional regularisation static is required. Following [44] and in common with several implementations of Doppler imaging techniques (see Sect. 12.1.2), we employ the maximum entropy method, using the MEMSYS package written by [97] to find the map of maximum entropy, subject to a χ^2 constraint. This method selects the reconstructed map which contains the least information with respect to a default map. Although we may assume some a priori information about surface intensity distribution, in general, a blank default map is adopted, where the pixels are set to the average value of the reconstructed map, and updated after each iteration. The technique is well tested, and artifacts are well characterised – see [107] for full details.

12.2.3 Roche Tomography: Early Maps

The early work using Roche tomography focused on using either single absorption or emission lines obtained using intermediate resolution spectrographs. The reconstructed maps of the secondaries in CVs such as AM Her, IP Peg, QQ Vul, and HU Aqr (maps 4–7 in Fig. 12.8) clearly showed the varying impact of irradiation from both the WD and accretion regions in these systems [108]. Indeed the asymmetric irradiation patterns found on the leading hemispheres in the polar-type CVs AM Her, QQ Vul and HU Aqr were most likely due to shielding by accretion streams.

The use of Roche tomography, in principle, allowed the most reliable mass determinations, since the intensity distribution across the surface was known. Adopting incorrect parameters introduced spurious artifacts during reconstruction, artificially increasing the information content of the map, thereby reducing its entropy. By assuming the correct binary parameters are those that minimise the number of artifacts in the image, the correct parameters could be determined by carrying out 'entropy landscapes' – reconstructing maps with many pairs of component masses and selecting the pair that produced the map of maximum entropy. In a similar fashion, the systemic velocity and inclination were obtained. These early maps (shown in Fig. 12.8) clearly demonstrated the potential of the technique in obtaining robust system parameters. However, single-line studies, combined with the inherent limit of signal-to-noise in these relatively faint and rapidly-rotating stars, only allowed for large scale features to be mapped. To map smaller-scale features, the technique of least-squares deconvolution (LSD) was adopted [24]. As explained in Sect. 12.1.3, this effectively stacks the thousands of stellar absorption lines observable in a single échelle spectrum to produce a 'mean' profile with greatly enhanced S/N (with a typical multiplex gain of $\sim$30). By adopting LSD, starspots (the most easily observed manifestation of stellar activity) could now be imaged, paving the way for the exploration of stellar magnetic activity on the secondaries in CVs.

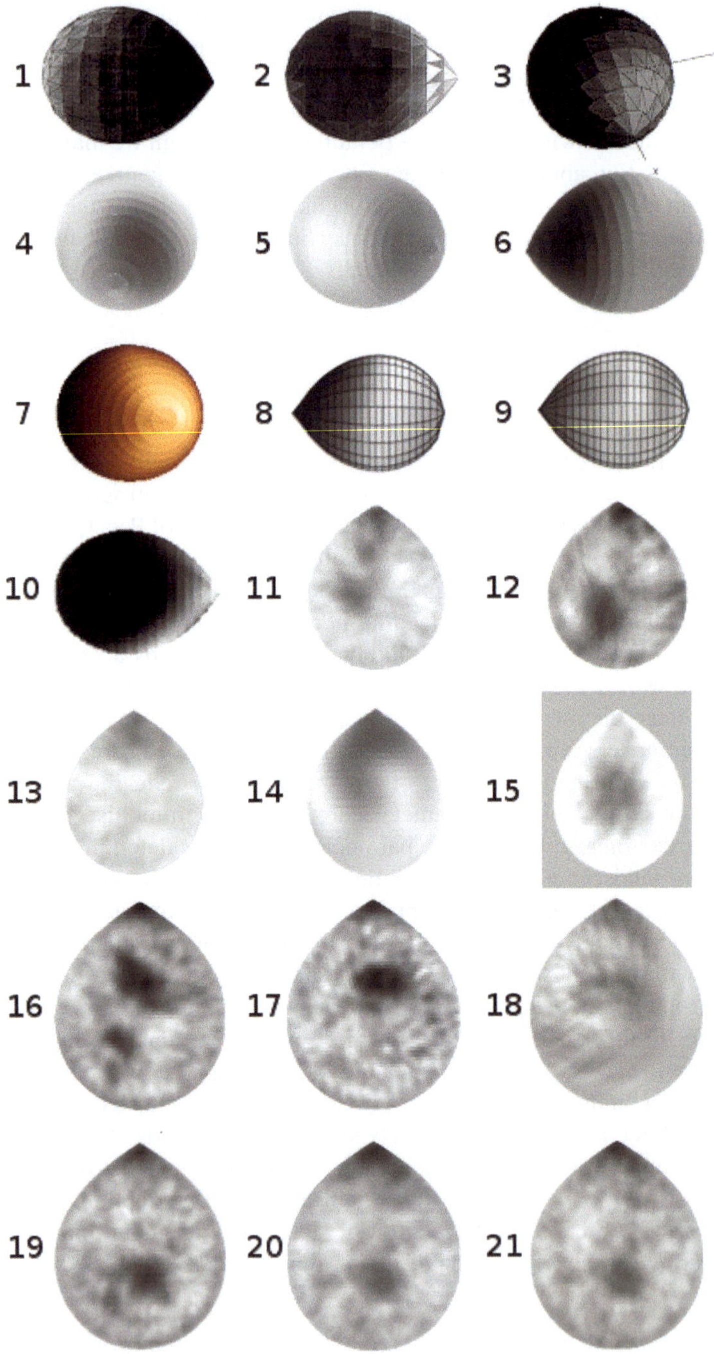

Fig. 12.8 Roche tomograms of the secondary stars in CVs that have been made over the last 20 years. The maps are numbered according to Table 12.1. All maps show irradiation and spots as *dark* features, apart from numbers 1, 2, 3, 7 & 10 where the scale is reversed. Not to scale

Table 12.1 Roche tomograms created over the last 20 years. Columns 1–5 list the map number as shown in Fig. 12.8, year of publication, the authors involved, the system being mapped, and the lines used in the reconstruction process, respectively. Column 6 lists the features that were mapped

Map	Year	Authors	System	Line used	Features
1	1994	Rutten and Dhillon	DW UMa	Balmer lines	Irr; land (un)
2	1996	Rutten and Dhillon	IP Peg	Na 1 doublet	Irr; land
3	1996	Davey and Smith	AM Her	Na 1 doublet	Irr (A); land
4	2003	Watson et al.	AM Her	Na 1 doublet	Irr (A); land
5	2003	Watson et al.	IP Peg	Na 1 doublet	Irr (A); land
6	2003	Watson et al.	QQ Vul	Na 1 doublet	Irr (A); land
7	2003	Watson et al.	HU Aqr	HeII 4686	Irr (A); land
8	2004	Staude et al.	QQ Vul	Na 1 doublet	Irr (A)
9	2004	Staude et al.	AM Her	Na 1 doublet	Irr (A)
10	2004	Potter et al.	V834 Cen	HeII 4686	Irr (A)
11	2006	Watson et al.	AE Aqr	LSD	Spots (HL); Irr; land
12	2007	Watson et al.	BV Cen	LSD	Spots (HL); Irr; Prom; land
13	2007	Watson et al.	V426 Oph	LSD	Spots; Irr; land
14	2012	Dunford et al.	RU Peg	LSD – low res.	Spot (HL); Irr; land
15	2014	Shahbaz et al.	Cen X-4	LSD	Spot (HL); Irr?; land
16–19	2015	Hill et al.	AE Aqr	LSD	Spots (HL); Irr; Land
20,21	2014	Hill et al.	AE Aqr	LSD	Spots (HL); Irr; DR; land

Irr irradiation, *A* asymmetric, *HL* high-latitude, *Land* entropy landscape was employed to find optimal parameters, *DR* differential rotation was measured, *Prom* a prominence was found

12.2.4 Probing Stellar Activity

Using Roche tomography (combined with LSD), the first ever starspots imaged on a donor star in a CV were found by Watson et al. [109] in their study of AE Aqr ($P_{rot} = 9.88$ h). The map clearly showed a large high-latitude spot, as well as a 'chain' of spots leading from the pole to the L_1 point, giving the first evidence that the emergence of spots may be influenced by tidal forces in these systems [43]. In later work, the longer period ($P_{rot} = 14.7$ h) CV, BV Cen, was successfully imaged, showing a similar distribution of spots to that of AE Aqr, as well as hosting a large prominence. A further map of V426 Oph supported the conclusions of the previous two maps – that the donor stars in CVs were highly magnetically active (with a large spot coverage), and that there was a clear indication of interaction between the WD primary and the emergence of magnetic flux on the secondary. Another notable first was achieved in more recent work, with the mapping of star spots on the secondary star in the X-ray binary, Cen X-4 [94]. Here, the low inclination of the system allowed a large polar-region feature to be imaged despite the poor S/N of the data. In further work, the long-term behaviour of the CV, AE Aqr, has been studied by comparing maps taken over an 8 year period (see maps 11 & 16–21 in Fig. 12.8). Here, the persistent band of spots $\sim 20 - 30°$ latitude, combined with the emergence of a band of spots $\sim 40 - 50°$ may indicate a magnetic activity cycle like that seen in the Sun.

Studies of magnetic activity are vital to better understand the behaviour of this class of binary over both long and short time periods. The distribution of spots reveals a wealth of information about the magnetic activity, and how it may be influenced by the underlying stellar dynamo. By studying stars with different fundamental parameters (e.g., spin, mass) we can learn which parameters are most influential on the stellar dynamo, providing tests that are simply impossible in isolated stars.

12.2.5 *Differential Rotation*

Many highly-active single stars display differential rotation (DR, where different stellar latitudes rotate with different velocities). This phenomenon is thought to play a crucial role in amplifying and transforming initially poloidal magnetic field into toroidal field through dynamo processes (the so-called Ω effect). However, theoretical work by Scharlemann [90] suggested that the tides raised on the secondary star by the WD act to suppress DR in tidally locked systems such as CVs, forcing the stellar envelope to co-rotate on average. This theory was put to the test when DR was measured for the first time in a mass-transferring secondary in the CV, AE Aqr [42]. Roche tomograms were created with data taken 9 days apart (see maps 20 & 21 in Fig. 12.8), allowing enough time for appreciable surface shear to build up, but short enough that spots did not significantly evolve. Strips of constant latitude on each map were cross-correlated, and the peaks were fit with a solar-like differential law, allowing a determination of the shear rate. A second method of calculating the shear rate was to minimise the residuals after applying the solar-like differential law to the first map, and subtracting the second map. Both techniques found an equator-pole lap time $\sim$270 d, showing that CV donors were not necessarily tidally locked – overturning the assumption that theoreticians and observers had held for decades. In comparison, the nearest comparable system, namely the pre-CV V471 Tau, was found to be rigidly rotating, despite its similar rotation period and spectral-type [47]. Furthermore, the determination of the co-rotation latitude of $\sim 40°$ in AE Aqr is something that is not possible for isolated stars.

It is clear that further observations of DR in tidally distorted systems are required to disentangle the most important parameters (period, stellar type, mass-ratio, Roche lobe filling factor) driving the stellar dynamo and influencing tidal dissipation efficiency. The discovery of DR on AE Aqr strongly suggests that our understanding of tidal dissipation in the convective envelopes of low-mass stars is lacking, and so continued studies of how tidal forces influence the magnetic activity on donor stars are crucial to our understanding of the stellar dynamo.

12.2.6 Future Prospects

Roche tomography celebrated its 20th anniversary of the first published map during the conference that led to this book [89]. The progress since its inception has been profound, with the level of detail in surface maps dramatically increasing over the years. In part this is due to the higher-quality of data from more modern instruments, but primarily the introduction of LSD has had the largest impact, clearly seen in maps 11–21 in Fig. 12.8. Roche tomography has clearly established its ability to determine robust system parameters in both CVs and X-ray binaries. Furthermore, it has allowed the detailed study of stellar magnetic activity in a unique testbed, with both rapid-rotation and tidal distortion. The discovery of differential rotation in a CV clearly demonstrates its potential in studies of surface rotation, and the level of detail achieved in maps shows its potential in studies of spot morphology, distribution, and lifetimes. The technique will be invaluable in the continuing study of stellar magnetic activity, and the underlying stellar dynamo.

12.2.6.1 Problems to be Solved

Despite the success of the technique, Roche tomography shows some problems which have yet to be resolved. Due to its lack of a two-temperature model, the growth of bright pixels is not thresholded, which means it is somewhat difficult to determine the absolute spot-filling factors for any given map. One may calculate the filling-factor in the same way, but maps may not be directly comparable, given that the number and size of features may be different. Furthermore, as with maps produced using Doppler imaging techniques, no formal uncertainties are given on either the map intensities or on any other system parameter. This problem can be mitigated somewhat by the use of bootstrapping (see [108]), giving a significance to each tile intensity, although the problem of uncertainty in system parameters remains. Over the years the treatment of limb-darkening has improved, in part due to the demands of higher-quality data. However, whilst using incorrect limb darkening coefficients lead to well characterised artifacts in the final maps, the limb darkening laws themselves are taken from spherical approximations, which is clearly not the case in a tidally-distorted star. A more thorough treatment of limb darkening would ensure that the optimal masses are more robust. Finally, we note that while the mass ratio is reliably determined, the system inclination (and thus component masses) are still somewhat poorly constrained – a disappointing fact considering the technique was initially developed to better determine these parameters.

12.2.6.2 Future Directions and Opportunities

Given the notable firsts of Roche tomography in mapping the first star spots on a CV and on an X-ray binary, and the recent measurement of differential rotation, the technique has shown its ability to reveal the magnetic activity in binaries. Furthermore, new developments may build on this success. The maximum-entropy regularisation statistic implemented in Roche tomography may need to be superseded by a new regularisation statistic, allowing small-scale features to be both more readily and more reliably mapped, and uncertainties to be more robustly quantified. In addition, the technique could be extended to include Zeeman Doppler imaging, which has been used to great effect to map magnetic fields in single stars and detached binary systems. The wealth of knowledge gained from such a technique would allow for a hitherto undiscovered insight into the nature and evolution of these systems.

References

1. Adelman, S.J., Gulliver, A.F., Kochukhov, O.P., et al.: The Variability of the Hg II λ3984 line of the mercury-manganese star α Andromedae. ApJ **575**, 449, (2002)
2. Anderson, R.I., Reiners, A., Solanki, S.K.: On detectability of Zeeman broadening in optical spectra of F- and G-dwarfs. A&A **522**, A81 (2010)
3. Babcock, H.W.: Zeeman effect in stellar spectra. ApJ **105**, 105 (1947)
4. Babcock, H.W.: The solar magnetograph. ApJ **118**, 387 (1953)
5. Babcock, H.W.: A catalog of magnetic stars. ApJS **3**, 141 (1958)
6. Bagnulo, S., Szeifert, T., Wade, G.A., et al.: Measuring magnetic fields of early-type stars with FORS1 at the VLT. A&A **389**, 191 (2002)
7. Baliunas, S.L., Donahue, R.A., Soon, W.H., et al.: Chromospheric variations in main-sequence stars. ApJ **438**, 269 (1995)
8. Berdyugina, S.V.: Starspots: a key to the stellar dynamo. Living Rev. Sol. Phys. **2**, 8 (2005)
9. Borra, E.F.: Polarimetry techniques applied to measurements of stellar magnetic fields. Ricerche Astronomiche **10**, 35 (1980)
10. Borra, E.F., Edwards, G., Mayor, M.: The magnetic fields of the late-type stars. ApJ **284**, 211 (1984)
11. Bouvier, J., Alencar, S.H.P., Harries, T.J., et al.: Magnetospheric accretion in classical T Tauri stars. In: Reipurth, B., Jewitt, D., Keil, K. (eds.) Protostars and Planets V, 951 pp. University of Arizona Press, Tucson (2007)
12. Bouvier, J., Matt, S.P., Mohanty, S., et al.: Angular momentum evolution of young low-mass stars and brown dwarfs: observations and theory. In: Beuther, H., Klessen, R.S., Dullemond, C.P., Henning, T. (eds.) Protostars and Planets VI, 914 pp. University of Arizona Press, Tucson (2014)
13. Braithwaite, J., Spruit, H.C.: A fossil origin for the magnetic field in A stars and white dwarfs. Nature **431**, 819 (2004)
14. Brown, S.F., Donati, J.-F., Rees, D.E., et al.: Zeeman-Doppler imaging of solar-type and AP stars. IV – maximum entropy reconstruction of 2D magnetic topologies. A&A **250**, 463 (1991)
15. Carroll, T.A., Strassmeier, K.G., Rice, J.B., et al.: The magnetic field topology of the weak-lined T Tauri star V410 Tauri. New strategies for Zeeman-Doppler imaging. A&A **548**, A95 (2012)
16. Charbonneau, P.: Stellar dynamos. Soc. Astron. Sci. Annu. Symp. **39**, 187 (2013)
17. Collier Cameron, A.: Modelling stellar photospheric spots using spectroscopy. In: Byrne, P.B., Mullan, D.J. (eds.) Surface Inhomogeneities on Late-Type Stars, Lecture Notes in Physics, vol. 397, p. 33. Springer, Berlin (1992)

18. Crossfield, I.J.M., Biller, B., Schlieder, J.E. et al.: A global cloud map of the nearest known brown dwarf. Nature **505**, 654 (2014)
19. del Toro Iniesta, J.C.: Introduction to Spectropolarimetry. Cambridge University Press, Cambridge (2003)
20. Delfosse, X., Donati, J.-F., Kouach, D., et al.: World-leading science with SPIRou – The nIR spectropolarimeter/high-precision velocimeter for CFHT. In: Cambresy, L., Martins, F., Nuss, E., Palacios, A. (eds.) SF2A-2013: Proceedings of the Annual meeting of the French Society of Astronomy and Astrophysics, p. 497 (2013)
21. Deutsch, A.J.: A method for mapping the surfaces of some magnetic stars. AJ **62**, 139 (1957)
22. Donati, J.-F., Brown, S.F.: Zeeman-Doppler imaging of active stars. V. Sensitivity of maximum entropy magnetic maps to field orientation. A&A **326**, 1135 (1997)
23. Donati, J.-F., Landstreet, J.D.: Magnetic fields of nondegenerate stars. ARA&A **47**, 333 (2009)
24. Donati, J.-F., Semel, M., Carter, B.D., et al.: Spectropolarimetric observations of active stars. MNRAS **291**, 658 (1997)
25. Donati, J.-F., Howarth, I.D., Jardine, M.M., et al.: The surprising magnetic topology of τ Sco: fossil remnant or dynamo output? MNRAS **370**, 629 (2006)
26. Donati, J.-F., Catala, C., Landstreet, J.D., et al.: ESPaDOnS: the new generation stellar spectro-polarimeter. Performances and first results. Astron. Soc. Pac. Conf. Ser. **358**, 362 (2006)
27. Donati, J.-F., Jardine, M.M., Gregory, S.G., et al.: Magnetic fields and accretion flows on the classical T Tauri star V2129 Oph. MNRAS **380**, 1297 (2007)
28. Donati, J.-F., Jardine, M.M., Gregory, S.G., et al.: Magnetospheric accretion on the T Tauri star BP Tauri. MNRAS **386**, 1234 (2008)
29. Donati, J.-F., Morin, J., Petit, P., et al.: Large-scale magnetic topologies of early M dwarfs. MNRAS **390**, 545 (2008)
30. Donati, J.-F., Skelly, M.B., Bouvier, J., et al.: Magnetospheric accretion and spin-down of the prototypical classical T Tauri star AA Tau. MNRAS **409**, 1347 (2010)
31. Donati, J.-F., Hébrard, E., Hussain, G., et al.: Modelling the magnetic activity and filtering radial velocity curves of young suns: the weak-line T Tauri star LkCa 4. MNRAS **444**, 3220 (2014)
32. Dorn, R.J., Anglada-Escude, G., Baade, D., et al.: CRIRES+: exploring the cold universe at high spectral resolution. Messenger **156**, 7 (2014)
33. Fares, R., Donati, J.-F., Moutou, C., et al.: Magnetic cycles of the planet-hosting star τ Bootis – II. A second magnetic polarity reversal. MNRAS **398**, 1383 (2009)
34. Fares, R., Moutou, C., Donati, J.-F. et al.: A small survey of the magnetic fields of planet-host stars. MNRAS **435**, 1451 (2013)
35. Gastine, T., Morin, J., Duarte, L., et al.: What controls the magnetic geometry of M dwarfs? A&A **549**, L5 (2013)
36. Goncharskij, A.V., Stepanov, V.V., Khokhlova, V.L., et al.: Mapping of chemical elements on the surfaces of AP stars. I. Solution of the inverse problem of finding local profiles of spectral lines. AZh **59**, 1146 (1982)
37. Gregory, S.G., Donati, J.-F., Morin, J., et al.: Can we predict the global magnetic topology of a pre-main-sequence star from its position in the Hertzsprung-Russell diagram? ApJ **755**, 97 (2012)
38. Hale, G.E.: On the probable existence of a magnetic field in sun-spots. ApJ **28**, 315 (1908)
39. Hall, J.C., Lockwood, G.W., Skiff, B.A.: The activity and variability of the sun and sun-like Stars. I. Synoptic Ca II H and K observations. AJ **133**, 862 (2007)
40. Hébrard, É.M., Donati, J.-F., Delfosse, X., et al.: Detecting planets around active stars: impact of magnetic fields on radial velocities and line bisectors. MNRAS **443**, 2599 (2014)
41. Hennebelle, P.: Formation of low-mass stars and brown dwarfs. EAS Publ. Ser. **57**, 91 (2012)
42. Hill, C.A., Watson, C.A., Shahbaz, T., et al.: Roche tomography of cataclysmic variables – VI. Differential rotation of AE Aqr – not tidally locked! MNRAS **444**, 192 (2014)

43. Holzwarth, V., Schüssler, M.: Dynamics of magnetic flux tubes in close binary stars. II. Nonlinear evolution and surface distributions. A&A **405**, 303 (2003)
44. Horne, K.: Images of accretion discs. I – the eclipse mapping method. MNRAS **213**, 129 (1985)
45. Hussain, G.A.J., Donati, J.-F., Collier Cameron, A., et al.: Comparisons of images derived from independent Zeeman Doppler imaging codes. MNRAS **318**, 961 (2000)
46. Hussain, G.A.J., Jardine, M., Collier Cameron, A.: Mapping potential fields on the surfaces of rapidly rotating stars. MNRAS **322**, 681 (2001)
47. Hussain, G.A.J., Allende Prieto, C., Saar, S.H., et al.: Spot patterns and differential rotation in the eclipsing pre-cataclysmic variable binary, V471 Tau. MNRAS **367**, 1699 (2006)
48. Hussain, G.A.J., Collier Cameron, A., Jardine, M.M., et al.: Surface magnetic fields on two accreting TTauri stars: CVCha and CRCha. MNRAS **398**, 189 (2009)
49. Jeffers, S.V., Barnes, J.R., Collier Cameron, A., et al.: Hubble space telescope observations of SV Cam – I. The importance of unresolved star-spot distributions in light-curve fitting. MNRAS **366**, 667 (2006)
50. Jeffers, S.V., Petit, P., Marsden, S.C., et al.: ϵ Eridani: an active K dwarf and a planet hosting star? The variability of its large-scale magnetic field topology. A&A **569**, A79 (2014)
51. Johns-Krull, C.M., Valenti, J.A., Koresko, C.: Measuring the magnetic field on the classical T Tauri star BP Tauri. ApJ **516**, 900 (1999)
52. Khokhlova, V.L.: Mapping of "spots" on the surface of Ap stars by means of line profiles. AZh **52**, 950 (1975)
53. Kochukhov, O., Bagnulo, S.: Evolutionary state of magnetic chemically peculiar stars. A&A **450**, 763 (2006)
54. Kochukhov, O., Piskunov, N.: Doppler Imaging of stellar magnetic fields. II. Numerical experiments. A&A **388**, 868 (2002)
55. Kochukhov, O., Wade, G.A.: Magnetic Doppler imaging of α^2 Canum Venaticorum in all four stokes parameters. Unveiling the hidden complexity of stellar magnetic fields. A&A **513**, A13 (2010)
56. Kochukhov, O., Makaganiuk, V., Piskunov, N.: Least-squares deconvolution of the stellar intensity and polarization spectra. A&A **524**, A5 (2010)
57. Kochukhov, O., Bagnulo, S., Wade, G.A., et al.: Magnetic Doppler imaging of 53 Camelopardalis in all four Stokes parameters. A&A **414**, 613 (2004)
58. Kochukhov, O., Drake, N.A., Piskunov, N., et al.: Multi-element abundance Doppler imaging of the rapidly oscillating Ap star HR 3831. A&A **424**, 935 (2004)
59. Kochukhov, O., Adelman, S.J., Gulliver, A.F., et al.: Weather in stellar atmosphere revealed by the dynamics of mercury clouds in α Andromedae. Nat. Phys. **3**, 526 (2007)
60. Lammer, H., Bredehöft, J.H., Coustenis, A., et al.: What makes a planet habitable? A&A Rev. **17**, 181 (2009)
61. Landstreet, J.D.: Magnetic fields at the surfaces of stars. A&A Rev. **4**, 35 (1992)
62. Landstreet, J.D.: Observing and modelling stellar magnetic fields 1. Basic physics and simple models. EAS Publ. Ser. **39**, 1 (2009)
63. Landstreet, J.D.: Observing and modelling stellar magnetic fields 2. models. EAS Publ. Ser. **39**, 21 (2009)
64. Makaganiuk, V., Kochukhov, O., Piskunov, N., et al.: The search for magnetic fields in mercury-manganese stars. A&A **525**, A97 (2011)
65. Mangeney, A., Praderie, F.: The influence of convection and rotation on X-ray emission in main sequence stars. A&A **130**, 143 (1984)
66. Martínez González, M.J., Asensio Ramos, A., Carroll, T.A., et al.: PCA detection and denoising of Zeeman signatures in polarised stellar spectra. A&A **486**, 637 (2008)
67. Mathis, S.: Transport processes in stellar interiors. In: Goupil, M., et al. (eds.) Studying Stellar Rotation and Convection, Lecture Notes in Physics, vol. 865, p. 23. Springer, Berlin (2013)
68. Mathys, G., Hubrig, S., Landstreet, J.D., et al.: The mean magnetic field modulus of AP stars. A&AS **123**, 353 (1997)

69. Morgenthaler, A., Petit, P., Morin, J., et al.: Direct observation of magnetic cycles in sun-like stars. Astronomische Nachrichten **332**, 866 (2011)
70. Morin, J., Donati, J.-F., Petit, P., et al.: Large-scale magnetic topologies of mid M dwarfs. MNRAS **390**, 567 (2008)
71. Morin, J., Donati, J.-F., Petit, P., et al.: Large-scale magnetic topologies of late M dwarfs. MNRAS **407**, 2269 (2010)
72. Morin, J., Dormy, E., Schrinner, M., et al.: Weak- and strong-field dynamos: from the Earth to the stars. MNRAS **418**, L133 (2011)
73. Morin, J., Jardine, M., Reiners, A., et al.: Multiple views of magnetism in cool stars. Astronomische Nachrichten **334**, 48 (2013)
74. Neiner, C., Alecian, E.: The BinaMIcS project: binarity and magnetism. EAS Publ. Ser. **64**, 75 (2013)
75. Noyes, R.W., Hartmann, L.W., Baliunas, S.L., et al.: Rotation, convection, and magnetic activity in lower main-sequence stars. ApJ **279**, 763 (1984)
76. O'Neal, D., Neff, J.E., Saar, S.H.: Measurements of starspot parameters on active stars using molecular bands in Echelle spectra. ApJ **507**, 919 (1998)
77. Petit, P., Dintrans, B., Solanki, S.K., et al.: Toroidal versus poloidal magnetic fields in Sun-like stars: a rotation threshold. MNRAS **388**, 80 (2008)
78. Petit, P., Dintrans, B., Morgenthaler, A., et al.: A polarity reversal in the large-scale magnetic field of the rapidly rotating sun HD 190771. A&A **508**, L9 (2009)
79. Petit, P., Donati, J.-F., Hébrard, E., et al.: Maximum entropy detection of planets around active stars. A&A (arXiv:1503.00180, in press)
80. Piskunov, N., Kochukhov, O.: Doppler imaging of stellar magnetic fields. I. Techniques. A&A **381**, 736 (2002)
81. Pizzolato, N., Maggio, A., Micela, G., et al.: The stellar activity-rotation relationship revisited: dependence of saturated and non-saturated X-ray emission regimes on stellar mass for late-type dwarfs. A&A **397**, 147 (2003)
82. Preston, G.W.: The chemically peculiar stars of the upper main sequence. ARA&A **12**, 257 (1974)
83. Reiners, A.: Observations of cool-star magnetic fields. Living Rev. Sol. Phys. **9**, 1 (2012)
84. Reiners, A., Basri, G.: On the magnetic topology of partially and fully convective stars. A&A **496**, 787 (2009)
85. Rice, J.B., Strassmeier, K.G.: Doppler imaging from artificial data. Testing the temperature inversion from spectral-line profiles. A&AS **147**, 151 (2000)
86. Rice, J.B., Wehlau, W.H., Khokhlova, V.L.: Mapping stellar surfaces by Doppler imaging – technique and application. A&A **208**, 179 (1989)
87. Romanova, M.M., Long, M., Lamb, F.K., et al.: Global 3D simulations of disc accretion on to the classical T Tauri star V2129 Oph. MNRAS **411**, 915 (2011)
88. Rosén, L., Kochukhov, O., Wade, G.A.: First Zeeman Doppler imaging of a cool star using all four stokes parameters, ApJ **805**, 169 (2015)
89. Rutten, R.G.M., Dhillon, V.S.: Roche tomography: imaging the stars in interacting binaries. A&A **288**, 773 (1994).
90. Scharlemann, E.T.: Tides in differentially rotating convective envelopes. II – the tidal coupling. ApJ **253**, 298 (1982)
91. Semel, M.: Zeeman-Doppler imaging of active stars. I – basic principles. A&A **225**, 456 (1989)
92. Semel, M., Ramírez Vélez, J.C., Martínez González, M.J., et al.: Multiline Zeeman signatures through line addition. A&A **504**, 1003 (2009)
93. Sennhauser, C., Berdyugina, S.V.: Zeeman component decomposition for recovering common profiles and magnetic fields. A&A **522**, A57 (2010)
94. Shahbaz, T., Watson, C.A., Dhillon, V.S.: The spotty donor star in the X-ray transient Cen X-4. MNRAS **440**, 504 (2014)
95. Shulyak, D., Ryabchikova, T., Mashonkina, L., et al.: Model atmospheres of chemically peculiar stars. Self-consistent empirical stratified model of HD 24712. A&A **499**, 879 (2009)

96. Shulyak, D., Reiners, A., Seemann, U., et al.: Exploring the magnetic field complexity in M dwarfs at the boundary to full convection. A&A **563**, A35 (2014)
97. Skilling, J., Bryan, R.K.: Maximum entropy image reconstruction – general algorithm. MNRAS **211**, 111 (1984)
98. Snik, F., Kochukhov, O., Piskunov, N., et al.: The HARPS polarimeter. In: Kuhn, J.R., Harrington, D.M., Lin, H., Berdyugina, S.V., Trujillo-Bueno, J., Keil, S.L., Rimmele, T. (eds.) Solar Polarization 6, Proceedings of a conference held in Maui, Hawaii, USA on May 30 to June 4, 2010, vol. 437, p. 237. Astronomical Society of the Pacific, San Francisco (2011)
99. Solanki, S.K., Unruh, Y.C.: Spot sizes on sun-like stars. MNRAS **348**, 307 (2004)
100. Strassmeier, K.G.: Starspots. A&A Rev. **17**, 251 (2009)
101. Tikhonov, A.N., Vitkevich, V.V., Artyukh, V.S., et al.: Reconstruction of the Radio-Brightness Distribution over a Source., AZh **46**, 472 (1969)
102. Vidotto, A.A., Gregory, S.G., Jardine, M., et al.: Stellar magnetism: empirical trends with age and rotation. MNRAS **441**, 2361 (2014)
103. Vidotto, A.A., Fares, R., Jardine, M., et al.: On the environment surrounding close-in exoplanets. MNRAS **449**, 4117 (2015)
104. Vogt, S.S., Penrod, G.D., Hatzes, A.P.: Doppler images of rotating stars using maximum entropy image reconstruction. ApJ **321**, 496 (1987)
105. Wade, R.A., Horne, K.: The radial velocity curve and peculiar TiO distribution of the red secondary star in Z Chamaeleontis. ApJ **324**, 411 (1988)
106. Wade, G.A., Donati, J.-F., Landstreet, J.D., et al.: Spectropolarimetric measurements of magnetic Ap and Bp stars in all four stokes parameters. MNRAS **313**, 823 (2000)
107. Watson, C.A., Dhillon, V.S.: Roche tomography of cataclysmic variables – I. Artefacts and techniques. MNRAS **326**, 67 (2001)
108. Watson, C.A., Dhillon, V.S., Rutten, R.G.M., et al.: Roche tomography of cataclysmic variables – II. Images of the secondary stars in AM Her, QQ Vul, IP Peg and HU Aqr. MNRAS **341**, 129 (2003)
109. Watson, C.A., Dhillon, V.S., Shahbaz, T.: Roche tomography of cataclysmic variables – III. Star-spots on AE Aqr. MNRAS **368**, 637 (2006)
110. Webb, N.A., Naylor, T., Jeffries, R.D.: Spectroscopic evidence for starspots on the secondary star of SS Cygni. ApJ **568**, L45 (2002)
111. Wehlau, W.H., Rice, J.B., Khokhlova, V.L.: The Surface of the magnetic Cp-Star 21 Per. Astron. Astrophys. Trans. **1**, 55 (1991)
112. Zeeman, P.: On the influence of magnetism on the nature of the light emitted by a substance. ApJ **5**, 332 (1897)

Chapter 13
AGN Reverberation Mapping

Misty C. Bentz

13.1 Introduction and Motivation

In the 25 years that the *Hubble Space Telescope* (*HST*) has been in low Earth orbit, there have been leaps and bounds in our understanding of many astrophysical phenomena, not the least of which are supermassive black holes. Dedicated ground-based programmes, coupled with the exquisite spatial resolution afforded by *HST*, have led to the common understanding that massive galaxies host supermassive black holes in their cores (see the review by Ferrarese and Ford [21]). In general, studies have shown that more massive galaxies host more massive black holes, which is interpreted as a symbiosis between galaxies and black holes in which they grow together and regulate each other's growth. However, recent studies have begun to find several galaxies that do not appear to follow this simple scaling relationship, and our picture of galaxy and black hole co-evolution throughout the Universe's history is becoming more complicated [50].

The active galactic nucleus, or AGN, phase is now understood to be a sporadic event in the life of a typical supermassive black hole, thought to be triggered by a merger or secular process in the host galaxy (see [32] and references therein). During this phase, the black hole is accreting at a relatively high rate, and the accretion process is releasing large amounts of energy across the electromagnetic spectrum. The lifetime of a typical AGN event for a typical black hole is small compared to the age of the Universe, and is generally thought to be on the order of $\sim 10^8$ years (e.g., [42]). Cosmic downsizing is observed in AGNs as it is in galaxies: the bright quasars we see in the early Universe are associated with massive black

M.C. Bentz (✉)
Department of Physics and Astronomy, Georgia State University, 25 Park Place Suite 605, Atlanta, GA 30303, USA
e-mail: bentz@astro.gsu.edu

H.M.J. Boffin et al. (eds.), *Astronomy at High Angular Resolution*, Astrophysics and Space Science Library 439, DOI 10.1007/978-3-319-39739-9_13

holes and large accretion rates, while today's active black holes tend to be fewer in number, less massive, and have smaller accretion rates (e.g., [38, 42, 70, 74, 75]).

Unfortunately, the rarity of bright AGNs compared to the plethora of galaxies hosting quiescent black holes today leads to the situation in which we find ourselves, where AGNs are generally distant and difficult to study even with the pristine spatial resolution afforded by *HST*. The techniques that have been developed to constrain the masses of inactive black holes in nearby galaxies rely on spatially resolving the innermost parsecs of a galactic nucleus, so they are limited to distances $\lesssim$100 Mpc and therefore not applicable to most AGNs. And yet AGNs act as beacons that shine across the entire observable Universe, tempting us to try to understand black hole and galaxy growth, and evolution out to $z = 7.1$ and perhaps beyond [54].

Luckily, AGNs are not only bright, they are also highly variable. We can, in effect, substitute time resolution for spatial resolution in a technique known as reverberation mapping to probe microarcsecond scales in the nuclei of even the most distant active galaxies [13, 64].

13.2 Reverberation Mapping Primer

Many independent studies have led to the general picture of AGN structure that we understand today (e.g., [1, 57, 65, 77] and references therein) as represented by the cartoon diagram in Fig. 13.1. In the centre is the supermassive black hole, with a mass in the range of 10^6–$10^{10}\,M_\odot$, and its associated accretion disc. The jet (if the AGN has one) is perpendicular to the accretion disc and highly collimated. On larger scales (corresponding to $\sim$0.01 pc for typical Seyferts or approximately the extent of the inner Oort Cloud in our own Solar System) lies a region of photoionized gas that radiates line emission. The location of this gas deep within the potential well of the black hole results in line-of-sight gas velocities that are quite large, causing the emission lines to appear Doppler broadened in the AGN spectrum by a few 1000 km s^{-1}. We imaginatively call this region of gas the broad line region (BLR). The outer edge of the BLR is most likely set by the dust sublimation radius (e.g., [25, 55, 58]), as the inclusion of dust in the BLR gas will extinguish the line emission. The dusty gas that exists outside this radius is generally referred to as the "torus", although the exact geometry of the region is not known. The dust torus causes the AGN system to have different spectral signatures when viewed at different orientations—a system that is close to face-on will have broad emission lines in its spectrum, while a system that is viewed edge-on will have the dust torus blocking the observer's line of sight to the BLR, so no broad emission lines will be seen. On even larger scales ($\sim$tens of pc for typical Seyferts), additional gas that is photoionized by the AGN system also exists, but the location of this gas on galactic scales results in line-of-sight velocities that are on the order of a few 100 km s^{-1}. We see the signature of this gas as narrow emission lines in the AGN spectrum, hence the name attributed to this region of gas is the narrow line region (NLR).

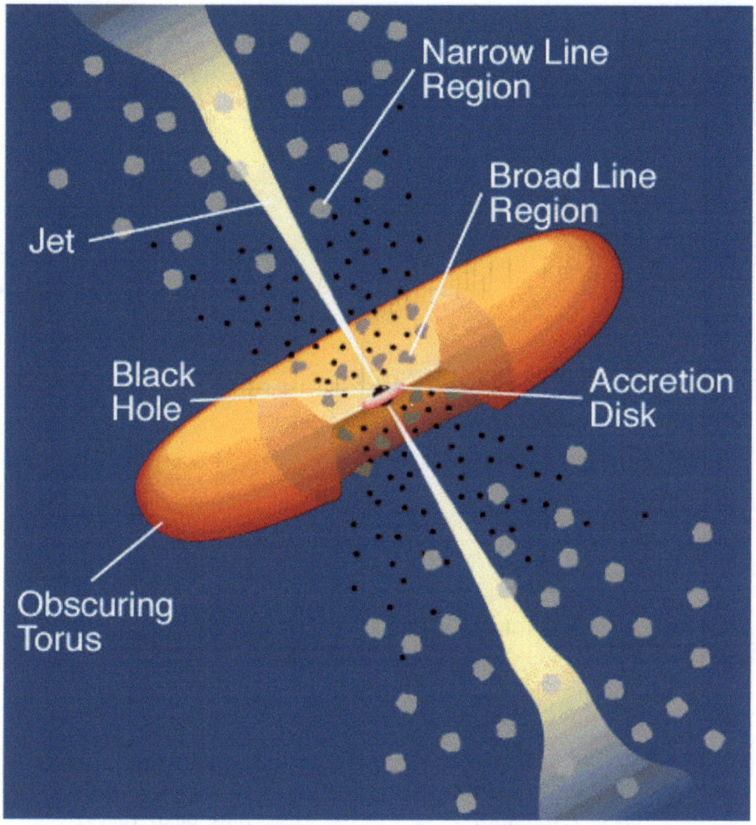

Fig. 13.1 Cartoon diagram of the typically-assumed structure of an AGN. Depending on the orientation at which this structure is viewed, different spectral signatures will be seen. In particular, the broad line region (BLR) is only visible to observers with a relatively face-on view. If viewed from the side, the torus blocks the BLR and only narrow emission lines will be seen in the AGN spectrum. (Reprinted with permission from Urry and Padovani [77])

Reverberation mapping (or "echo mapping") measures the light travel time between different regions in an AGN system. The continuum emission is expected to arise from the accretion disc, and in the ultraviolet, optical, and near-IR it is observed to vary on timescales of hours to days. The source of this variability is not yet understood: one possible explanation is magnetic recombination in the accretion disc, e.g., [41]. Whatever the cause, the variations that are observed in the continuum emission are seen echoed at a later time in the fluxes of the broad emission lines (see Fig. 13.2 for example light curves for the nearby Seyfert galaxy NGC 4151). The time delay between the variations in the continuum and the echo of those variations in an emission line is simply the average light travel time from the accretion disc to the BLR. The accretion disc is generally assumed to be very compact, and so the time delay can be interpreted as the average radius of the BLR in the AGN

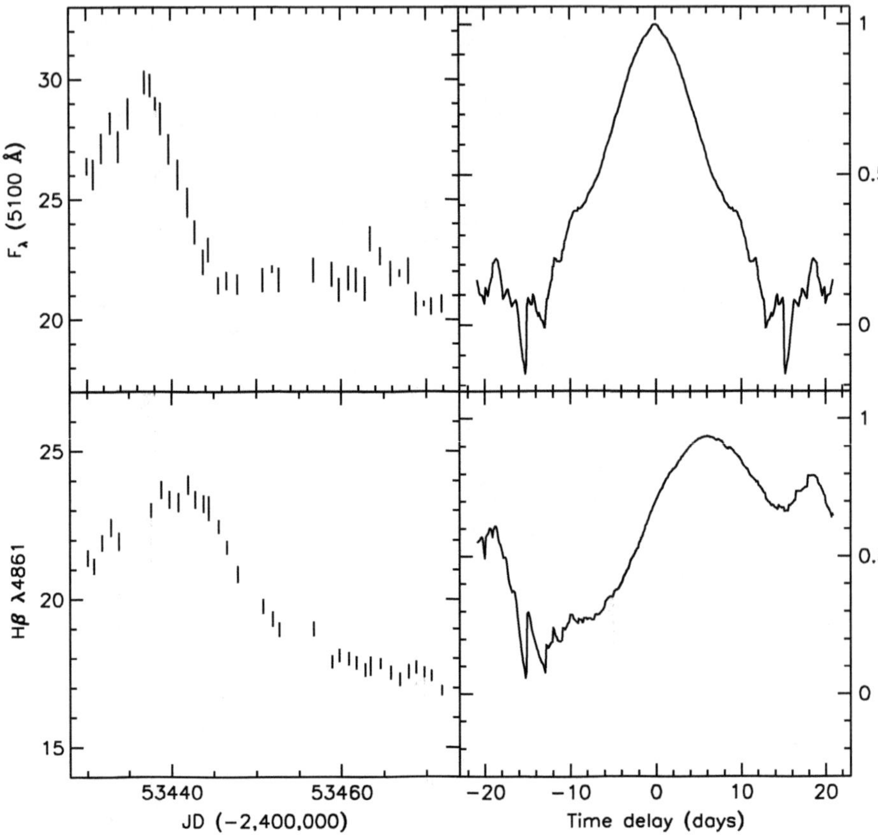

Fig. 13.2 Continuum and Hβ light curves for the Seyfert galaxy NGC 4151 (*left*), autocorrelation function of the continuum light curve (*top right*), and cross-correlation of Hβ relative to the continuum (*bottom right*). The emission-line light curve is visibly delayed from that of the continuum and smoothed in time, both signatures that are obvious to the eye in the light curves and the cross-correlation function, and indicative of the extended nature of the Hβ-emitting BLR gas. (Reprinted with permission from Bentz et al. [4])

system. By definition, reverberation mapping requires a line of sight that permits the observer to view the broad emission lines in the AGN spectrum, so it is only applicable to AGNs with a relatively face-on orientation.

Within the BLR, different emission lines are observed to respond to continuum variations with different time delays, such that species with higher ionization potentials, like C IV λ1549, respond with a shorter time delay than those with lower ionization potentials, like Hβ (e.g., [12, 46, 69]). This behaviour points to ionization stratification within the BLR—more highly ionized line emission is radiated from a smaller radius within the BLR, while emission from more neutral gas occurs at larger radii, farther from the central ionizing source. Photoionization modeling constraints agree with this interpretation: the photoionized gas in the BLR preferentially emits line emission wherever the temperature and density are most favorable for a specific atomic transition (e.g., [2, 48, 49]).

With reverberation mapping, the radius we measure is the responsivity-weighted average radius, which may not exactly coincide with the emissivity-weighted average radius of the gas for that particular line emission. But the general behaviour that we observe is the same: in a single AGN, if the central luminosity increases, then the time delay we measure for any specific emission line becomes larger (e.g., [5, 43, 66]). It is important to keep in mind that we are not measuring a geometrical radius, such as an inner or outer boundary of the BLR, with emission-line reverberation mapping.

Even for a relatively nearby AGN, the size of the region probed by reverberation mapping is quite compact and not resolvable with current imaging technology or that which is likely to be developed in the foreseeable future. For a typical Seyfert galaxy at a distance of 40 Mpc, the time delay expected for the Hβ λ4861 emission line is $\sim$10 light-days, which projects to an angular radius of $\sim$50 microarcsec.

In practice, reverberation mapping relies on dense spectrophotometric monitoring of an AGN system over an extended period of time. The basic requirements for a successful monitoring programme are the following:

1. total campaign length of at least three times the longest expected time delay to maximize the probability that variability of a large enough amplitude will occur during the campaign [35];
2. sampling cadence that is sufficiently dense to resolve the variability and expected time delays (e.g., Nyquist sampled in time);
3. exposure times that yield a signal-to-noise >50 in the continuum and substantially higher in the emission lines, where the amplitude of variability is generally only a few percent;
4. spectral resolution that is sufficiently high to distinguish broad emission lines from each other and from overlapping or nearby narrow emission lines (generally *R* of a few thousand);
5. flux calibration that is good to 2 % or better from observation to observation throughout the campaign [67];
6. strong nerves and a healthy dose of good luck.

Meeting all of these constraints with a ground-based telescope is observationally quite challenging. The weather especially can cause an otherwise well-planned reverberation campaign to not live up to its potential or to fail outright. Added to this uncertainty is the fact that AGN variations are stochastic and not guaranteed to occur during the course of a monitoring campaign for any particular AGN of interest (cf. the case of Mrk 290 which was monitored in 2007 and showed strong variations [18], but showed little to no variation when it was monitored again in 2008 [11]). The typical warning for the stock market also applies here: past performance is no guarantee of future behaviour.

To date, successful reverberation campaigns have been carried out for $\sim$60 different AGNs [9]. Constraints on readily-available resources have generally limited the size of the telescope used in a reverberation campaign to 1.0–4.0-m class telescopes. Coupled with the need for high signal-to-noise ratio spectra in each visit, most of these 60 AGNs are apparently bright and reside within the nearby Universe ($z < 0.1$). Reverberation mapping is not fundamentally limited to nearby objects, but the high luminosities necessary for high signal-to-noise spectra of $z = 2 - 3$ quasars directly translates to a long emission-line time delay, which is further stretched through cosmological dilation (cf. [39] and their monitoring campaign length of 10 years). Ongoing efforts to multiplex reverberation mapping with the Sloan Digital Sky Survey multi-object spectrograph may soon increase the sample size by a substantial fraction and push the median of the sample to somewhat larger redshifts [72], but large statistical samples of reverberation results at multiple redshifts spanning the course of cosmic history are unlikely for the forseeable future. Luckily, the results we have in hand for the current reverberation sample are able to provide us with a foothold for investigating cosmological black hole growth.

13.3 Reverberation Mapping Products

Substantial progress over the last 10 years especially has led to several valuable and widely used reverberation-mapping products, including direct black hole mass measurements, black hole scaling relationships that can be used to quickly estimate large numbers of black hole masses, and detailed information on the geometry and kinematics of the BLR gas. As we describe in the following sections, the first two items are fairly well-developed at this time, while we are just now beginning to truly exploit the third.

13.3.1 Black Hole Masses

In order to directly constrain the mass of a black hole, a luminous tracer (usually gas or stars) must be used to probe the local gravitational potential. In the case of reverberation mapping, the photoionized BLR gas is deep within the potential well of the black hole where its motion is expected to be dominated by gravity in the absence of strong radiation pressure. While the effect of radiation pressure is still debatable for lines such as C IV, these conditions are most likely to be satisfied for Hβ (e.g., [52, 56, 59]) which is expected to arise from gas that is well-shielded from the central ionizing source. The Doppler-broadened width of the Hβ emission line is therefore a measure of the line-of-sight velocity of the gas within the BLR. And, as described in Sect. 13.2, the time delay in the Hβ emission line is a measure of the radius of the BLR for that same line-emitting gas. Coupling these two measurements together through the virial theorem allows a direct constraint on the mass of the central black hole, modulo a scaling factor to account for the detailed geometry (including inclination) and kinematics (whether rotation, infall, or outflow) of the gas.

Given the orientation-dependent picture of AGNs described above, and our inability to directly resolve the BLRs of even nearby AGNs with current technology, it is necessary to constrain the black hole mass scaling factor through some means. This is typically accomplished in an indirect way by comparing the relationship between black hole mass and host-galaxy bulge stellar velocity dispersion, the $M_{\rm BH}-\sigma_\star$ relationship, for nearby quiescent galaxies with stellar- and gas-dynamical modeling-based black hole masses to the relationship for AGNs with as-yet-unscaled reverberation masses, i.e., virial products. The average multiplicative factor that must be applied to the AGN virial products to bring the AGN relationship into agreement with that of the quescent galaxies is found to be $\sim$4$-$5 (most recently 4.3 ± 1.1 [27], when all AGNs are treated equally and local high-luminosity AGNs are included to extend the range of the $M_{\rm BH} - \sigma_\star$ relationship).

Certainly, there are many assumptions involved in determining this average scaling factor, and it is important to keep in mind when applying the scaling factor that it is a population average and therefore likely to be uncertain by a factor of 2–3 for any individual AGN. Nonetheless, several lines of independent evidence point to this value of the mass scaling factor being in the right ballpark.

The first is to simply assume that the main contribution to the scale factor is the inclination of the system, or the $\sin i$ term in the gas velocity. A scale factor of 4.3 would imply a typical broad-lined AGN inclination of $\sim$29°, which is well in line with expectations based on our current understanding of AGN structure. Furthermore, this average inclination agrees well with the geometric inclinations derived for the extended and resolved narrow-line region structure of several nearby AGNs [23].

A more rigorous test is to compare the black hole masses derived from reverberation mapping with those derived from stellar- or gas-dynamical modeling in the same objects. As previously discussed, most AGNs are too far away for dynamical modeling techniques to be applied, but a few very nearby AGNs can be examined in this way. Direct comparisons have been carried out for two AGNs to date—NGC 3227 [15, 18] and NGC 4151 [4, 61] (see Fig. 13.3). So far, the resultant black hole masses agree remarkably well between such disparate measurement techniques, each derived from independent observations and each with their own independent set of assumptions and biases. While the agreement is reassuring for these two objects, a sample of two is hardly definitive. Two more AGNs with reverberation masses—NGC 6814 [11] and NGC 5273 [7]—are in various stages of the dynamical modeling process, and a handful of other AGNs are being targeted for reverberation-mapping with the hope of dynamical modeling to follow.

One current complication to this test, however, is the effect of bars on the central stellar dynamics, and therefore the derived black hole masses (e.g., [14, 31]). Even for the relatively face-on galaxy, NGC 4151, a significant bias was found to be induced in the best-fit black hole mass from the weak galaxy scale bar [61]. NGC 5273 will be an especially interesting case for testing black hole masses from reverberation mapping versus stellar dynamical modeling given its unbarred S0 morphological type. Furthermore, in the next several years, it is likely that JWST will allow some advancement in the numbers of AGNs that can provide direct mass

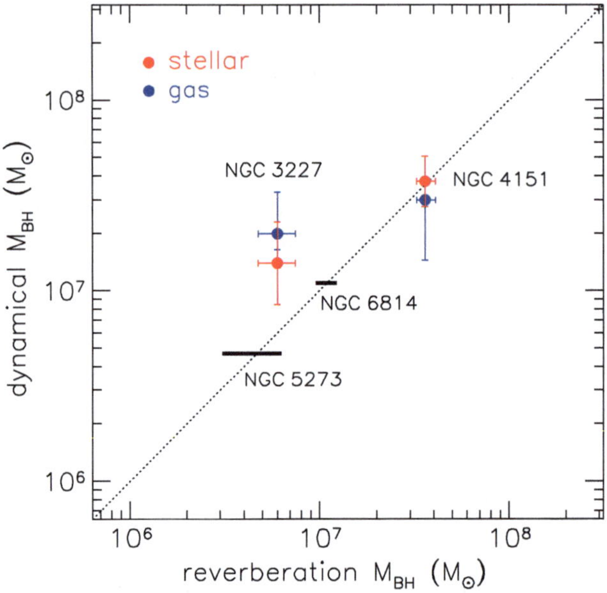

Fig. 13.3 Comparison of black hole masses derived from reverberation mapping with an average mass scale factor applied, and from stellar dynamical modeling and gas dynamical modeling which do not rely on a scale factor. Stellar dynamical modeling is in progress for NGC 6814 and NGC 5273, but the expected location based on the reverberation mass for each is denoted in the figure

comparisons across techniques. Gultekin et al. argue that dynamical modeling can still place strong constraints on black hole mass even if the radius of influence of the black hole is not strictly resolved in the observations [30]. JWST will provide a comparable spatial resolution for studies of host-galaxy stellar absorption features to that which is currently achieved with ground-based observatories and adaptive optics, but the advantages of JWST include a stable PSF, a significantly higher Strehl ratio, and very low backgrounds, all of which are important for deriving tight dynamical constraints on the black hole mass.

Finally, as we discuss in Sect. 13.3.3, it is possible to directly constrain the black hole mass, without needing to resort to the use of a scaling factor, from the reverberation-mapping data itself. Data quality concerns have generally not allowed this goal to be met in the past, but recent progress is encouraging, and the resultant mass constraints generally agree with our expectations based on the arguments above.

13.3.2 Black Hole Scaling Relationships

One of the most useful scaling relationships to arise from the compendium of reverberation-mapping measurements is the $R_{\rm BLR} - L_{\rm AGN}$ relationship: the relationship between the time delay, or average radius of emission, for a specific emission line and the luminosity of the central AGN at some particular wavelength. This particular scaling relationship was expected from simple photoionization arguments and looked for in the early days of reverberation mapping when the number of successful monitoring campaigns and reverberation measurements was still very small [47]. The addition of several high-luminosity local quasars from the Palomar-Green Bright Quasar Survey to the reverberation sample led to the first well-defined functional form of the $R_{\rm BLR} - L_{\rm AGN}$ relationship [40]. Figure 13.4 shows the most current calibration of the $R_{\rm BLR} - L_{\rm AGN}$ relationship between Hβ and the AGN continuum luminosity at rest-frame 5100 Å, where all luminosities have been carefully corrected for the contamination of host-galaxy starlight using high-resolution *HST* images and two-dimensional decompositions of the images to separate the galaxy starlight from the AGN emission [6].

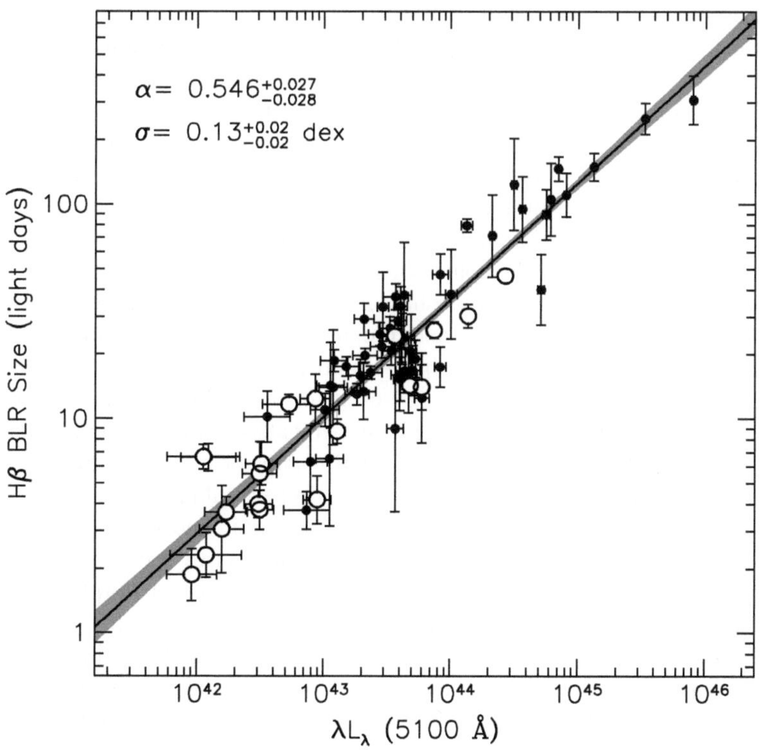

Fig. 13.4 The relationship between the Hβ time delay and the specific luminosity of the AGN at 5100 Å, the $R_{\rm BLR} - L_{\rm AGN}$ relationship. (Reprinted with permission from Bentz et al. [6])

Rather than carrying out a long-term monitoring campaign for any AGN of interest, the R_{BLR}–L_{AGN} relationship allows a single spectrum to provide an estimate of the black hole mass through two simple measurements: the width of the broad emission line as a proxy for the gas velocity, and the continuum luminosity of the AGN as a proxy for the time delay expected in the emission line. This handy shortcut provides a means for taking large spectroscopic surveys and producing catalogs full of black hole mass estimates (e.g., [71]). Of course, as one might expect, the devil is in the details.

Currently, Hβ is the only emission line for which a well-calibrated $R_{BLR} - L_{AGN}$ relationship exists. Unfortunately, Hβ shifts out of the observed-frame optical bandpass at redshifts of only $z = 0.7 - 0.8$. The Mg II and C IV emission lines in the ultraviolet rest-frame are therefore more accessible for most quasar surveys conducted from the ground. But at this time, only a handful of measurements of C IV reverberation time delays exist (see [39] for a first attempt to constrain a C IV $R_{BLR} - L_{AGN}$ relationship), and even fewer reverberation measurements exist for Mg II. Quasar black hole masses therefore require bootstrapping the estimates into the UV using the Hβ $R_{BLR} - L_{AGN}$ relationship for local AGNs as the cornerstone (e.g., [78]). Furthermore, the exact prescription for turning two simple spectral measurements into an unbiased mass estimate is still highly debatable (see [17, 19] for discussions of specific details that can cause biases).

Interestingly, the small scatter in the $R_{BLR} - L_{AGN}$ relationship has led to the proposal that it may be used to turn AGNs into standardizable candles for investigating cosmological expansion [45, 82]. Quasars are easily observed well beyond $z \approx 1$, where Type Ia supernovae become rare and difficult to find and where the differences in cosmological models are more apparent. One of the largest practical difficulties in turning this idea into reality, however, is again the long time delays involved in monitoring high redshift quasars and the necessity of high signal-to-noise spectroscopy over the course of such a monitoring campaign.

Other black hole scaling relationships include the aforementioned AGN $M_{BH} - \sigma_\star$ relationship [26, 27, 60, 63] and the relationship between the AGN black hole mass and the host-galaxy bulge luminosity, the $M_{BH} - L_{bulge}$ relationship [3, 10, 80]. While the more commonly-used forms of these relationships tend to be those derived for black holes with dynamically-modeled black hole masses (e.g., [22, 24, 30, 50, 51, 53]), the AGN relationships provide a useful counterpoint given the differences between the two samples. In particular, the AGN reverberation sample has a large percentage of late-type, disc-dominated galaxies, whereas the quiescent galaxy sample with dynamical black hole masses is comprised mainly of early-type galaxies. It is not at all clear that galaxies of different morphological types should all follow the same scaling relationships (see the review by Kormendy and Ho [50]). Furthermore, active galaxies may not follow the same scaling relationships as quiescent galaxies (e.g., [79]), but most studies of galaxy and black hole co-evolution at cosmological distances are necessarily limited to active galaxies.

Even despite these differences between the samples, it appears that our current constraints on the general forms of the $M_{BH} - L_{bulge}$ and $M_{BH} - \sigma_\star$ relationships are consistent for AGNs with reverberation masses and for quiescent galaxies with dynamical black hole masses. Several ongoing studies aim to refine and more accurately constrain these relationships, and part of this effort is devoted to tackling the key observational uncertainties that remain—such as determining accurate distances to the AGN host galaxies, and replacing stellar velocity dispersion measurements from long-slit spectra with those obtained from integral field spectroscopy of the host galaxies. Here again, barred galaxies (and galaxies with "discy" bulges) are a source of confusion. While they are seen to be outliers in the quiescent $M_{\mathrm{BH}} - \sigma_{\star}$ relationship [26, 37, 50], there is no such offset seen in the AGN $M_{\mathrm{BH}} - \sigma_{\star}$ relationship [27] unless it is artificially inserted by scaling the black hole masses in those galaxies by a different value [33]. Comparisons between the different assumptions and biases in the AGN $M_{\mathrm{BH}} - \sigma_{\star}$ relationship versus the quiescent galaxy relationship will therefore help to shed light on the underlying causes for such puzzles. And while it has been the focus of less intense study in the last decade or so, an accurately calibrated $M_{\mathrm{BH}} - L_{\mathrm{bulge}}$ relationship will be especially necessary for constraining galaxy evolution through upcoming deep all-sky photometric surveys with no spectroscopic component, like LSST.

13.3.3 BLR Geometry and Kinematics

Most of the progress in reverberation mapping has focused on the lowest-order measurement that can be made, namely the average time delay of an emission line, because this is what is required to make a dynamical mass measurement. Historically, it was also the only measurement that could be recovered from the marginally-sampled light curves from early reverberation campaigns. However, there is much more information encoded in densely-sampled light curves. In particular, the emission-line light curve is a convolution of the continuum variations and the extended response of the BLR gas at different line-of-sight velocities and light travel times relative to the observer. Resolving the time delays as a function of velocity across an emission-line profile can therefore give constraints on the detailed geometry and kinematics of the BLR gas. In Fig. 13.5, we show three examples of emission-line response given one fairly simple model for the BLR and three different possible kinematics of the gas: rotation, infall, and outflow. The differences between the three are apparent in a full deconvolution of the emission-line response (Fig. 13.5, shaded regions), or a first-order analysis in which the mean time delays are computed for velocity bins across the line profile (Fig. 13.5, error bars).

While the wealth of information that is potentially available from reverberation-mapping datasets has been understood for quite some time, the practical difficulties involved in deconvolving a faint signal from sparsely- and irregularly-sampled noisy data have limited much progress in this area. Notable early attempts include [20, 76, 81], but recovered maps of velocity-resolved results, reminiscent of those shown in Fig. 13.5, were ambiguous at best. In the last few years, however, reverberation programmes have enjoyed much more success given the careful experimental setup and the resultant high quality of the data [8, 11, 16, 18, 29].

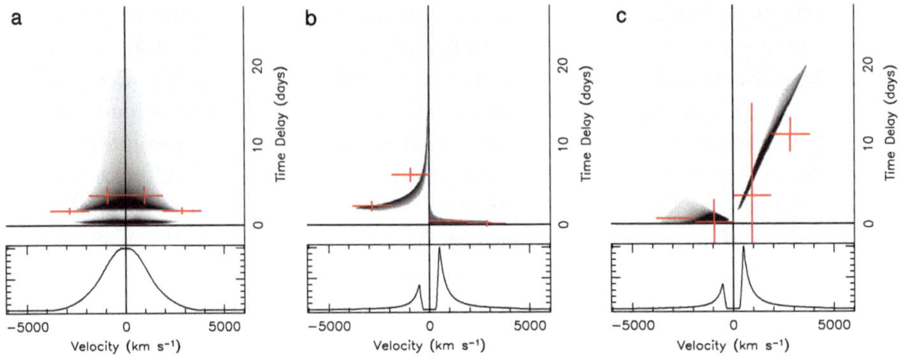

Fig. 13.5 The expected emission-line response for a toy model BLR with three different possible kinematics: (**a**) rotation, (**b**) infall, and (**c**) outflow. For simplicity, the geometry is kept the same for all three cases—the line emission is restricted to a bicone with a semi-opening angle of 30° and the model is inclined at 20° so that the observer is inside the beam. The radiation structure within the BLR clouds is set so that the emission is enhanced for clouds at smaller radii, and the line emission is partially anisotropic, such that the emission is enhanced in the direction of the illuminating source. The gray-scale images show the full two-dimensional structure in time lag versus line-of-sight velocity, while the *vertical red error bars* show the weighted mean and standard deviation of the time lag within discrete velocity bins that are represented by the *horizontal red error bars*. The overall shape is different for each of the three models: a symmetric structure around zero velocity for circular Keplerian orbits, longer lags in the blueshifted emission for infall, and longer lags in the redshifted emission for outflow. (Reprinted with permission from Bentz et al. [11])

With such high quality data now in hand, there are two general methods for extracting the reverberation signal. Deconvolution techniques can be used to produce a model-independent, but potentially difficult-to-interpret, velocity delay map like the models displayed in Fig. 13.5. Direct modelling of the spectroscopic data, on the other hand, produces easily-interpreted constraints on different possible physical models, but is by definition model-dependent and limited by human imagination.

The most widely employed deconvolution algorithm to date is the MEMECHO code [34, 36], which uses maximum entropy balanced by χ^2 to find the simplest possible solution that fits the data. Each pixel in the AGN spectrum can be treated as a separate velocity bin for which an entire light curve exists throughout the monitoring campaign. The code solves for the time delay response function in all the individual light curves for each of the pixels in the spectrum. From the many response functions, a map of time delay as a function of velocity is reconstructed.

Figure 13.6 shows an example of the deconvolved response of the broad optical recombination lines in the spectrum of the Seyfert galaxy Arp 151 [8]. The differences in expected mean time delays for the lines is immediately apparent, and also interesting is the strong prompt response in the red wings of the Balmer lines compared to the lack of prompt response in the blue wings of those same emission lines. An asymmetric response such as this could be produced by either rotating gas with enhanced emission in one location (such as a hot spot or warp), or by infalling

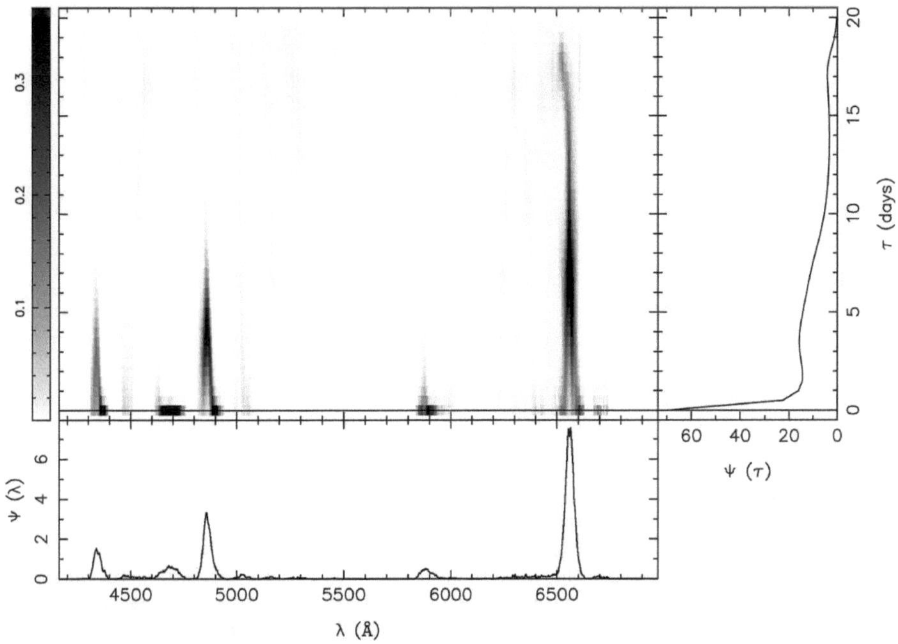

Fig. 13.6 The deconvolved emission-line response as a function of velocity for the broad optical recombination lines in the spectrum of Arp 151. (Reprinted with permission from Bentz et al. [8])

gas, or some combination of these simplistic models. Similar asymmetries are also seen in the deconvolved responses of a handful of additional AGNs [28].

Direct modeling, on the other hand, tests the data against specific geometric and kinematic models to constrain the family of models that best represent the observations. For the Arp 151 dataset above, direct modeling results prefer a thick disc BLR geometry, inclined at $\sim 25°$ to the observer's line of sight, and the kinematics are dominated by inflow with some contribution from rotation [62], in general agreement with the interpretation of the deconvolution results. Pancoast et al. also find similar results for a few additional AGNs [62].

Furthermore, direct modeling is able to determine the *individual* scaling factor that would need to be applied to the reverberation-based black hole mass as described in Sect. 13.3.1 above. For the handful of AGNs with successful dynamical models, we can see that the scale factor indeed varies from object to object, as expected for a population of objects with random inclinations within some limited range (see Fig. 13.7). The average of these individual scaling factors also agrees quite well with the population average derived above through use of the $M_{\mathrm{BH}} - \sigma_{\star}$ relationship (see Sect. 13.3.1).

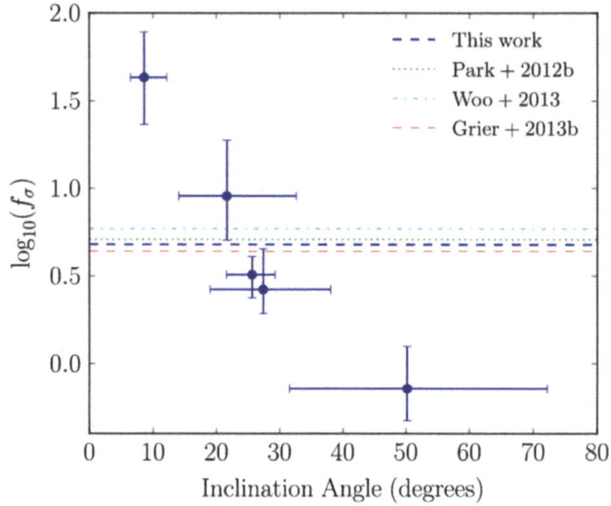

Fig. 13.7 The individual mass scale factor for 5 AGNs with direct modeling of their velocity-resolved responses. The average value for this small sample is $f \approx 4.8$ (denoted by the *horizontal blue dashed line*), in good agreement with values determined by comparison of the active galaxy and quiescent galaxy $M_{BH} - \sigma_\star$ relationships (denoted by the other *horizontal lines*). (Reprinted with permission from Pancoast et al. [62])

13.4 Looking Ahead

The future is looking promising for applications of reverberation mapping. In addition to the many ongoing areas of study summarized in the previous pages, several recent or ongoing programmes, such as the multi-object SDSS reverberation programme [72] and the massive HST plus ground-based reverberation programme for NGC 5548 [16], are just starting to report results that will lead to both new insights and new puzzles in the area of AGN physics. Additionally, the upcoming OzDES programme [44] will help push to higher redshifts, providing a more stable anchor for black hole mass estimates of high-z quasars.

The flurry of recent activity in velocity-resolved reverberation mapping is unlikely to abate any time soon, and here we may hope to unlock many of the secrets surrounding AGN feeding and feedback. New codes to deal with velocity-resolved reverberation mapping data are currently being developed ([73], Anderson et al. in prep), and new features are being added to currently-existing codes (Pancoast, *private communication*).

With UV astronomy currently dependent on the continuation of *HST*, it is certainly conceivable to think that the *Kronos* spacecraft [68], or a similar instrument, may again make an appearance in proposal form. It is an exciting time for supermassive black holes, reverberation mapping, and AGN physics!

Acknowledgements This work is supported by NSF CAREER grant AST-1253702.

References

1. Antonucci, R.: Unified models for active galactic nuclei and quasars. ARA&A **31**, 473 (1993)
2. Baldwin, J., Ferland, G., Korista, K., Verner, D.: Locally optimally emitting clouds and the origin of quasar emission lines. ApJ **455**, L119 (1995)
3. Bentz, M.: Black hole scaling relationships and NLS1s. In: Narrow-Line Seyfert 1 Galaxies and their Place in the Universe, vol. 33. PoS, Trieste (2011)
4. Bentz, M.C., Denney, K.D., Cackett, E.M., et al.: A reverberation-based mass for the central black hole in NGC 4151, ApJ **651**, 775 (2006)
5. Bentz, M.C., Denney, K.D., Cackett, E.M., et al.: NGC 5548 in a low-luminosity state: implications for the broad-line region. ApJ **662**, 205 (2007)
6. Bentz, M.C., Denney, K.D., Grier, C.J., et al.: The low-luminosity end of the radius-luminosity relationship for active galactic nuclei. ApJ **767**, 149 (2013)
7. Bentz, M.C., Horenstein, D., Bazhaw, C., et al.: The mass of the central black hole in the nearby Seyfert galaxy NGC 5273. ApJ **796**, 8 (2014)
8. Bentz, M.C., Horne, K., Barth, A.J., et al.: The Lick AGN monitoring project: velocity-delay maps from the maximum-entropy method for Arp 151. ApJ **720**, L46 (2010)
9. Bentz, M.C., Katz, S.: The AGN black hole mass database. PASP **127**, 67 (2015)
10. Bentz, M.C., Peterson, B.M., Pogge, R.W., et al.: The black hole mass-bulge luminosity relationship for active galactic nuclei from reverberation mapping and Hubble Space Telescope imaging. ApJ **694**, L166 (2009)
11. Bentz, M.C., Walsh, J.L., Barth, A.J., et al.: The Lick AGN monitoring project: broad-line region radii and black hole masses from reverberation mapping of Hβ. ApJ **705**, 199 (2009)
12. Bentz, M.C., Walsh, J.L., Barth, A.J., et al.: The Lick AGN monitoring project: reverberation mapping of optical hydrogen and helium recombination lines. ApJ **716**, 993 (2010)
13. Blandford, R.D., McKee, C.F.: Reverberation mapping of the emission line regions of Seyfert galaxies and quasars. ApJ **255**, 419 (1982)
14. Brown, J.S., Valluri, M., Shen, J., Debattista, V.P.: On the offset of barred galaxies from the black hole $M_{BH} - \sigma$ relationship. ApJ **778**, 151 (2013)
15. Davies, R.I., Thomas, J., Genzel, R., et al.: The star-forming torus and stellar dynamical black hole mass in the Seyfert 1 nucleus of NGC 3227. ApJ **646**, 754 (2006)
16. De Rosa, G., Peterson, B.M., Ely, J., et al.: Space telescope and optical reverberation mapping project.I. Ultraviolet observations of the Seyfert 1 galaxy NGC 5548 with the cosmic origins spectrograph on Hubble Space Telescope, ApJ **806**, 128 (2015)
17. Denney, K.D., Peterson, B.M., Dietrich, M., Vestergaard, M., Bentz, M.C.: Systematic uncertainties in black hole masses determined from single-epoch spectra. ApJ **692**, 246 (2009)
18. Denney, K.D., Peterson, B.M., Pogge, R.W., et al.: Reverberation mapping measurements of black hole masses in six local Seyfert galaxies. ApJ **721**, 715 (2010)
19. Denney, K.D., Pogge, R.W., Assef, R.J., Kochanek, C.S., Peterson, B.M., Vestergaard, M.: C IV line-width anomalies: the perils of low signal-to-noise spectra. ApJ **775**, 60 (2013)
20. Done, C., Krolik, J.H.: Kinematics of the broad emission line region in NGC 5548. ApJ **463**, 144 (1996)
21. Ferrarese, L., Ford, H.: Supermassive black holes in galactic nuclei: past, present and future research. Space Sci. Rev. **116**, 523 (2005)
22. Ferrarese, L., Merritt, D.: A fundamental relation between supermassive black holes and their host galaxies. ApJ **539**, L9 (2000)
23. Fischer, T.C., Crenshaw, D.M., Kraemer, S.B., Schmitt, H.R.: Determining inclinations of active galactic nuclei via their narrow-line region kinematics. I. Observational results. ApJS **209**, 1 (2013)
24. Gebhardt, K., Bender, R., Bower, G., et al.: A relationship between nuclear black hole mass and galaxy velocity dispersion. ApJ **539**, L13 (2000)
25. Goad, M.R., Korista, K.T., Ruff, A.J.: The broad emission-line region: the confluence of the outer accretion disc with the inner edge of the dusty torus. MNRAS **426**, 3086 (2012)

26. Graham, A.W., Onken, C.A., Athanassoula, E., Combes, F.: An expanded $M_{BH} - \sigma$ diagram, and a new calibration of active galactic nuclei masses. MNRAS **412**, 2211 (2011)
27. Grier, C.J., Martini, P., Watson, L.C., et al.: Stellar velocity dispersion measurements in high-luminosity quasar hosts and implications for the AGN black hole mass scale. ApJ **773**, 90 (2013)
28. Grier, C.J., Peterson, B.M., Horne, K., et al.: The structure of the broad-line region in active galactic nuclei. I. Reconstructed velocity-delay maps. ApJ **764**, 47 (2013)
29. Grier, C.J., Peterson, B.M., Pogge, R.W., et al.: Reverberation mapping results for five Seyfert 1 galaxies. ApJ **755**, 60 (2012)
30. Gültekin, K., Richstone, D.O., Gebhardt, K., et al.: The M−σ and M−L relations in galactic bulges, and determinations of their intrinsic scatter. ApJ **698**, 198 (2009)
31. Hartmann, M., Debattista, V.P., Cole, D.R., et al.: The effect of bars on the $M_* - \sigma_e$ relation: offset, scatter and residuals correlations. MNRAS **441**, 1243 (2014)
32. Heckman, T.M., Best, P.N.: The coevolution of galaxies and supermassive black holes: insights from surveys of the contemporary universe. ARA&A **52**, 589 (2014)
33. Ho, L.C., Kim, M.: The black hole mass scale of classical and pseudo bulges in active galaxies. ApJ **789**, 17 (2014)
34. Horne, K.: Echo mapping problems maximum entropy solutions. In: Astronomical Society of the Pacific Conference Series, Reverberation Mapping of the Broad-Line Region in Active Galactic Nuclei, vol. 69, p. 23. Astronomical Society of the Pacific, San Francisco (1994)
35. Horne, K., Peterson, B.M., Collier, S.J., Netzer, H.: Observational requirements for high-fidelity reverberation mapping. PASP **116**, 465 (2004)
36. Horne, K., Welsh, W.F., Peterson, B.M.: Echo mapping of broad H-beta emission in NGC 5548. ApJ **367**, L5 (1991)
37. Hu, J.: The black hole mass-stellar velocity dispersion correlation: bulges versus pseudo-bulges. MNRAS **386**, 2242 (2008)
38. Kalfountzou, E., Civano, F., Elvis, M., Trichas, M., Green, P.: The largest X-ray-selected sample of z>3 AGNs: C-COSMOS and ChaMP. MNRAS **445**, 1430 (2014)
39. Kaspi, S., Brandt, W.N., Maoz, D., et al.: Reverberation mapping of high-luminosity quasars: first results. ApJ **659**, 997 (2007)
40. Kaspi, S., Smith, P.S., Netzer, H., et al.: Reverberation measurements for 17 quasars and the size-mass-luminosity relations in active galactic nuclei. ApJ **533**, 631 (2000)
41. Kawaguchi, T., Mineshige, S., Machida, M., Matsumoto, R., Shibata, K.: Temporal $1/f^{\alpha}$ fluctuations from fractal magnetic fields in black-hole accretion flow. PASJ **52**, L1 (2000)
42. Kelly, B.C., Vestergaard, M., Fan, X., et al.: Constraints on black hole growth, quasar lifetimes, and eddington ratio distributions from the SDSS broad-line quasar black hole mass function. ApJ **719**, 1315 (2010)
43. Kilerci Eser, E., Vestergaard, M., Peterson, B.M., Denney, K.D., Bentz, M.C.: On the scatter in the radius-luminosity relationship for active galactic nuclei. ApJ **801**, 8 (2015)
44. King, A.L., Martini, P., Davis, T.M., et al.: Simulations of the OzDES AGN reverberation mapping project. MNRAS **453**, 1701 (2015)
45. King, A.L., Davis, T.M., Denney, K.D., Vestergaard, M., Watson, D.: High-redshift standard candles: predicted cosmological constraints. MNRAS **441**, 3454 (2014)
46. Kollatschny, W., Bischoff, K., Robinson, E.L., Welsh, W.F., Hill, G.J.: Short-term emission line and continuum variations in Mrk 110. A&A **379**, 125 (2001)
47. Koratkar, A.P., Gaskell, C.M.: Radius-luminosity and mass-luminosity relationships for active galactic nuclei. ApJ **370**, L61 (1991)
48. Korista, K., Ferland, G., Baldwin, J.: Do the broad emission line clouds see the same continuum that we see? ApJ **487**, 555 (1997)
49. Korista, K.T., Goad, M.R.: What the optical recombination lines can tell us about the broad-line regions of active galactic nuclei. ApJ **606**, 749 (2004)
50. Kormendy, J., Ho, L.C.: Coevolution (or not) of supermassive black holes and host galaxies. ARA&A **51**, 511 (2013)

51. Magorrian, J., Tremaine, S., Richstone, D., et al.: The demography of massive dark objects in galaxy centers. AJ **115**, 2285 (1998)
52. Marconi, A., Axon, D.J., Maiolino, R., et al.: The effect of radiation pressure on virial black hole mass estimates and the case of narrow-line Seyfert 1 galaxies. ApJ **678**, 693 (2008)
53. McConnell, N.J., Ma, C.-P.: Revisiting the scaling relations of black hole masses and host galaxy properties. ApJ **764**, 184 (2013)
54. Momjian, E., Carilli, C.L., Walter, F., Venemans, B.: The highest redshift quasar at z = 7.085: a radio-quiet source. AJ **147**, 6 (2014)
55. Nenkova, M., Sirocky, M.M., Nikutta, R., Ivezić, Ž., Elitzur, M.: AGN dusty tori. II. Observational implications of clumpiness. ApJ **685**, 160 (2008)
56. Netzer, H.: Radiation pressure force and black hole mass determination in low-redshift type-I and type-II active galactic nuclei. ApJ **695**, 793 (2009)
57. Netzer, H.: Revisiting the unified model of active galactic nuclei. ARA&A **53**, 365 (2015)
58. Netzer, H., Laor, A.: Dust in the narrow-line region of active galactic nuclei. ApJ **404**, L51 (1993)
59. Netzer, H., Marziani, P.: The effect of radiation pressure on emission-line profiles and black hole mass determination in active galactic nuclei. ApJ **724**, 318 (2010)
60. Onken, C.A., Ferrarese, L., Merritt, D., et al.: Supermassive black holes in active galactic nuclei. II. Calibration of the black hole mass-velocity dispersion relationship for active galactic nuclei. ApJ **615**, 645 (2004)
61. Onken, C.A., Valluri, M., Brown, J.S., et al.: The black hole mass of NGC 4151. II. Stellar dynamical measurement from near-infrared integral field spectroscopy. ApJ **791**, 37 (2014)
62. Pancoast, A., Brewer, B.J., Treu, T., et al.: Modelling reverberation mapping data – II. Dynamical modelling of the Lick AGN monitoring project 2008 data set. MNRAS **445**, 3073 (2014)
63. Park, D., Kelly, B.C., Woo, J.-H., Treu, T.: Recalibration of the virial factor and $M_{BH} - \sigma_*$ relation for local active galaxies. ApJS **203**, 6 (2012)
64. Peterson, B.M.: Reverberation mapping of active galactic nuclei. PASP **105**, 247 (1993)
65. Peterson, B.M.: An Introduction to Active Galactic Nuclei. Cambridge University Press, Cambridge (1997)
66. Peterson, B.M., Berlind, P., Bertram, R., et al.: Steps toward determination of the size and structure of the broad-line region in active galactic nuclei. XVI. A 13 year study of spectral variability in NGC 5548. ApJ **581**, 197 (2002)
67. Peterson, B.M., Ferrarese, L., Gilbert, K.M., et al.: Central masses and broad-line region sizes of active galactic nuclei. II. A homogeneous analysis of a large reverberation-mapping database. ApJ **613**, 682 (2004)
68. Peterson, B.M., Polidan, R.S., Robinson, E.L.: Kronos: a multiwavelength observatory for mapping accretion-driven sources. In: Blades, J.C., Siegmund, O.H.W. (eds.) Society of Photo-Optical Instrumentation Engineers (SPIE) Conference Series, Future EUV/UV and Visible Space Astrophysics Missions and Instrumentation, vol. 4854 p.311. SPIE, Bellingham (2003)
69. Peterson, B.M., Wandel, A.: Evidence for supermassive black holes in active galactic nuclei from emission-line reverberation. ApJ **540**, L13 (2000)
70. Shankar, F., Weinberg, D.H., Miralda-Escudé, J.: Self-consistent models of the AGN and black hole populations: duty cycles, accretion rates, and the mean radiative efficiency. ApJ **690**, 20 (2009)
71. Shen, Y., Richards, G.T., Strauss, M.A., et al.: A catalog of quasar properties from Sloan Digital Sky Survey data release 7. ApJS **194**, 45 (2011)
72. Shen, Y., Brandt, W.N., Dawson, K.S., et al.: The Sloan Digital Sky Survey reverberation mapping project: technical overview. ApJS **216**, 4 (2015)
73. Skielboe, A., Pancoast, A., Treu, T., et al.: Constraints on the broad line region from regularized linear inversion: velocity-delay maps for five nearby active galactic nuclei. MNRAS **454**, 144 (2015)

74. Ueda, Y., Akiyama, M., Hasinger, G., Miyaji, T., Watson, M.G.: Toward the standard population synthesis model of the X-Ray background: evolution of X-ray luminosity and absorption functions of active galactic nuclei including compton-thick populations. ApJ **786**, 104 (2014)
75. Ueda, Y., Akiyama, M., Ohta, K., Miyaji, T.: Cosmological evolution of the hard X-ray active galactic nucleus luminosity function and the origin of the hard X-Ray background. ApJ **598**, 886 (2003)
76. Ulrich, M.-H. Horne, K.: A month in the life of NGC 4151: velocity-delay maps of the broad-line region. MNRAS **283**, 748 (1996)
77. Urry, C.M., Padovani, P.: Unified schemes for radio-loud active galactic nuclei. PASP **107**, 803 (1995)
78. Vestergaard, M., Peterson, B.M.: Determining central black hole masses in distant active galaxies and quasars. II. Improved optical and UV scaling relationships. ApJ **641**, 689 (2006)
79. Wandel, A.: The black hole-to-bulge mass relation in active galactic nuclei. ApJ **519**, L39 (1999)
80. Wandel, A.: Black holes of active and quiescent galaxies. I. The black hole-bulge relation revisited. ApJ **565**, 762 (2002)
81. Wanders, I., Goad, M.R., Korista, K.T., et al.: The geometry and kinematics of the broad-line region in NGC 5548 from HST and IUE observations. ApJ **453**, L87 (1995)
82. Watson, D., Denney, K.D., Vestergaard, M., Davis, T.M.: A new cosmological distance measure using active galactic nuclei. ApJ **740**, L49 (2011)

Index

H.M.J. Boffin et al. (eds.), *Astronomy at High Angular Resolution*, Astrophysics and Space Science Library 439, DOI 10.1007/978-3-319-39739-9

MIX
Papier aus verantwortungsvollen Quellen
Paper from responsible sources
FSC® C105338

If you have any concerns about our products,
you can contact us on
ProductSafety@springernature.com

In case Publisher is established outside the EU,
the EU authorized representative is:
Springer Nature Customer Service Center GmbH
Europaplatz 3, 69115 Heidelberg, Germany

Printed by Libri Plureos GmbH
in Hamburg, Germany